Phospholipids and Cellular Regulation

Volume I

Editor

J. F. Kuo
Department of Pharmacology
Emory University School of Medicine
Atlanta, Georgia

CRC Press, Inc.
Boca Raton, Florida

Library of Congress Cataloging in Publication Data
Main entry under title:

Phospholipids and cellular regulation.

Bibliography: p.
Includes index.
1. Cellular control mechanisms. 2. Phospholipids.
3. Cell membranes. I. Kuo, J. F. (Jyh-Fa), 1933-
QH604.P46 1985 574.87′5 84-28569
ISBN 0-8493-5537-0 (v. 1)
ISBN 0-8493-5538-9 (v. 2)

This book represents information obtained from authentic and highly regarded sources. Reprinted material is quoted with permission, and sources are indicated. A wide variety of references are listed. Every reasonable effort has been made to give reliable data and information, but the author and the publisher cannot assume responsibility for the validity of all materials or for the consequences of their use.

Direct all inquiries to CRC Press, Inc., 2000 Corporate Blvd., N.W., Boca Raton, Florida, 33431.

Second Printing, 1986

International Standard Book Number 0-8493-5537-0 (Volume I)
International Standard Book Number 0-8493-5538-9 (Volume II)

Library of Congress Card Number 84-28569
Printed in the United States

PREFACE

The importance of phospholipids in biological systems has become increasingly recognized and appreciated in recent years. These two volumes are intended to give a comprehensive and critical coverage of the progress made in certain major areas of research directly or indirectly related to phospholipids. Volume I begins with chapters dealing with the structural and functional roles of phospholipids in biological membranes and techniques for studying phospholipid membranes. More dynamic aspects of phospholipids follow, which include their roles in the modifications of the functional properties of membranes, in intracellular mediator systems, and in human disease. Volume II covers two biologically important phospholipids (i.e., platelet activating factor and antitumor agent alkyllysophospholipid), some major aspects of the newly discovered (thanks to Professor Nishizuka and his able colleagues in Japan) phospholipid/Ca^{2+}-dependent protein kinase (protein kinase C) system, including its role in transduction of membrane signals, and finally, regulation of oxidative enzymes by phospholipids and membrane. My deep gratitude is extended to many leading investigators in the fields, who have kindly contributed their chapters. I hope that the volumes will serve as a comprehensive and valuable source of references, thus fostering a further advancement in this ever rapidly expanding research on the roles of phospholipids in cellular function and regulation.

J. F. Kuo

THE EDITOR

J. F. Kuo holds the Ph.D. in Biochemistry from the University of Illinois (Urbana) and is Professor of Pharmacology at Emory University School of Medicine. His previous positions include Research Biochemist at Lederle Laboratories and Assistant and Associate Professor of Pharmacology at Yale University School of Medicine. He was a Visiting Professor of the Swedish Medical Research Council at Linköping University, Sweden, where he was awarded the D. Med. (hon.) in 1980. Dr. Kuo's research covers the biochemical, pharmacological, immunological, and pathophysiological aspects of membrane receptors, cyclic nucleotides, calcium, phospholipids, and protein phosphorylation systems. He has published extensively in these areas of investigation.

CONTRIBUTORS

Fulton T. Crews
Assistant Professor
Department of Pharmacology
University of Florida College of Medicine
J. Hills Miller Health Center
Gainesville, Florida

Pieter R. Cullis
Associate Professor
Department of Biochemistry
University of British Columbia
Vancouver, British Columbia
Canada

John N. Fain
Professor and Chairman
Section of Biochemistry
Brown University
Providence, Rhode Island

Robert V. Farese
Professor of Medicine
Director, Division of Endocrinology and Metabolism
Department of Internal Medicine
University of South Florida College of Medicine
and
Associate Chief of Staff
Research and Development
James A. Haley Veterans Hospital
Tampa, Florida

R. Adron Harris
Associate Professor
Department of Pharmacology
University of Colorado
Denver, Colorado

Robert J. Hitzemann
Associate Professor
Departments of Pharmacology and Psychiatry
University of Cincinnati
Cincinnati, Ohio

M. J. Hope
Research Associate
Department of Biochemistry
University of British Columbia
Vancouver, British Columbia
Canada

Karl Y. Hostetler
Associate Professor of Medicine
Department of Medicine
University of California at San Diego
San Diego, California

B. de Kruijff
Professor
Institute of Molecular Biology
State University of Utrecht
Padualaan, Utrecht
The Netherlands

Irene Litosch
Research Associate
Division of Biology and Medicine
Section of Biochemistry
Brown University
Providence, Rhode Island

Horace H. Loh
Professor
Departments of Pharmacology and Psychiatry
University of California
San Francisco, California

Alfred H. Merrill
Assistant Professor
Department of Biochemistry
Emory University School of Medicine
Atlanta, Georgia

J. Wylie Nichols
Assistant Professor
Departments of Biochemistry and Physiology
Emory University School of Medicine
Atlanta, Georgia

C. P. S. Tilcock
Research Associate
Department of Biochemistry
University of British Columbia
Vancouver, British Columbia
Canada

A. J. Verkleij
Associate Professor
Institute of Molecular Biology
State University of Utrecht
Padualaan, Utrecht
The Netherlands

TABLE OF CONTENTS

Chapter 1

STRUCTURAL PROPERTIES AND FUNCTIONAL ROLES OF PHOSPHOLIPIDS IN BIOLOGICAL MEMBRANES

P. R. Cullis, M. J. Hope, B. de Kruijff, A. J. Verkleij, and C. P. S. Tilcock

TABLE OF CONTENTS

I. INTRODUCTION

A. Scope

Membranes contain an astonishing variety of lipids. The development of this diversity must result in significant evolutionary advantages to the uni- or multicellular organism in which they reside. This suggests particular functional roles for each component; however, clarification of these functional roles for individual lipid species has proven difficult. In this work we present an overview of the physical properties of phospholipids as demonstrated in model membrane systems and their possible functional roles in membrane-mediated phenomena.

The literature in this area is large and is growing rapidly. In order to achieve a basic coherence and form, we have largely restricted the scope of this review to the lipids of eukaryotic membrane systems. This has two major consequences. First, there is now a large body of evidence to suggest that most, if not all, eukaryotic membrane phospholipids are in the fluid liquid crystalline state at physiological temperatures. As a result, this work is largely restricted to the properties of liquid crystalline lipid systems. The gel-liquid crystalline characteristics of membrane lipids, which have been extensively reviewed elsewhere,[1] are only mentioned where germane. Second, where possible, we have concentrated on studies revealing the properties of lipids actually found in membranes. As a result, the literature on synthetic lipid systems is only included to the extent that it contributes to an understanding of the behavior of naturally occurring lipid species.

B. Roles of Lipids in Membranes: An Overview

Development of an understanding of the functional roles of lipids in membranes began with the early experiments of Gorter and Grendel,[2] who came to the conclusion that the erythrocyte contained sufficient lipid to provide a bilayer lipid matrix surrounding the red cell. The resulting concept that the major functional role of lipids is to provide a bilayer permeability barrier between external and internal compartments has remained a dominant theme in our understanding of biological membranes. Subsequent observations that membranes are fluid, allowing rapid lateral diffusion in the plane of the membrane,[3] and that membrane proteins are often inserted into and through the lipid bilayer have further contributed to our present understanding of membranes, resulting in the Singer and Nicholson[4] "fluid mosaic model". A refined version of this model is shown in Figure 1, which emphasizes the observations that membranes exhibit asymmetric transbilayer distributions of lipid and protein components. In the red cell membrane, for example, phosphatidylserine and phosphatidylethanolamine are largely localized to the interior monolayer,[5] whereas phosphatidyl choline, sphingomyelin, and the carbohydrate-containing moieties of glycolipids and glycoproteins are found on the membrane exterior. For completeness, the membrane potential, $\Delta\psi$, arising from transbilayer chemical gradients of ions such as Na^+, K^+, and H^+ is also indicated.

Within the terms of this model, the functional roles of lipids fall into categories related to an ability to self-assemble into bilayer structures on hydration, thus providing a permeability barrier as well as a matrix with which functional membrane proteins can be associated. Roles of individual lipid components may therefore concern establishing the bilayer structure itself, establishing appropriate permeability characteristics, satisfying insertion and packing requirements in the region of integral proteins, as well as allowing the surface association of peripheral protein via electrostatic interactions. All of these demands are clearly critical. An intact permeability barrier to small ions such as Na^+, K^+, and H^+, for example, is vital to establishing the electrochemical gradients. These give rise to a membrane potential which drives other membrane-mediated transport processes, whereas the lipid in the region of membrane protein must seal the protein into the bilayer so as to both prevent nonspecific leakage and to provide an environment appropriate to a functional protein conformation.

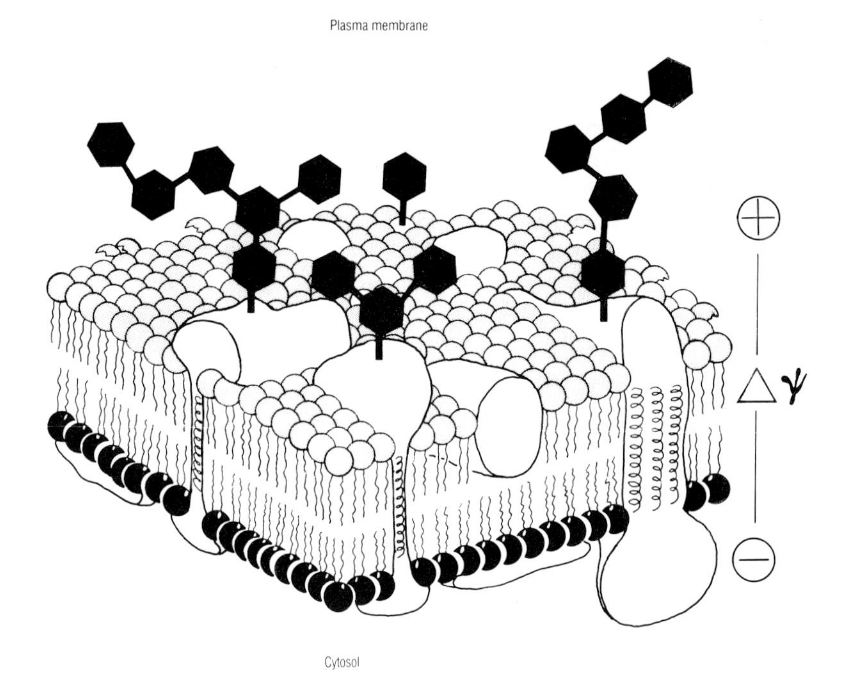

FIGURE 1. A refined fluid mosaic model of biological membranes. The transbilayer asymmetry of lipids and carbohydrates is emphasized as is the membrane potential $\Delta\psi$ arising from transbilayer electrochemical gradients of various ions.

C. Organization

The organization of this chapter stems from a model building approach to the understanding of the functional roles of individual lipid components of membranes. Briefly, the physical properties and roles of individual lipids are exceedingly difficult to ascertain in an intact biological membrane due to the complex lipid composition. In order to gain insight into the roles of individual components, much simpler "model membrane" systems are required which contain the lipid species of interest. The simplest of these model systems consists of the isolated lipid dispersed in an aqueous buffer, which can be used to determine structural and motional properties. At the next level of sophistication, unilamellar model systems are required to examine properties such as permeability. Subsequent, more sophisticated unilamellar models will then include reconstituted proteins if an understanding of lipid-protein interactions or the influence of a given lipid on protein function is required. This step-by-step development of increasingly sophisticated well-defined systems which model aspects of the composition and function of biological membranes should allow the properties and roles of individual components to be elucidated in an unambiguous manner.

This model approach dictates the development we follow here. First, some overview of methods of generating model membrane systems is required, as presented in the next section. Subsequently, the structural properties of lipids as elucidated in lipid-water dispersions are discussed. The fact that membrane lipids can adopt a variety of structures in addition to the bilayer organization leads to the possibility that the roles of certain lipids may be related to an ability to form nonbilayer structures. Membrane fusion is a particularly important example as indicated in Section V. The permeability properties of lipids and the roles of lipids in protein function as indicated by studies of reconstituted lipid-protein systems are reviewed in the following two sections. Finally, these observations are related to observations on intact biological membranes themselves in Section VIII.

II. MODEL MEMBRANE SYSTEMS

A. Dispersions of Lipids in Water

As indicated above, the simplest model system is obtained by depositing the lipid of interest as a film (normally by evaporation from chloroform), which is subsequently hydrated by mechanical agitation (e.g., vortexing), in the presence of aqueous buffer. In the case of lipids which adopt bilayer structure on hydration, the result is a milky suspension containing large "multilamellar vesicles" (MLVs) or liposomes.[6] These systems commonly range in size from 0.5 to 10 μm and consist of concentric lipid bilayers in an onionskin configuration. As little as 10% of the total lipid is contained in the outermost bilayer.

Lipids that adopt the hexagonal H_{II} phase (see Section IV) are usually much more difficult to disperse in water and form a fine particulate suspension which is easy to distinguish from MLV systems. The particles basically consist of a hydrocarbon matrix penetrated by aqueous channels (~20 Å diameter) which is surrounded by a monolayer of lipid with polar head groups oriented to the external aqueous medium.

B. Small Unilamellar Vesicles

The simple lipid-water dispersions are most useful for studies on the structural preferences of lipids and factors which modulate these preferences; however, they are clearly not accurate models of biological membranes in that MLVs contain internal bilayers, whereas H_{II} dispersions do not exhibit a well-defined permeability barrier between external and internal environments. Unilamellar systems provide more representative model systems. The simplest techniques for producing such systems generate "small unilamellar vesicles" (SUVs) which exhibit diameters in the range of 25 to 40 nm. Note that only lipid or lipid mixtures which form bilayer structure on hydration (e.g., MLVs on mechanical agitation) can form stable SUVs. The major techniques for making SUVs involve sonication[7] or passage through a French press. The small size of the vesicles produced has certain advantages. For example, high resolution nuclear magnetic resonance (NMR) spectra can be obtained from such systems, allowing outside-inside lipid distributions to be determined,[9] as well as details of acyl chain mobility;[10] however, these advantages are outweighed by the disadvantages concerning the very small radius of curvature which can perturb the physical properties of lipid components.[10] Related limitations concern the small interior aqueous volume, which limits the amount of material that can be trapped inside. For example, a vesicle of 25 nm diameter can contain only ~1.6×10^3 atoms of a solute trapped at a nominal 1 m*M* concentration, assuming a bilayer thickness of 4 nm. Parameters such as internal pH are difficult to define, as illustrated by the observation that at pH = 7 only 1 in approximately 10 of these vesicles contained a free proton.

C. Large Unilamellar Vesicles

The difficulties encountered with the SUV systems have led to a substantial effort to obtain large unilamellar vesicles, which may be generally defined as unilamellar systems with diameters greater than 50 nm. Most of these procedures have been well reviewed elsewhere.[11] Briefly, the LUV techniques can be divided into three general classes consisting of vesicles prepared by detergent dialysis,[12] vesicles prepared employing organic solvents,[13-15] and LUVs prepared directly from MLV dispersions.[7,16] The LUVs thus produced range in size from 50 to 200 nm and each technique has various advantages and limitations. The detergent dialysis techniques involve solubilizing the lipid in the detergent of choice which is followed by dialysis to remove the detergent. This leaves unilamellar vesicles. Difficulties encountered concern lipid species which are difficult to solubilize, limited trapping efficiences (defined as the percentage of available solute which is trapped), and the presence of residual detergent even after extended dialysis or gel filtration procedures.

Advantages include the gentle nature of the procedure and the fact that solubilized membrane proteins can be reconstituted into the vesicles during the dialysis procedure.[17]

Techniques involving organic solvents have become increasingly popular. Most of these procedures involve the solubilization of the lipid in an organic solvent (e.g., ether, ethanol) which is subsequently injected into an aqueous buffer. The solvent may be removed by incubation of the buffer at a temperature above the boiling point of the organic solvent or by subsequent dialysis. A modification known as the reverse phase evaporation (REV) procedure[15] involves making an emulsion of lipid (dissolved in ether) with aqueous buffer. The organic solvent is removed under vacuum, giving rise to hydrated lipid in the form of a thick gel. This can subsequently be diluted and sized under low (≤80 psi) pressure through polycarbonate filters to give LUVs of a defined size. An advantage of this technique is the high trapping efficiencies available (~35%). A general disadvantage of LUV preparations involving organic solvent concerns the differing solubilities of various species of lipid. This often necessitates different organic solvent mixtures according to the lipid employed.

Recently,[16] it has been shown that LUVs can be produced directly from MLVs by repetitively extruding MLVs through polycarbonate filters (with 100-nm pore size or less) under moderate pressures (100 to 700 psi). This technique has many advantages due to the absence of organic solvents or detergents, the straightforward and rapid protocol involved, the high trapping efficiencies available (~30%), and the generality of the technique. In particular, all lipid systems which can be dispersed in MLV form can be subsequently converted to LUV form by the extrusion procedure.

III. MEMBRANE FLUIDITY

Before discussing the properties of lipid in model systems of varying complexity, it is appropriate to indicate the problems associated with current characterizations of the properties of lipids in membranes. Chief among these is the concept of membrane fluidity, which can be most misleading. For example, it is commonly assumed that more saturated lipids or the presence of cholesterol makes membranes less "fluid". This is not necessarily the case. Strictly speaking, the fluidity parameter is the reciprocal of the membrane viscosity, which in turn is inversely proportional to the rotational and lateral diffusion rates (D_R and D_T, respectively) of membrane components.[18] Thus, a linear relation between membrane fluidity and D_R and D_T would be expected, which is not observed. Incorporation of cholesterol into phosphatidylcholine model membranes (at temperatures above the acyl chain gel to liquid crystalline transition temperature) has little or no influence on the lateral diffusion rates observed,[19,20] and can actually increase the rotational diffusion rates.[21] The major influence of cholesterol or decreased unsaturation is to increase the order in the hydrocarbon matrix. It is this increase or decrease in order, which is a measurable quantity expressed by NMR or ESR "order parameters",[22] for example, that should be correlated with such changes as decreased or increased membrane permeabilities.

IV. STRUCTURAL PROPERTIES OF LIPIDS

A. Introduction

The structural properties of phospholipids are most conveniently characterized in the simplest model systems consisting of a dispersion of the lipid in an aqueous buffer. Dispersions of individual species of liquid crystalline phospholipids of biological origin adopt either of the three structures shown in Figure 2. These include the micellar phase, which is preferred by minority lipid components such as lysophospholipids, the familiar bilayer phase, or the hexagonal H_{II} phase. The hexagonal H_{II} phase is composed of hexagonally packed cylinders of lipid where the cylinders are composed of a central aqueous channel toward

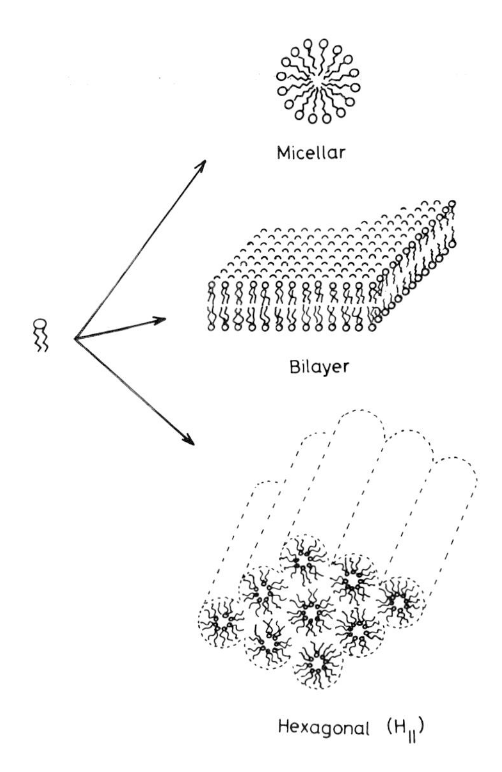

FIGURE 2. Structural preferences of liquid crystalline lipids dispersed in water at concentrations above the critical micellar concentration.

which the polar head groups are oriented. Mixed lipid systems can adopt a variety of other phases, such as those exhibiting cubic symmetry; however, the structures of these three dimensional entities are difficult to determine unequivocally. Within the context of this chapter the ability of lipids to adopt these liquid crystalline phase structures, particularly the bilayer or H_{II} arrangements or variations thereof, will be referred to as "lipid polymorphism". The literature characterizing these polymorphic properties of lipids is considerable. Here we present a synopsis of the techniques employed to monitor lipid polymorphism, analyses of the polymorphic properties of individual membrane phospholipids, and the properties of mixed lipid systems.

B. Techniques

The polymorphism of membrane lipids has been investigated by a variety of diffraction, spectroscopic, calorimetric, and other techniques. The major techniques are X-ray, NMR, and freeze-fracture; the ^{31}P NMR as well as the freeze-fracture characteristics of various lipid phases are illustrated in Figure 3. Here we briefly outline the utility and limitation of these various protocols.

X-ray and neutron diffraction are the only techniques that provide, in principle, unambiguous information on the macroscopic structure of lipid phases. A detailed discussion of their application to the study of lipid polymorphism and membrane structure is given elsewhere.[23-26] The utility of the X-ray technique for the determination of lipid phase structure

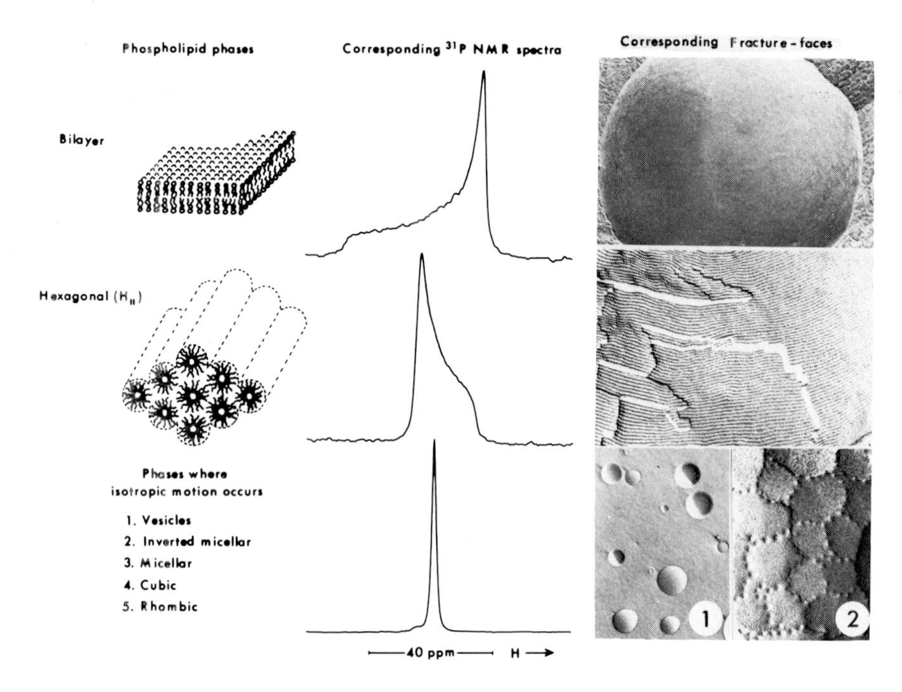

FIGURE 3. ^{31}P NMR and freeze-fracture characteristics of phospholipids in various phases. The bilayer ^{31}P NMR spectrum was obtained from aqueous dispersions of egg yolk phosphatidylcholine, and the hexagonal (H_{II}) phase spectrum from phosphatidylethanolamine (prepared from soybean phosphatidylcholine). The ^{31}P NMR spectrum representing isotropic motion was obtained from a mixture of 70 mol% soya phosphatidylethanolamine and 30% egg yolk phosphatidylcholine after heating to 90°C for 15 min. All preparations were hydrated in 10 m*M* Tris-acetic acid (pH 7.0) containing 100 m*M* NaCl and the ^{31}P NMR spectra were recorded at 30°C in the presence of proton decoupling. The freeze-fracture micrographs represent typical fracture faces obtained from bilayer and H_{II} phase systems as well as structures giving rise to isotropic motional averaging. The bilayer configuration (total erythrocyte lipids) gives rise to a smooth fracture face, whereas the hexagonal (H_{II}) configuration is characterized by ridges displaying a periodicity of 6 to 15 mm. Common conformations that give rise to isotropic motion are represented in the bottom micrograph: (1) bilayer vesicles (~100 nm diameter) of egg phosphatidylcholine prepared by extrusion techniques and (2) large lipid structures containing lipidic particles. This latter system was generated by fusing SUVs composed of egg phosphatidylethanolamine and 20 mol% egg phosphatidylserine which were prepared at pH 7 and then incubated at pH 4 for 15 min to induce fusion.

lies in the fact that lipids in phases possessing long-range order, such as the lamellar or hexagonal H_{II} phases, exhibit characteristic small-angle diffraction patterns. Lipids in a lamellar phase give rise to X-ray diffraction patterns with long spacing of the first and higher orders in the ratio 1:1/2:1/3 etc., whereas for lipids in a hexagonal phase, long spacings occur in the ratio $1:1/\sqrt{3}:1/2:1/\sqrt{7}$, etc.[23,27-31] While small-angle X-ray diffraction usually yields unambiguous information about the structure of liquid crystalline lipid phases such as the lamellar or hexagonal phases, it is generally more difficult to obtain definitive characterizations of other lipid phases such as cubic phases (although there are exceptions[32]). This is because cubic phases tend to give an insufficient number of reflections to permit unambiguous lattice assignments.[24,33,34]

Systems in which there is little long-range order exhibit diffraction profiles which are more difficult to interpret. It is difficult to detect and quantify a minority component in two or more coexisting phases, particularly when one component lacks long-range order. Further, if sophisticated detectors[35] are not employed, long exposure times (hours to days) are required; this is a problem for labile lipids. In addition, sample preparation is generally not as straightforward as with other spectroscopic or calorimetric techniques. In order to maximize signal intensities it is advantageous to prepare the sample as oriented, closely opposed multilayers

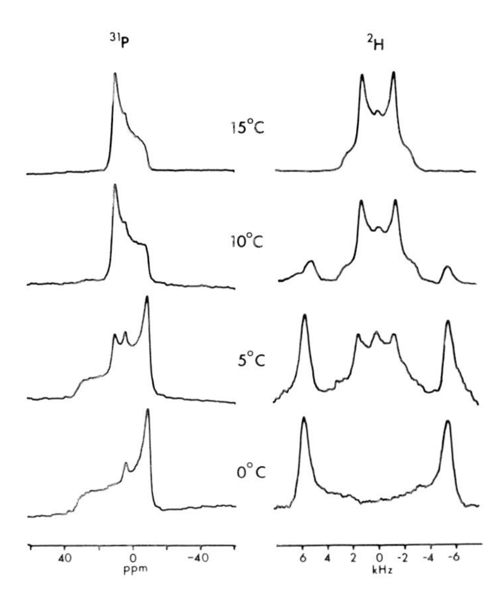

FIGURE 4. ^{31}P and ^{2}H NMR spectra as a function of temperature of fully hydrated dioleoylphosphatidylethanolamine (DOPE), which is ^{2}H labeled at the C_{11} position of the acyl chains ($[C_{11}-{}^{2}H_2]$DOPE). The ^{31}P NMR spectra were obtained at 81.0 MHz in the presence of proton decoupling, whereas the ^{2}H NMR spectra were obtained at 30.4 MHz. (From Tilcock, C. P. S. et al., *Biochemistry*; 21, 4596, 1982. With permission.)

(where the phase properties allow); while this is a relatively straightforward procedure for pure lipid systems, it is not so readily achieved with biological membranes.[24] These disadvantages aside, X-ray continues to be the method of choice for the unambiguous determination of lipid phase structure. Studies with neutron diffraction have been limited by the somewhat expensive hardware involved.

More recently, the techniques of ^{31}P NMR and ^{2}H NMR have been applied to the study of lipid polymorphism. Detection of a bilayer-H_{II} transition in DOPE ^{2}H labeled in the acyl chains ($[C_{11}-{}^{2}H_2]$-DOPE) by ^{31}P and ^{2}H NMR is illustrated in Figure 4. ^{31}P NMR has proven particularly useful and relies on the following factors. The electron distribution surrounding a phospholipid phosphorus nucleus is anisotropic and thus in a magnetic field the phosphorus experiences an orientation-dependent "shielding" termed the "chemical shift anisotropy" (CSA). For a dry lipid powder the resulting ^{31}P NMR spectrum represents a superposition of resonances from all orientations. For lipids in large (radius $\geqslant$200 nm) liquid crystalline systems the CSA is partially averaged on the NMR timescale by the rotational axial motion of the molecule. In the presence of proton decoupling, this results in a characteristic asymmetric ^{31}P NMR lineshape with a low-field shoulder and high-field peak, exhibiting an effective CSA (measured from the low field shoulder to the high field peak) of approximately -40 to -50 ppm, depending upon the lipid species, temperature, or other factors.[8,36-39] That this lineshape also represents a superposition over all orientations may be seen on examining the ^{31}P NMR spectrum from bilayers oriented between glass plates.[39] Different orientations of the plates with respect to the applied field results in individual narrow resonances with an angular-dependent chemical shift. The resonance position of the high-field peak of the unoriented system corresponds to an orientation of the lipids with their long axes normal to the applied field, while the resonance position of the shoulder corresponds to an orientation parallel to the field. The asymmetry of the lineshape arising from the unoriented system arises from the fact that there is a greater probability of finding a lipid with its long axis perpendicular, rather than parallel to, the applied field.

For a lipid in the hexagonal H_{II} phase additional motional averaging is experienced due to diffusion of the lipids around the cylinders. This results in a two-fold reduction in the effective CSA and a reversed asymmetry.[8,36,38] In particular, there is now a smaller probability of finding a cylinder of lipid parallel to the field (in which all the long axes of the lipid will be perpendicular to the field) than finding a cylinder of lipid perpendicular to the field. Thus, the relative intensities of the low- and high-field shoulders are reversed compared to the lamellar system. It should be recognized that if the rate of diffusion of the lipid molecules around the cylinder is slow relative to the spectroscopic timescale ($\sim 10^{-5}$ sec) then no additional averaging effects will be observed. This is part of the reason why ESR spin-probes are not well suited to monitoring lamellar-H_{II} transitions.[40] In systems where molecular reorientation occurs over all possible orientations (isotropic motion) in times on the order of the NMR timescale or less, motional averaging results in a single narrow ^{31}P NMR resonance. This occurs for lipids in small vesicles, micelles, inverted micelles, or other phases such as the cubic.

The reliability of ^{31}P NMR phase identifications has been questioned as theoretical considerations show that changes in the conformation of the phosphoryl segment of the lipid headgroups could result in the observed spectral differences without any change in phase structure.[41,42] However, numerous experiments have failed to demonstrate any such conformational changes, and the assignments of phase structure based on ^{31}P NMR data have been fully corroborated by X-ray diffraction for various phosphodiester lipids.[31,34,43-45] This supports the general utility of the ^{31}P NMR method. It should, however, be pointed out that ^{31}P NMR (and ^{2}H NMR) does not provide structural information per se, but tells of the motional averaging properties of the ensemble. ^{31}P and ^{2}H NMR are thus used in an extrapolative manner, the interpretation being based on prior structural assignments by X-ray or neutron diffraction.

An advantage of ^{31}P NMR over X-ray diffraction is that it is technically easier to apply, particularly to complex biological systems. Since ^{31}P is the naturally occurring isotope, no isotopic enrichment is required; as there is only one phosphorus per phospholipid, interpretation is straightforward. The principal disadvantages of the technique are that it can only be used for lipids that contain phosphorus and that it is generally not possible to examine the phase behavior of a particular lipid within a mixture of phospholipids. Exceptions include situations where the lipid of interest has a considerably different CSA[45,46] or a different spin-lattice relaxation time.[47] Finally, compared to X-ray, ^{31}P NMR phase determinations require a large amount of material (20 μmol phospholipid or more) if excessive accumulation times are to be avoided.

^{2}H NMR techniques can also be employed to identify lipid phase structure (Figure 4) Such determinations are based upon very similar principles to ^{31}P NMR. In particular, the quadrupolar moment of each deuterium gives rise to a doublet, whose splitting is dependent on the orientation of the magnetic field with respect to the ^{2}H nucleus. In non-oriented lamellar systems this gives rise to a "powder pattern" where the separation between the two major peaks is known as the quadrupolar splitting. In the absence of additional local motion, lateral diffusion around the cylinders of H_{II} phase systems results in a reduction in the quadrupole splitting by a factor of 2, whereas for isotropic motion, the ^{2}H NMR spectrum consists of a single line. Clearly, for studies of lipid polymorphism, it is possible to have only one lipid in a multicomponent system labeled with deuterium and thus be able to examine its particular phase behavior. The primary disadvantage of the application of ^{2}H NMR to the study of lipid polymorphism and membranes in general is the requirement for organic syntheses of specifically deuterated molecules; however, methods for the incorporation of deuterium at many positions within a lipid molecule are now well established.[22,63] In addition, the development of "de-Paking" techniques for the study of ^{2}H NMR spectra avoids some of these difficulties.[48] The application of ^{2}H NMR to the study of lipid and membrane structure and dynamics is well reviewed in the papers of Seelig and Seelig[22] and Davis.[48]

Freeze-fracture electron microscopy has proven to be an immensely useful technique for visualizing lipid polymorphism. Unlike spectroscopic techniques which provide ensemble averaged information, freeze-fracture electron microscopy allows the observation of local structure. This allows the direct visualization of structures in model lipid systems. With the exception of certain phosphatidylcholines that can exhibit a rippled (gel state) $P\beta'$ phase,[28] lamellar phases give rise to extended smooth fracture faces, whereas hexagonal H_{II} phases exhibit fracture planes with a corrugated appearance[49] as illustrated in Figure 3. Lipidic particle structures, apparently corresponding to interbilayer inverted micellar structures, are often found in mixed lipid systems in which one or more of the lipids prefers the hexagonal H_{II} phase in isolation. These have been suggested to be intermediate structures between the lamellar and hexagonal H_{II} phases,[50,51] and are further discussed in Section IV.G.

Freeze-fracture has potential artifacts due to the need to fix (freeze) the sample and to inhibit formation of ice crystals.[52] By way of illustration, phospholipids can undergo gel to liquid-crystal or lamellar to nonlamellar phase transitions as the temperature is raised. Thus, it is essential that the rate of cooling of the samples is as high as possible in order to "trap" the high temperature event. For example, when a sample of dimyristoylphosphatidylcholine is dip cooled (cooling rates up to 1000°C/sec) above its gel to liquid-crystal transition temperature (~22°C) a transition to the $P\beta$ gel phase occurs during cooling, giving rise to a characteristic rippled pattern. However, if the sample is quenched using ultra-rapid spray freezing techniques, the fractured lipid vesicles exhibit smooth fracture faces characteristic of liquid crystalline lamellae.[53] To inhibit the formation of ice crystals, cryoprotectants such as glycerol, ethylene glycol, polyethyleneglycol, or dimethylsulfoxide (DMSO) have been used. Such cyroprotectants can induce morphological changes in tissue[54] and lipid dispersions[55] and also affect the thermotropic properties of lipids.[56] In addition to these technical problems, some structural reorganization of lipids may occur so quickly as to escape detection by freeze-fracture.[57]

Other techniques used to monitor the lipid phase state include differential scanning calorimetry (DSC), which detects lamellar-micellar transitions in dispersions of lysophosphatidylcholines.[58] Lamellar-H_{II} transitions are also detectable by DSC for a variety of phospholipids.[59] Although limited, the data indicates that the lamellar to hexagonal H_{II} transition is characterized by an enthalpy which is approximately an order of magnitude less than that for the gel to liquid-crystal transition. Alternatively, Fourier transform infrared spectroscopy can be employed to monitor variations in certain absorption bands occurring during structural transitions. Thus, changes in the absorption bands for the chain methylene or carbonyl stretching vibrations have been used to monitor lamellar-H_{II} transitions.[60]

C. Terminology

The terminology used to describe the macroscopic structure of liquid crystals can be confusing. Here we clarify some of the more common terms pertinent to lipid polymorphism.

Liquid crystals are "mesophases" (middle states) that occur between the crystalline and fully melted or solution states. Substances that undergo transitions between mesophases as the temperature varies are said to exhibit thermotropic mesomorphism. Similarly, substances that can exist in different mesophases as the hydration is varied exhibit "lyotropic" mesomorphism. The lyotropic phases formed by soaps or phospholipids also exhibit thermotropic mesomorphism where the phase state depends upon both water content and temperature.

Two terms often encountered in the earlier literature on lipid polymorphism are "smectic" and "middle". The smectic phase is a turbid, fairly viscous phase with rheological properties similar to soaps. The essential feature of a smectic phase is that the molecules are arranged in lamellae with their long axes approximately normal to the layer plane. The structural analogy with multilamellar lipid bilayer assemblies is clear. "Middle" refers to structures that have a bulk solvent (aqueous) phase containing amphiphiles arranged in parallel tubes, packed into a hexagonal array where the polar groups are oriented outwards. This is commonly known as "hexagonal H_I". The term "reversed middle" describes the similar structure where the polar groups are oriented inwards, which is usually referred to as "hexagonal H_{II}".

Two further terms commonly encountered are "fluid isotropic" and "viscous isotropic". The term "isotropic" indicates that the phase is not optically active, giving rise to translucent dispersions. Viscous isotropic states, in general, occur in systems where the lipid is fully hydrated; the cubic phases such as those exhibited by aqueous dispersions of monoglycerides are prime examples.[61] Shipley[62] presents a summary of the major features displayed by various mesophases.

D. Polymorphic Phase Properties of Individual Lipid Species

The polymorphic phase that a given liquid crystalline lipid species adopts on hydration depends upon intrinsic factors such as the nature of the head group or acyl chains, or extrinsic factors such as hydration, temperature, pH, ionic strength, or the presence of divalent cations, other lipids, or proteins. In this section we describe the phase properties of various lipids in isolation, considering the effects of hydration, temperature, pH, or the effect of divalent cations on their phase properties where such data is available. Subsequently, we discuss the phase preferences of lipid mixtures and the regulation of phase behavior by a variety of factors. By way of introduction, Table 1 provides a summary of the phase behavior of various classes of naturally occurring membrane lipids under conditions of full hydration over temperatures within the physiological regime.

1. Phosphatidylcholine

The phase behavior of PCs has been extensively investigated.[23,36,38,64-68] At temperatures below the gel to liquid crystal transition temperature, T_c, all PCs adopt a lamellar, gel ($L\beta$) phase where the acyl chains are extended in an all-*trans* configuration. Above T_c the phase is also lamellar, but liquid crystalline ($L\alpha$) where the acyl chains are disordered. For saturated PCs with chain length greater than 14 carbons, the $L\beta$ to $L\alpha$ transition occurs via a (lamellar) intermediate $P\beta'$ phase associated with the onset of chain tilt.[28,64,65] The associated transition is referred to as the "pretransition". At low hydration ($<5\%$ water) and at elevated temperatures, an unsaturated PC such as egg PC can adopt a variety of nonlamellar phases including body-centered cubic, hexagonal H_{II}, and rhombohedral.[66-68]

2. Phosphatidylsulfocholine

Phosphatidylsulfocholine, in which the trimethyl ammonium function of the choline moiety of PC is replaced by a dimethylsulfonium group, possesses thermotropic properties intermediate between PCs and phosphatidylethanolamines. Phosphatidylsulfocholines exhibit gel to liquid crystal transitions at higher temperatures than corresponding PCs and show similar pretransitional behavior. Available evidence indicates that phosphatidylsulfocholines adopt a lamellar phase under physiological conditions.[69]

3. Phosphonolipids

Phosphonolipids have been isolated from various species of Tetrahymena. NMR evidence suggests a lamellar structure for the isolated lipids from *T. spyroformis* between -20 and 20°C, with indications of hexagonal H_{II} structure at temperatures above approximately 45°C.[70,71]

4. Sphingomyelin

X-Ray[72-75] and NMR[76] data show that hydrated sphingomyelin, either derived from bovine

Table 1
POLYMORPHIC PHASE PREFERENCES OF LIQUID CRYSTALLINE, UNSATURATED LIPIDS

	Phase preferences	
Lipid	Physiological conditions	Other conditions
Phosphatidylcholine	L	H_{II}, low hydration, and high temp
Sphingomyelin	L	
Phosphatidylethanolamine	H_{II}	L, pH $\geq$ 8.5, low temp
Phosphatidylserine	L	H_{II}, pH $\leq$ 3.5
Phosphatidylglycerol	L	H_{II}, high temp, high salt conc.
Phosphatidylinositol	L	
Cardiolipin	L	H_{II}, divalent cations, pH $\leq$ 3, high salt
Phosphatidic acid	L	H_{II}, divalent cations, pH $\leq$ 3.5, high salt
Monoglucosyldiglyceride	H_{II}	
Diglucosyldiglyceride	L	
Monogalactosyldiglyceride	H_{II}	
Digalactosyldiglyceride	L	
Cerebroside	L	
Cerebroside sulfate	L	
Ganglioside	M	
Lysophosphatidylcholine	M	
Cholesterol		Induces H_{II} phase in mixed lipid systems
Unsaturated fatty acids		Induce H_{II} phase

Note: L: Lamellar; H_{II} hexagonal H_{II}; M: micellar.

brain or the synthetic palmitoyl species, adopts only the lamellar organization over a wide range of temperatures and hydration. Among naturally occurring phospholipids, sphingomyelin exhibits the highest gel-liquid crystalline transition temperature (e.g., T_c ~35°C for erythrocyte sphingomyelin), consistent with a relatively saturated acyl chain composition.

5. Phosphatidylethanolamine

PE is the major phospholipid of eukaryotic systems which spontaneously adopts H_{II} phase structure in the presence of excess aqueous buffer at physiological temperatures. However, fully hydrated PEs adopt either lamellar or hexagonal H_{II} phase organization dependent on temperature, hydration,[27,31,38,43,59,60,67,77,78] ionic strength,[77,79] pH,[27,38,59,79,80] acyl chain unsaturation,[38,59,77,78] or the presence of branched acyl chains.[81,82] Unsaturated PEs of both synthetic and natural origin undergo a transition between the lamellar and H_{II} phases with increasing temperature as illustrated in Figure 5 for various species of naturally occurring PEs. The temperature at which this transition occurs decreases with increasing unsaturation as summarized in Table 2 for synthetic species of PE. For a homologous series of saturated diacyl or diakyl PEs, the lamellar-H_{II} transition temperature (T_{BH}) decreases with increasing chain length.[79] The T_{BH} for diakyl derivatives is lower than for the corresponding diacyl species.[79] For dilauroyl (dialkyl) PE, dehydration results in an increase in the temperature of the gel to liquid-crystal transition, but a decrease in the temperature of the lamellar-H_{II} transition.[79] A similar effect is induced by high salt concentrations.[31,77,79] The lamellar-H_{II} transition occurs either above or concomitant[9] with the gel to liquid crystal transition.

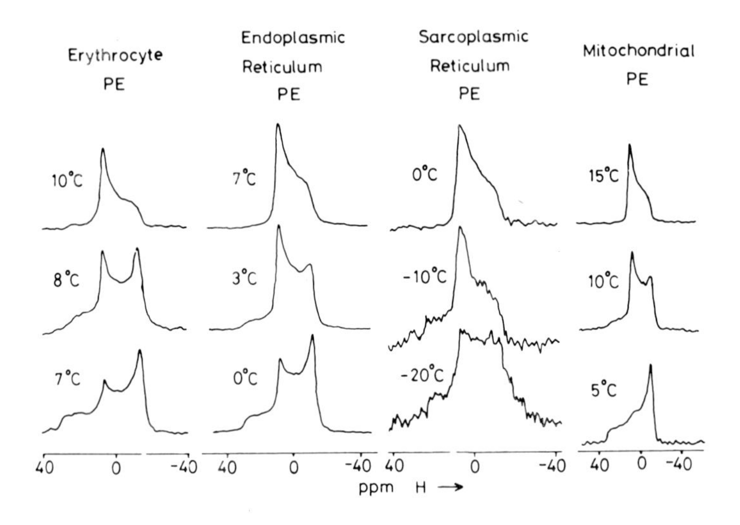

FIGURE 5. ^{31}P NMR spectra at selected temperatures illustrating bilayer to H_{II} transitions for various species of naturally occurring phosphatidylethanolamine hydrated in excess aqueous buffer. For experimental conditions and other details on erythrocyte PE, see Cullis and de Kruijff;[59] endoplasmic reticulum PE, see de Kruijff et al.,[250] and mitochondrial PE, see Cullis et al.[120] Sarcoplasmic reticulum PE was isolated from rabbit muscle sarcoplasmic reticulum. (From Cullis, P. R. et al., *Membrane Fluidity in Biology*, Vol. 1, Aloia, R. C., Ed., Academic Press, N.Y., 1983, 39. With permission.)

Table 2
BILAYER TO HEXAGONAL (H_{II}) TRANSITION TEMPERATURES (T_{BH}) FOR PHOSPHATIDYLETHANOLAMINES

Molecular species	T_{BH} (°C)	Ref.
14:0/14:0 PE	>90	78
18:0/18:0 PE	>100	77
16:0/18:1$_c$ PE	75	80
18:1$_t$/18:1$_t$ PE	55—65	21, 59, 78
18:1$_c$/18:1$_c$ PE	10	8, 59, 78
16:1$_c$/16:1$_c$ PE	0	273
18:2$_c$/18:2$_c$ PE	< − 15	78
18:3$_c$/18:3$_c$ PE	<0	273
20:4$_c$/20:4$_c$ PE	< − 30	273
22:6$_c$/22:6$_c$ PE	< − 30	273
Egg PE	25—35	59, 80, 81
PE from egg PC	40—55	60, 81
E. coli PE (wild type)	55—60	59
Erythrocyte PE (human)	8	59
Endoplasmic reticulum PE (rat liver)	7	250
Sarcoplasmic reticulum PE (rabbit)	− 10	274
Inner mitochondrial membrane PE (rat liver)	10	120
Soya PE	− 10	136
PE from soya PC	0—20	76

NMR[84,85] and FT-IR studies[60] indicate that the acyl chains of PEs are relatively more disordered in the hexagonal H_{II} phase compared to the liquid-crystalline lamellar phase. Lamellar -H_{II} transitions for PE are pH sensitive. For example, egg PE is stabilized in the lamellar phase at pH ≥8.5,[59] a finding supported by Seddon and co-workers[79] for didodecyl

PE in the presence of high salt concentrations. This corresponds to deprotonation of the primary amine of PE and thus a net negative polar head group charge.

6. Mono- and Dimethyl Phosphatidylethanolamine

Successive methylation of diacyl PE yields derivatives with characteristics intermediate between PE and PC. With successive methylation the gel to liquid-crystal transition temperature decreases for dipalmitoyl species.[86] For dioleoylmonomethyl PE, the lamellar-H_{II} transition temperature is increased more than 40°C compared to dioleoyl PE, whereas for dioleoyldimethyl PE bilayer structure is maintained up to at least 80°C.[87]

7. Phosphatidylserine

At neutral pH, where phosphatidylserine (PS) has a net negative charge, disaturated species of PS exhibit gel to liquid-crystal transition temperatures which are higher than for the corresponding PC, phosphatidylglycerol (PG), or cardiolipin (CL) with similar fatty acid compositions, but lower than for the corresponding PE. No pretransitional behavior is observed.[88,89] X-Ray and spectroscopic data all indicate that at neutral pH aqueous dispersions of saturated and unsaturated PS adopt a lamellar phase.[90-92] Protonation of the serine carboxyl function results in an increase in the gel-liquid crystal transition temperature and at pH values below the pK of the carboxyl group (~3.5), unsaturated PS adopts the hexagonal H_{II} phase.[91] The interaction of divalent cations with PS is of major interest in relation to calcium-induced lateral phase separations and membrane fusion phenomena. Addition of calcium to both saturated and unsaturated PS results in apparent dehydration of the head group region[88,91] with formation of a calcium-PS complex[88,91,92] This results in the precipitation of the lipid and the formation of distinctive cochleate lipid cylinders as visualized by freeze-fracture electron microscopy.[88,93] For the Ca^{2+}-PS complex both Raman[94] and NMR[91] data indicate that the acyl chains and PS head group are rigid and immobilized, consistent with the elevated transition temperatures observed by DSC.[88-90]

8. Phosphatidylglycerol

At neutral pH, disaturated PGs are negatively charged and exhibit gel to liquid crystal transition temperatures with associated enthalpies similar to PCs. Addition of Ca^{2+} can give rise to cochleate crystalline forms with elevated gel to liquid crystal transition temperatures.[95-98] In the liquid crystal state both X-ray[98,99] and NMR[100] evidence indicates that a variety of saturated and unsaturated PGs adopt the lamellar phase. It has been reported that dimyristoyl PG at high temperature (90°C) and in the presence of 1 *M* $CaCl_2$ will form a hexagonal H_{II} phase.[98]

9. Phosphatidic Acid

Phosphatidic acid is a central intermediate in phospholipid metabolism. Its turnover is closely correlated with stimulus-secretion coupling and the PI response.[101] It differs from the majority of the naturally occurring phospholipids in that it is a monoester rather than a diester and thus possesses two dissociable protons on the phosphate group with pKs of approximately 3.5 and 8. The thermotropic properties of saturated PAs, as a function of pH, have been extensively investigated, primarily by calorimetric techniques.[77,88,89,102-105] Disaturated PAs exhibit gel to liquid crystal transitions at temperatures higher than even the corresponding PEs, which has been attributed to intermolecular hydrogen bonding. Both saturated and unsaturated PAs adopt the lamellar phase above pH 4. In the case of saturated species this organization is maintained for the free acid form or the Ca^{2+}, Mg^{2+}, or Ba^{2+} salts.[102-109] These latter forms are usually in the gel state. Unsaturated PA adopts the H_{II} organization below pH 4, and at pH 6 the H_{II} phase can be induced by Ca^{2+}.[108,109] The H_{II} organization has been observed for a saturated PA species at high temperatures and salt concentrations.[77]

10. Cardiolipin

Cardiolipin (CL) has two negative charges associated with the head group. Saturated CLs exhibit gel to liquid crystal transition temperatures and enthalpies similar to the corresponding phosphatidylglycerols which increase monotonically with chain length.[110] At neutral pH and in the absence of divalent cations, saturated and unsaturated cardiolipins adopt a lamellar phase, whereas in the presence of calcium at concentrations above 1 m*M* or under conditions of low hydration unsaturated (naturally occurring) cardiolipin precipitates and forms a hexagonal H_{II} phase.[49,111-115,120] The temperature of the lamellar-hexagonal H_{II} transition for cardiolipins varies with both acyl chain unsaturation and salt form.[111] Hexagonal H_{II} phase structure may also be induced for bovine heart CL by decreasing the pH below 3 or, at neutral pH, by increasing the salt concentration above 1.5 *M*.[115]

11. Phosphatidylinositol

While it is a minority component of eukaryote membranes, the phosphatidylinositol (PI) effect[101] accompanying receptor-mediated phenomena suggests an important role for this lipid in cellular responses to external stimuli. ^{31}P NMR and freeze-fracture studies show that unsaturated varieties of this negatively charged lipid adopt a liquid crystalline lamellar phase upon hydration both in the absence and presence of calcium or magnesium;[116] however, in the presence of divalent cations there is some reduction in the local motion in the phosphate region.

12. Cerebrosides and Gangliosides

While the primary structure, chemistry, metabolism, and to some extent function of various glycosphingolipids is known,[117,118] relatively little is known about their phase behavior. Bovine brain cerebrosides and palmitoylgalactosylsphingosine adopt a lamellar gel phase in aqueous dispersion at physiological temperatures and lamellar liquid-crystal phases at higher temperatures.[67,119,121,122] Cerebrosides, uncharged monogalactosyl ceramides, only form the lamellar phase in water; however, human brain sulfatide, a sulfated negatively charged cerebroside, will apparently adopt lamellar and micellar phases with increasing water content.[119] With regard to gangliosides (glycosphingolipids that contain sialic acid), it is known that bovine brain gangliosides adopt the micellar phase at higher water content.[123]

13. Lysophospholipids

There have been relatively few studies of monoacyl phospholipids. Lysophosphatidylcholines, like diacyl PCs, exhibit cooperative chain melting transitions upon heating as detected by DSC.[124-127] At temperatures below the chain melting temperature, lysolecithins form a lamellar gel phase with interdigitated acyl chains[127] and then form micelles concomitant with the chain melt. NMR data[128] suggest that compared to diacyl PCs, the head group of lysolecithins experiences greater motional freedom although it is unclear whether this is due to increased motion of the molecule as a whole or greater local motional freedom about the C1 and C2 carbon of the glyceryl backbone.[128]

14. Glycolipids of Plants and Prokaryotes

Mono- and digalactosyldiglycerides (MGalDG, DGalDG) are the two major polar lipid species of photosynthetic membranes. They exhibit complex thermotropic phase behavior with evidence of metastable gel states.[129,130] Early studies by Rivas and Luzatti[131] indicated that the total galactolipids from maize chloroplasts could adopt a number of phases depending upon temperature and hydration. This included hexagonal H_{II} between -20 and 100°C at water contents of less than 10%, a cubic phase between 60 and 100°C at water contents between 10 and 20%, and a lamellar phase between 0 and 40°C for samples with more than 20% water. Similar studies upon the galactolipids isolated from pelargonium leaves[132] showed that over all temperatures and hydrations in the physiological range, MGalDG adopts a hexagonal H_{II} phase whereas digalactosyldiglyceride adopts a lamellar configuration. Studies using synthetic lipids of known acyl chain composition shown that MGalDG adopts lamellar and hexagonal H_{II} phases dependent upon the acyl chain unsaturation. The dilinolenoyl derivative spontaneously adopts a hexagonal H_{II} phase in the liquid crystal state, whereas the saturated distearoyl derivative adopts the lamellar phase. By comparison, DGalDGs adopt a lamellar phase irrespective of the degree of unsaturation of the acyl chains. Studies upon the closely related materials mono- and diglucosyldiglyceride (MGluDG, DGluDG) isolated from Acholeplasma[133,134] indicate similar phase behavior where the monoglucosy-diglyceride spontaneously adopts a hexagonal H_{II} phase while the diglucosyl derivative forms a lamellar phase. Mixtures of mono- and diglucosyldiglycerides can form cubic mesophases[135] dependent upon the MGluDG/DGluDG ratio and the acyl chain unsaturation.

E. Polymorphic Phase Behavior of Mixed Lipid Systems

Characterization of the polymorphic phase behavior of individual lipid species found in membranes leads to two important conclusions. First, all biological membranes appear to contain appreciable proportions of lipids, such as PE, which prefer the H_{II} organization in isolation under physiological conditions. Second, under appropriate conditions of ionic strength, pH, temperature, hydration, and divalent cation concentration, nearly all of the major lipid species found in membranes adopt both liquid crystalline bilayer or H_{II} phase structure. This shows that the structural preferences of lipids can be isothermally regulated by physiologically relevant factors. In this section we review the polymorphic behavior of mixed lipid systems with the following focus: there is now overwhelming evidence that the vast majority of lipids in membranes are in the bilayer organization. Any physiologically relevant departures from lamellar structure in response to various stimuli must therefore involve a relatively minor proportion of endogenous lipids. Thus, the features of particular interest are how nonbilayer lipids are stabilized into a bilayer organization in mixed lipid systems, and how sensitive this structure is to variables such as ionic strength, pH, and divalent cations.

1. Mixtures of Phosphatidylcholine and Phosphatidylethanolamine

Phosphatidylcholine can stabilize unsaturated PEs into a lamellar configuration in a manner which is sensitive to the acyl chain unsaturation of the lipids, the temperature, and the ratio of PC to PE.[34,38,43,85,136] In general, increased unsaturation of the PE or PC component, high temperatures, or a low molar ratio of PC to PE are all factors which favor destabilization of lamellar structure and formation of hexagonal H_{II} phase, intermediate cubic, or inverted micellar structure. Approximately 20 to 25 mol% PC in a given PC-PE mixture is usually sufficient to stabilize a lamellar phase, although it should be noted that the lamellar phase may be metastable.[85] The ability of egg PC to stabilize soya PE in a lamellar phase is shown in Figure 6. For the equimolar mixture, ^{31}P NMR indicates a lamellar structure for the combined lipids. At intermediate PC to PE ratios a resonance indicative of isotropic motional averaging is observed which may arise from inverted micellar "intermediate" structures.

This ability of PC to stabilize unsaturated PEs into a bilayer appears to be a general consequence of the bilayer-preferring properties of the PC component, rather than due to any specific interaction. In mixtures of (liquid crystalline) saturated and unsaturated PEs, for example, the saturated (bilayer preferring) dimyristoyl PE (DMPE) species can stabilize an overall bilayer structure.[78] It is interesting to note that at lower temperatures (where DMPE adopts the gel phase) lateral segregation of the gel state PE removes this bilayer stabilization capacity, allowing the unsaturated dioleoyl PE (DOPE) species to revert to the H_{II} organization it prefers in isolation.

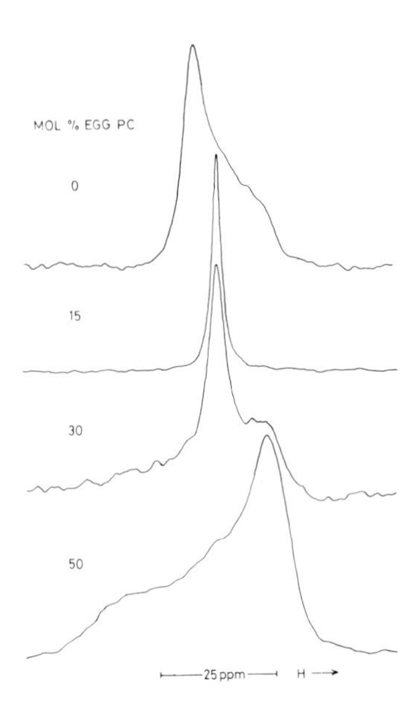

FIGURE 6. 36.4 MHz ^{31}P NMR spectra of aqueous dispersions of soya phosphatidylethanolamine/egg yolk phosphatidylcholine mixtures at 30°C. The mol% phosphatidylcholine (with respect to total phospholipid) contained in the various samples are (A) 0 mol%, (B) 15 mol%, (C) 35 mol%, and (D) 50 mol%. All dispersions contained 25 m*M* Tris/acetic acid (pH 7.0) and 2 m*M* EDTA. (From Cullis, P. R. and de Kruijff, B., *Biochim. Biophys. Acta*, 507, 207, 1978. With permission.)

2. Mixtures of Phosphatidylethanolamine with Acidic Phospholipids

As previously indicated, the structural preferences of acidic (negatively charged) lipids are sensitive to variables such as ionic strength, pH, and divalent cations. Given that negatively charged phospholipids adopt the bilayer phase at neutral pH in the absence of divalent cations, it may be expected that they can stabilize bilayer structure, for instance, in mixtures with unsaturated PE. It may be equally expected that factors tending to segregate the acidic lipids, or which reduce their stabilizing ability, induce H_{II} structure. That this is the case is illustrated in Figure 7 for mixtures of soya PE with unsaturated varieties of PS, PI, PA, or CL, where 15 to 30% of the acidic lipid is sufficient to stabilize an overall lamellar organization. The subsequent introduction of Ca^{2+} triggers H_{II} phase formation in all these mixed systems. The mechanism involved differs. In the PE-PS systems Ca^{2+} can induce lateral segregation of the PS into crystalline cochleate domains, thus allowing the PE component to revert to the H_{II} organization.[45,47,137] Alternatively, for mixtures of PE with CL or PA the addition of Ca^{2+} converts the bilayer stabilizing species to a H_{II} species, which permits the whole system to undergo a bilayer-H_{II} transition. In the case of the PG-PE system, Ca^{2+} appears to reduce the bilayer stabilizing capacity of PG (possibly by reducing electrostatic repulsion effects), again allowing the whole mixture to adopt H_{II} structure.[100] For PI-PE systems results are more equivocal, and the mechanism could correspond to that observed for PG or a partial lateral segregation of PI.[116] It is interesting to note that PI is the most effective acidic lipid for stabilizing bilayer structure, which may be related to the large PI head group.

The properties of unsaturated PE-PS systems are of particular interest given that they are major lipid components of systems such as the inner monolayer of the erythrocyte.[5] PS is usually found to be more unsaturated than other phospholipids in eukaryotic membranes and it is therefore important to characterize the ability of Ca^{2+} to sequester more unsaturated

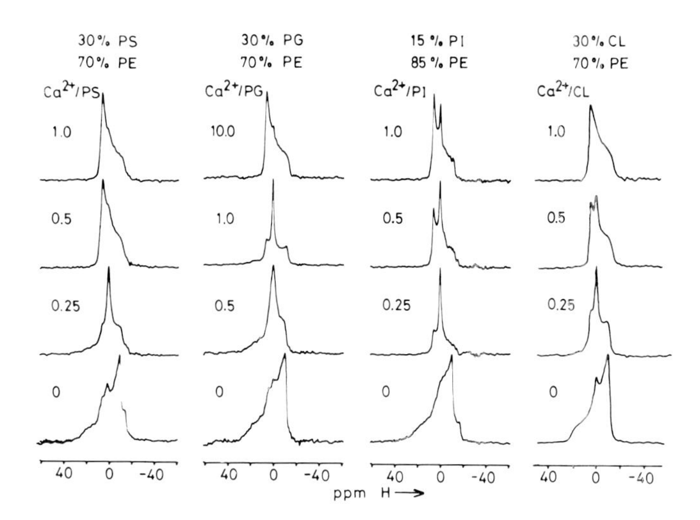

FIGURE 7. ^{31}P NMR spectra arising from mixtures of acidic phospholipids with soya phosphatidylethanolamine (a polyunsaturated PE derived from soya PC[76]) in the presence of various molar ratios of Ca^{2+}. All samples were prepared from 50 μmol total phospholipid hydrated in MLV form by vortex mixing. The Ca^{2+} was added as aliquots from a 100 m*M* stock solution. For further details regarding (egg) PS-PE, see Tilcock and Cullis;[47] (egg) PG-PE, see Farren and Cullis;[100] (soya) PI-PE, see Nayar et al.,[116] and (beef heart) CL-PE, see deKruijff and Cullis.[233] (From Cullis, P. R. et al., *Membrane.Fluidity in Biology*, Vol. 1, Aloia, R. C., Ed., Academic Press, New York, 1983, 39. With permission.)

varieties of PS in mixed lipid systems. Whereas Ca^{2+} can segregate DOPS in DOPS-DOPE (1:1) systems, such effects are not observable in equimolar mixtures of $18:2_c/18:2_c$ (DLPS) with DOPE.[45] The corresponding inability of Ca^{2+} to induce H_{II} phase structure in DOPE-DLPS systems is illustrated in Figure 8. This suggests that Ca^{2+}-induced segregation of the unsaturated varieties of PS found in biological membranes may not be readily achievable.

It is important to note that any factor reducing the bilayer stabilizing capacity of the acidic lipid will induce H_{II} phase structure. Thus at lower pH values where unsaturated PS tends to adopt H_{II} organization, PE-PS systems would be expected to adopt H_{II} organization. This is illustrated in Figure 9, and similar effects would be expected in unsaturated mixtures containing PA and other acidic lipids.

3. Influence of Cholesterol and Other Sterols

In bilayer systems cholesterol exhibits a well-characterized ability to inhibit formation of the crystalline (gel) state and to decrease the permeability of liquid crystalline lamellar systems.[138] In unsaturated PE systems equimolar cholesterol levels either reduce or have little effect on bilayer-H_{II} transition temperatures.[59] However, in unsaturated PE-PC systems cholesterol exhibits a remarkable ability to destabilize bilayer structure and induce the hexagonal H_{II} phase.[38,85,139,142,145] This is illustrated in Figure 10 for equimolar DOPE-DOPC mixtures, where the presence of equimolar cholesterol (with respect to phospholipid) results in complete H_{II} organization. In mixtures stabilized by less PC, less cholesterol is required for destabilization. In a DOPE-DOPC (4:1) mixture, for example, 20 mol% cholesterol induces complete H_{II} phase structure, and as little as 2 mol% can inhibit lipid structures

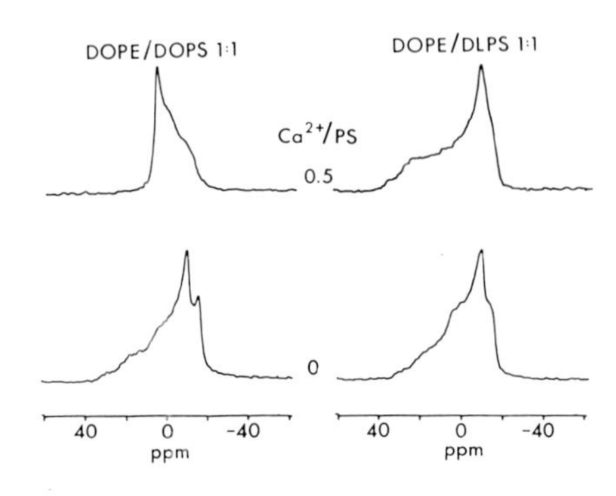

FIGURE 8. 81.0 MHz ^{31}P NMR spectra obtained at 30°C from aqueous DOPE-DOPS (1:1) and DOPE-DLPS (1:1) dispersions in the absence and presence of Ca^{2+}. The Ca^{2+}/PS ratio refers to the molar ratio of Ca^{2+} to phosphatidylserine.

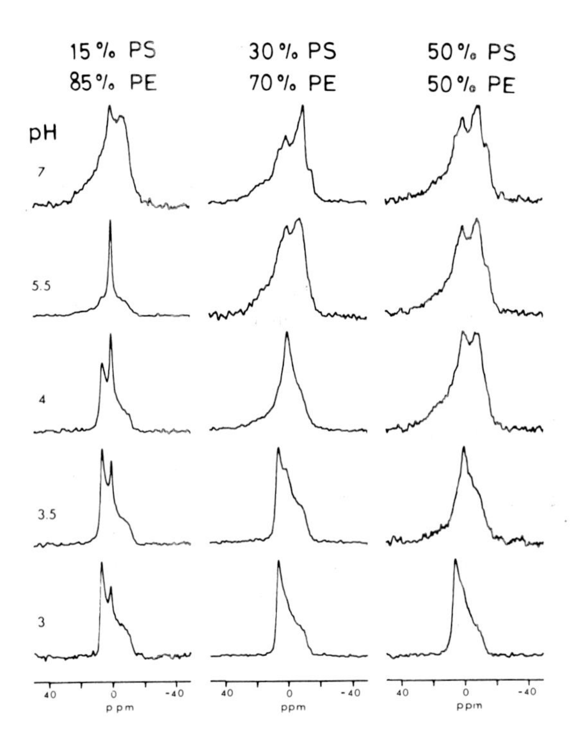

FIGURE 9. 81.0 MHz ^{31}P NMR spectra at 30°C obtained from aqueous dispersions of soya PE containing (a) 15% soya PS, (b) 30% soya PS, and (c) 50% soya PS at pH values 7.0, 5.5, 4, 3.5 and 3. (From Tilcock, C. P. S. and Cullis, P. R., *Biochim. Biophys. Acta*, 641, 189, 1981. With permission.)

giving rise to an isotropic resonance and produce detectable H_{II} phase.[85] It may be noted that sphingomyelin is a much more effective bilayer stabilizing agent than PC when cholesterol is present.[76]

Other sterols such as coprostanol, epicoprostanol, stigmasterol, ergosterol, and androstanol can also induce hexagonal H_{II} structure for PC-PE mixtures,[277] indicating that the conformation of the 3β-hydroxyl is not important for such an ability. Further, neither a coplanar

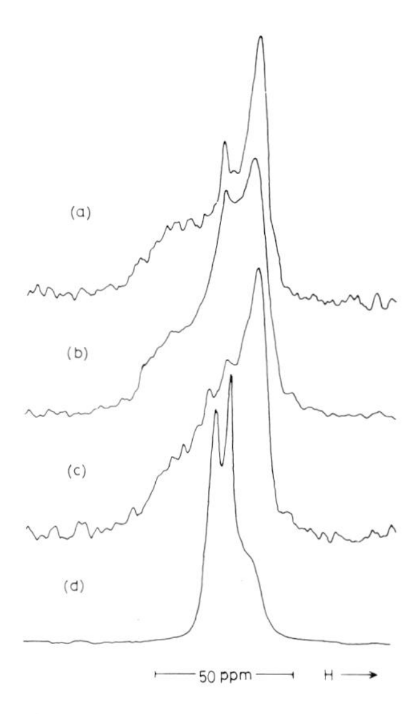

FIGURE 10. 36.4 MHz ^{31}P NMR spectra obtained at 30°C from equimolar mixtures of $18:1_c/18:1_c$ phosphatidylethanolamine (DOPE) with $18:1_c/18:1_c$ phosphatidylcholine (DOPC) in the presence of (a) 0%, (b) 15%, (c) 30%, and (d) 50 mol% cholesterol. All dispersions contained 10 m*M* Tris-acetic acid (pH 7.0) and 2 m*M* EDTA. (From Cullis, P. R. et al., *Biochim. Biophys. Acta*, 513, 21, 1978. With permission.)

ring system nor the presence of a side chain is required for the destabilizing effects of the sterols. It has been noted[140] that there is a direct relationship between the lamellar-hexagonal H_{II} transition temperature and the mean molecular areas occupied by the sterol.

These bilayer destabilizing effects of cholesterol are quite general in that corresponding effects are observed for unsaturated PE-PS systems. Thus, equimolar cholesterol can induce essentially complete H_{II} phase structure in soya PE-soya PS (2:1) systems.[137] However, the most interesting feature concerns the enhanced sensitivity of the net structure to various cations and the influence of cholesterol on PS segregation. A particularly important point concerns the inhibition of Ca^{2+}-induced segregation of PS into crystalline cochleate domains. In DOPE-DOPS (1:1) systems, for example, the presence of 50 mol% cholesterol (with respect to total phospholipid) completely inhibits segregation of PS by Ca^{2+}.[45] In particular, addition of Ca^{2+} then triggers H_{II} phase organization where the PE and the PS are present in the H_{II} phase matrix. This conclusion is supported by a great deal of evidence[45] including the fact that ^2H NMR studies of DOPE-DOPS-cholesterol (1:1:1) systems containing ^2H-labeled DOPS reveal a factor of 2 decrease in the quadrupolar splitting on addition of Ca^{2+}; this indicates H_{II} phase structure for the PS component (Figure 11).

In mixtures containing more unsaturated varieties of PS ($18:2_c/18:2_c$ PS, DLPS) the presence of cholesterol results in an ability of Ca^{2+} to trigger bilayer-H_{II} transitions as illustrated in Figure 12. This again proceeds by a direct incorporation of the DLPS into the H_{II} phase, and contrasts with the situation in the absence of cholesterol (Figure 8). The presence of cholesterol also results in an enhanced structural sensitivity to other cations. For example, Mg^{2+} cannot usually trigger bilayer-H_{II} transitions in unsaturated PE-PS systems. However, in analogous systems containing cholesterol,[45,137] such effects are observed. Further, Mg^{2+} can then act as an adjunct for Ca^{2+}, reducing the Ca^{2+} concentrations required for stimulating H_{II} structures from 2 m*M* to 200 μ*M* or less.[137] Similarly, whereas high salt concentrations do not normally influence PE-PS phase behavior, the presence of 1 *M* NaCl results in H_{II} structure for soya PE-soya PS-cholesterol (1:1:2) systems.[137]

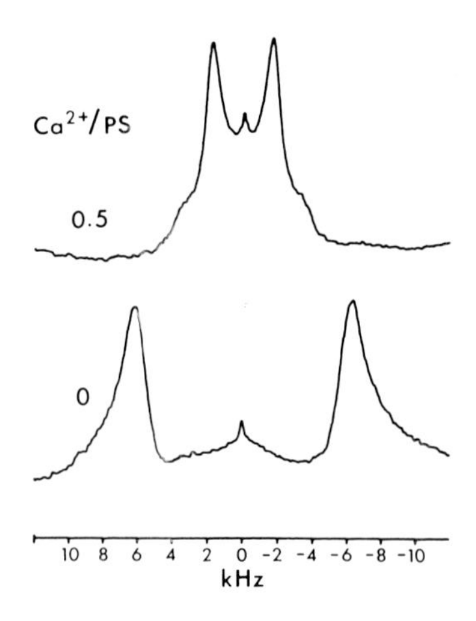

FIGURE 11. 30.7 MHz ^{2}H NMR spectra obtained at 30°C for a DOPE-[$C_{11} - {}^2H_2$]DOPS-cholesterol (1:1:1) mixture in the absence and presence of Ca^{2+} (Ca^{2+}/DOPS molar ratio = 0.5). (From Tilcock, C. P. S. et al., *Biochemistry*, 23, 2696, 1984. With permission.)

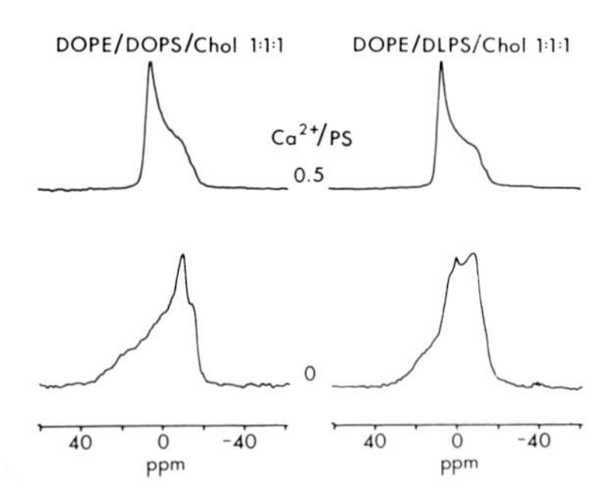

FIGURE 12. 81.0 MHz ^{31}P NMR spectra obtained at 30°C from aqueous DOPE-DOPS-CHOL (1:1:1) and DOPE-DLPS-CHOL (1:1:1) dispersions in the presence of varying amounts of Ca^{2+}. The ratio Ca^{2+}/PS refers to the molar ratio of Ca^{2+} to phosphatidylserine. (From Tilcock, C. P. S. et al., *Biochemistry*, 23, 2696, 1984. With permission.)

4. Influence of Anesthetics and Other Lipophilic Compounds

Anesthetics constitute a chemically diverse group of lipophilic compounds which are able to inhibit function of the Na^+ channel of excitable membranes in a nonspecific manner. The close correlation between anesthetic potency and solubility in a hydrocarbon environment strongly suggests that these agents act by perturbing some property of the lipid domain.[141] The local (positively charged) amine anesthetics exhibit diverse effects on the polymorphism of lipid systems containing negatively charged lipids, where dibucaine and chlorpromazine, for example, induce H_{II} organization in unsaturated CL and PA[112,142] systems. Alternatively, in unsaturated PE-PS systems, dibucaine stabilizes the bilayer organization and inhibits the ability of Ca^{2+} to trigger H_{II} organization.[143] In contrast, neutral anesthetics such as the longer chain ($C \geq 6$) *n*-alcohols and alkanes promote H_{II} organization in egg PE systems.[144,145]

These observations lead to the general conclusion that any lipophilic molecule will affect the polymorphic preferences of appropriate lipid systems. Thus, long-chain unsaturated fatty

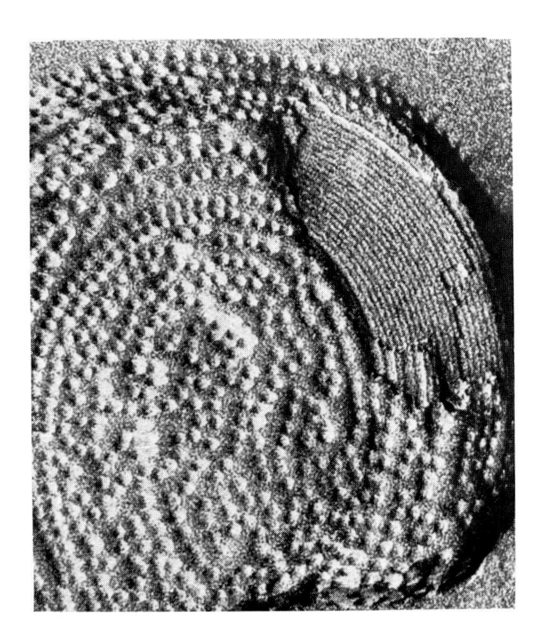

FIGURE 15. Freeze-fraction micrograph of a cardiolipin-egg phosphatidylcholine (1:1) dispersion incubated in the presence of Mn^{2+}. (From Verkleij, A. J. et al., *Biochim. Biophys. Acta*, 600, 620, 1980. With permission.)

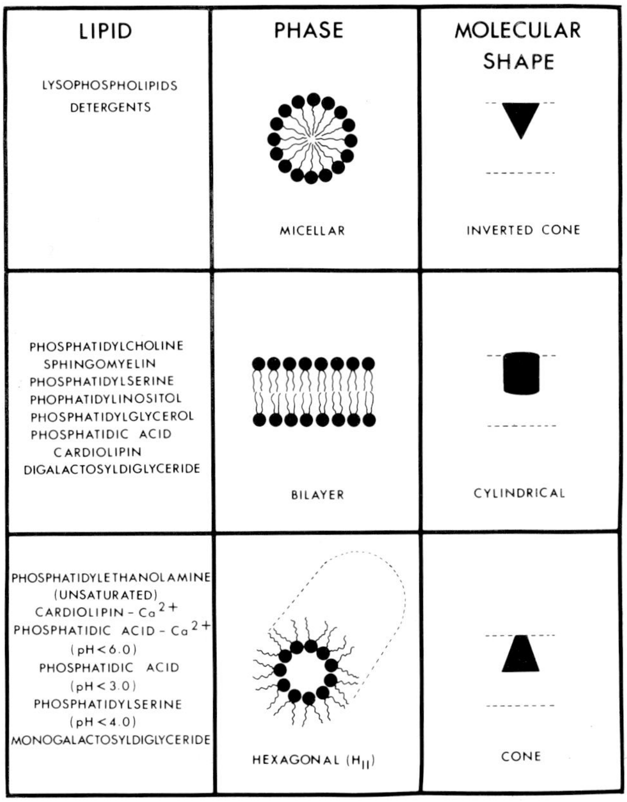

LIPID	PHASE	MOLECULAR SHAPE
LYSOPHOSPHOLIPIDS DETERGENTS	MICELLAR	INVERTED CONE
PHOSPHATIDYLCHOLINE SPHINGOMYELIN PHOSPHATIDYLSERINE PHOPHATIDYLINOSITOL PHOSPHATIDYLGLYCEROL PHOSPHATIDIC ACID CARDIOLIPIN DIGALACTOSYLDIGLYCERIDE	BILAYER	CYLINDRICAL
PHOSPHATIDYLETHANOLAMINE (UNSATURATED) CARDIOLIPIN - Ca^{2+} PHOSPHATIDIC ACID - Ca^{2+} (pH < 6.0) PHOSPHATIDIC ACID (pH < 3.0) PHOSPHATIDYLSERINE (pH < 4.0) MONOGALACTOSYLDIGLYCERIDE	HEXAGONAL (H_{II})	CONE

FIGURE 16. Polymorphic phases and corresponding dynamic molecular shapes of component lipids.

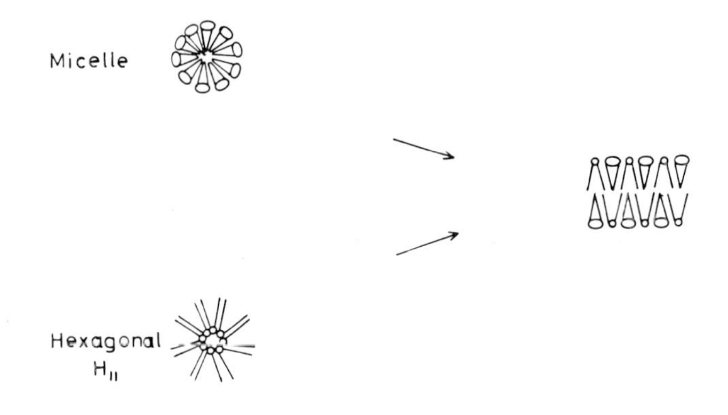

FIGURE 17. A net bilayer structure arising from mixtures of cone-shaped (H_{II} phase) lipids and inverted cone (micellar) lipids due to shape complementarity effects.

It should be recognized that "shape" is an inclusive term, reflecting the effects of size and motion of polar and apolar regions, head group hydration and charge, hydrogen bonding processes and effects of counterions, among other possibilities. The cone shape of (unsaturated) PE, for example, can be ascribed to a smaller, less hydrated head group (in comparison to PC) and possibilities of intermolecular hydrogen bonding,[148] which would further reduce the area per molecule in the head group region. Similarly, the smaller head groups of monoglucosyl and monogalactosyldiglycerides in comparison to bilayer-forming diglucosyl and digalactosyldiglycerides can be considered to promote H_{II} organization. Alternatively, increased temperatures or unsaturation, which increase acyl chain motion, may be considered to increase the hydrophobic volume and thus engender H_{II} organization as is observed for PEs. Conversely, factors which reduce the effective head group area (without inducing compensatory changes in the hydrophobic volume v by crystallization of the acyl chains) would be expected to favor H_{II} phase structure. This rationalizes such phenomena as Ca^{2+}-induced bilayer H_{II} transitions in liquid crystalline CL or PA systems, for example, where Ca^{2+} reduces inter-head group electrostatic repulsion, and thus reduces the effective a_o leading to the phase transition. In PE systems stabilized by PS, similar effects are predicted (and observed) when lateral segregation of PS is inhibited by factors such as cholesterol. Factors leading to protonation of the PS or PA head groups reduce the net head group charge, thus favoring the H_{II} organization observed at low pH.

An interesting prediction of the shape hypothesis is that appropriate proportions of inverted cone (micellar) lipid and cone-shaped (H_{II} phase) lipid in mixed systems should result in a net lamellar structure due to the shape complementarity considerations illustrated in Figure 17. That this is the case is indicated by the observation that egg PE can be stabilized in a bilayer organization by varying concentrations of detergents such as octylglucoside, Triton® X-100, and deoxycholate.[149] Similar considerations explain the net lamellar structure observed for mixtures of lysolecithin with cholesterol[150] and fatty acids.[151]

V. LIPID POLYMORPHISM AND MEMBRANE FUSION

A. Introduction

The most remarkable aspect of the results summarized in the previous section is that apparently all membrane lipids can adopt nonbilayer structure, either in isolation or in mixed systems, in situations which are not unduly removed from physiological conditions. The further observation that this polymorphism can be related to a generalized "shape" property of lipids leads to two classes of functional roles for liquid crystalline lipids in membranes.

The first concerns the ability to adopt nonbilayer structure per se, whereas the second concerns roles that lipids with differing shapes may satisfy in an intact bilayer membrane.

In this section we discuss membrane events which may rely on the ability of endogenous lipids to adopt transitory or long-lived nonbilayer arrangements. The primary focus will concern membrane fusion processes and related events. There is suggestive evidence that nonbilayer structure may play direct roles in protein insertion[152] and certain membrane transport processes;[153] however, these possibilities remain speculative. Evidence in favor of a role of nonlamellar intermediates in fusion, on the other hand, is strong. This is intuitively reasonable, as it is clearly a topological impossibility for two membrane-bound systems to fuse in order to achieve mixing of internal compartments without a transitory departure from bilayer structure at the fusion interface. Nonbilayer lipid structures provide attractive possibilities as transitory intermediates in the fusion event.

B. Background

Two models which have been previously proposed to explain fusion events are indicated in Figure 18. Here we briefly discuss the evidence leading to these models and their various limitations.

1. Lipid Disorder and Membrane Fusion

The "lipid disorder" model (Figure 18) was largely based on studies of cellular fusion induced by lipid soluble "fusogens", which are mainly unsaturated long-chain carboxylic acids and their esters.[154] Such fusogenic lipids partition into the cell membranes, resulting in an increased disorder in the acyl chain region.[155,156] It was therefore proposed that fusion events in vivo rely on segregation of protein from the fusion interface, which exposed closely apposed lipid bilayers. Local segregation or production of factors which increase lipid "disorder" (e.g., lysolipids, fatty acids) were then thought to allow the bilayer to intermix, resulting in fusion.

Certain aspects of this model were based on freeze-fracture observations that intramembranous particles (presumed to represent integral proteins) appear to be segregated from the fusion interface;[156] however, studies employing ultrafast freezing procedures show little or no change in particle distributions.[157] More importantly, as indicated later in this section, lipid-soluble fusogens are able to induce H_{II} phase organization in model and biological membrane systems.[158,159] This suggests that the "mixing" event of Figure 18 actually reflects formation of nonlamellar intermediate structure (see Section V.C).

2. Lateral Phase Separation and Membrane Fusion

Evidence supporting this mechanism has been reviewed elsewhere.[160,161] Briefly, the lateral phase separation model was stimulated by observations that SUVs composed of synthetic, saturated varieties of PC fuse to form larger systems when incubated near the lipid gel-liquid crystal transition temperature, T_c.[162] This is accompanied by increased membrane permeability, which is attributed to packing defects arising from the coexistence of gel and liquid crystalline domains ("lateral phase separation").

These observations led to the possibility that lateral phase separation of a particular lipid component into crystalline domains is a general requirement for fusion events. Attempts to correlate this with in vivo fusion events involved the observation that Ca^{2+} is often required for fusion, and that Ca^{2+} can induce isothermal gel to liquid crystalline phase transitions for certain acidic phospholipids such as PS. As summarized previously, X-ray analyses and ^{31}P NMR studies indicate that the crystalline PS-Ca^{2+} complex exhibits properties similar to anhydrous PS. This strong and specific interaction of PS with Ca^{2+} can result in Ca^{2+}-induced lateral phase separation of PS in mixed lipid systems.[160,162] Papahadjopoulos and collaborators have demonstrated that PC-PS vesicles increase in size in the presence of Ca^{2+}

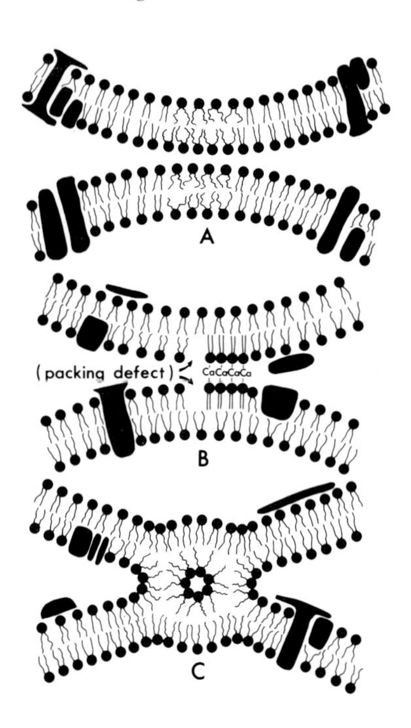

FIGURE 18. Intermediate structures postulated for membrane fusion models relying on (A) lipid "disordering" effects, (B) lateral phase separation effects, and (C) nonbilayer lipid structures.

for PS contents greater than 50 mol%. Calorimetric studies indicated that the mechanism involved is fusion rather than monomolecular exchange,[162] and that Ca^{2+}-induced gel phase domains are necessary for fusion to proceed.

These studies led to the fusion model shown in Figure 18, where Ca^{2+} induces a phase separation of acidic lipids resulting in "structural defects" exposing hydrocarbon chains to water. These "high energy" contact points between water and the bilayer interior are then thought to promote fusion.

Difficulties with this model include, first, the fact that in binary mixtures of PS with PC or PE, the levels of Ca^{2+} required to induce lateral segregation and crystallization of PS are 2 m*M* or higher.[45] These levels are higher than are likely to occur in the cell cytoplasm. Second, as indicated in the previous section, in more complex systems containing more unsaturated species of PS (naturally occurring PS is usually very unsaturated) Ca^{2+} is unable to induce lateral segregation of PS even at levels in excess of 20 m*M*.[45] Finally, the mechanism lacks generality. Calcium-induced crystallization or lateral phase segregation of naturally occurring phospholipids other than PS has not been observed, restricting the mechanism to PS containing biological membranes. Other biological stimuli for fusion, such as changes in pH, do not appear to result in lateral segregation events.

C. Fusion of Membranes by Lipid-Soluble Fusogens

The possibility that fusion processes in vivo proceed via nonbilayer inverted micelle or inverted cylinder (H_{II} phase) intermediates gained initial credence with the observation[158] that oleic acid (a member of the group of lipid-soluble fusogens identified by Ahkong et al.[154] induces H_{II} phase structure in erythrocyte (ghost) membranes at membrane concentrations close to those required to induce erythrocyte fusion in vitro. Further comparative studies[159] of the influence of lipid-soluble fusogens and chemically related nonfusogens revealed that fusogenic compounds induced H_{II} organization in total (isolated) erythrocyte

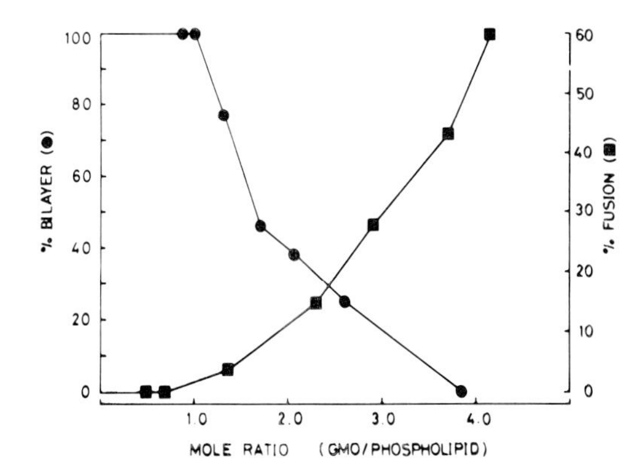

FIGURE 19. A comparison of the extent of fusion between erythrocytes and the amount of phospholipid remaining in the bilayer phase in erythrocyte (ghost) membranes at various membrane concentrations of glycerol monoleate (GMO) ● The percentage of membrane phospholipids in extended bilayers as indicated by ^{31}P NMR. ■ The percentage of fusion between erythrocytes following incubation for 2 hr with various concentrations of GMO which resulted in the indicated membrane-associated amounts of GMO. (From Hope, M. J. and Cullis, P. R., *Biochim. Biophys. Acta*, 640, 82, 1981. With permission.)

lipids and the ghost membrane, whereas the nonfusogens did not. Further, as illustrated in Figure 19, a close correlation is observed between the extent of fusion induced by the fusogen glycerol monooleate between erythrocytes and the amount of nonbilayer organization observed in erythrocyte ghost membranes containing similar membrane levels of fusogen.

These studies provided strong support for the fusion mechanism involving nonbilayer inverted intermediates illustrated in Figure 20. Briefly, once close intermembrane apposition has been achieved, the two outer monolayers of the adjoining membrane combine to form the "inverted" structure. This type of structural transition is analogous to the transition from bilayer to inverted cylinder structure in bilayer H_{II} transitions. Subsequently, a reversion to net bilayer structure can result in the completed fusion event. The inverted intermediate structure could be a short inverted cylinder or an inverted micelle (lipidic particle). The fact that bilayer inverted micelle or bilayer inverted cylinder (H_{II} structure) transitions can be induced in a wide variety of lipid mixtures by various physiologically relevant factors suggests that this mechanism may apply with some generality to fusion events in vivo.

D. Lipid Polymorphism and Fusion of Model Membranes

Biological membranes do not appear to rely on the presence of exogenous fusogens to initiate fusion events. However, in Section IV it was demonstrated that PE, which prefers the H_{II} organization in isolation, can be stabilized in the bilayer organization by all the major acidic phospholipids common to biological membranes, including PS, PI, PA, PG, and CL. The subsequent introduction of Ca^{2+} induces an isothermal bilayer to hexagonal H_{II} phase transition. A potential relation between such phase modulation and fusion was demonstrated for LUVs consisting of equimolar CL and PC.[50] On addition of Ca^{2+} these vesicles fused to form larger systems, an event accompanied by the appearance of lipidic particles which were often located at the fusion interface. It was concluded that fusion occurred via an intermediary formation of nonlamellar structure (possibly inverted micellar) between closely apposed vesicles. This is consistent with the fusogen effects and suggests that any factor which enhances the ability of lipids to adopt hexagonal H_{II} organzation will also promote fusion.

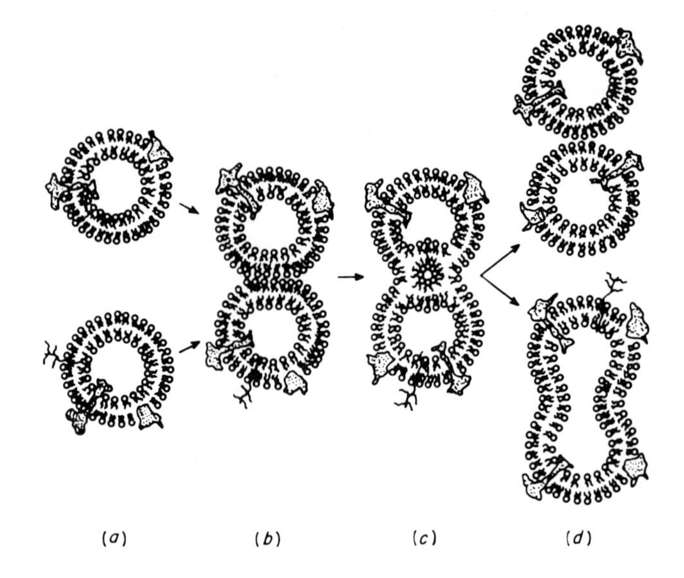

FIGURE 20. Proposed mechanism of membrane fusion proceeding via an inverted cylinder or inverted micellar intermediate. The process whereby the membranes come into close apposition (a—b) is possibly protein mediated, whereas the fusion event itself (b—c) is proposed to involve formation of an inverted lipid intermediate.

This hypothesis is further supported[163] by the micrographs shown in Figure 21, which illustrate the behavior of SUV systems composed of PE-PS, PE-PA, PE-PG, and PE-CL when dialyzed against a Ca^{2+} buffer. All these SUV systems fuse to form larger systems, with the concomitant appearance of lipidic particles which are often localized to the fusion interface. The similar morphology exhibited by these fusing systems indicates a common fusion mechanism. However, the system containing PS is the only one for which a Ca^{2+}-induced crystallization of the acidic lipid species can occur and is thus the only one for which the phase separation mechanism could apply. Thus the fact that all these systems undergo fusion by an apparently similar process, with the common feature that Ca^{2+} facilitates H_{II} phase formation, suggests that factors conducive to H_{II} phase formation are of more basic importance. This conclusion is also supported by the results of Figure 22 where incubation of PA-PS and PE-PS SUV systems at pH 4 induces fusion which again appears to proceed via a similar mechanism. This can be attributed to the abilities of lower pH values to enhance the ability of PS and PA to adopt H_{II} structures (Section IV.D). Similar conclusions apply to the temperature-induced fusion of PC-PE-cholesterol LUVs, where incubation at higher temperatures, which enhances formation of nonbilayer structure, also results in fusion.

It is logical to suppose that these Ca^{2+}, pH-, or temperature-induced fusion events involve some local enrichment of the nonbilayer-preferring species (PE) at the fusion interface. This would be expected to facilitate formation of the inverted nonlamellar intermediate. This may not be the case, however, as indicated by the ideal mixing properties exhibited by liquid crystalline PC-PE and PE-PS-cholesterol systems summarized in the previous section. It appears more likely that the factors initiating fusion reduce the overall ability of the bilayer-stabilizing species to maintain lamellar structure and thus enhance the probability of inverted structure formation (and subsequent fusion) at intermembrane contact points without segregating particular lipids into local domains.

A further point concerns the absolute concentrations of Ca^{2+} required to induce nonlamellar structure and fusion. In the absence of other factors, the Ca^{2+} concentrations required to induce nonlamellar structure in systems such as those of Figure 21 are 1 m*M* or higher.

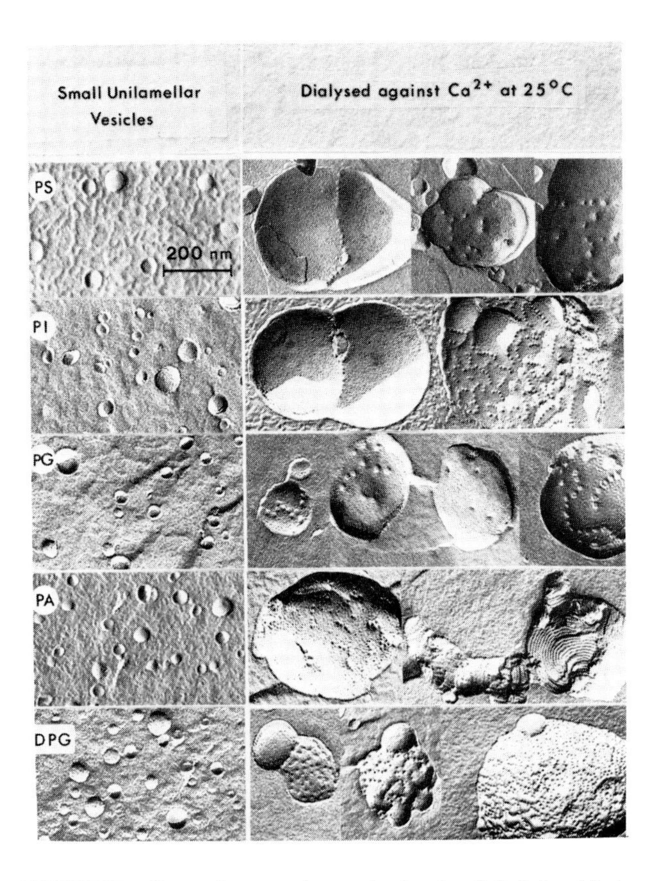

FIGURE 21. Freeze-fracture micrographs showing Ca^{2+}-induced fusion of unilamellar vesicles consisting of soya phosphatidylethanolamine and a negatively charged phospholipid in the molar ratio of 4:1. Vesicles were prepared by sonication in the absence of Ca^{2+} and subsequently dialyzed at 25°C against a buffer containing 10 m*M* Ca^{2+}. Samples were removed at various time intervals and frozen in the presence of glycerol (25% v/v) by plunging into a liquid-solid freon slush. Replicas were prepared employing standard procedures. The acidic phospholipids are PS (egg phosphatidylserine), PI (soya phosphatidylinositol), PG (egg phosphatidylglycerol), PA (dioleoylphosphatidic acid), and DPG (beef heart cardiolipin). (From Hope, M. J. et al., *Biochem. Biophys. Res. Commun.*, 110, 15, 1983. With permission.)

However, it has been shown[137] that lower Ca^{2+} concentrations (>200 μ*M*) can trigger H_{II} phase structure in PS-PE containing cholesterol when in the presence of 2 m*M* Mg^{2+}. With regard to in vivo fusion process such as those involved in exocytosis, such Ca^{2+} concentrations are higher than those usually encountered in the cytosol. However, local transitory Ca^{2+} concentrations on this order are potentially possible following inward Ca^{2+} fluxes triggering the exocytotic event.

In summary, factors which promote formation of the H_{II} phase structure in various multilamellar systems initially result in fusion of corresponding SUV and LUV systems. The observation that this fusion is accompanied by formation of the lipidic particle structure may be employed to suggest that lipid particles, interpreted as inverted micelles, play intermediary roles in the fusion event. In this regard recent experiments[164] employing fast-freezing techniques suggest that lipidic particles appear subsequent to the fusion process. It is, however, quite possible that inverted micelles occurring as intermediaries in fusion exhibit a sufficiently

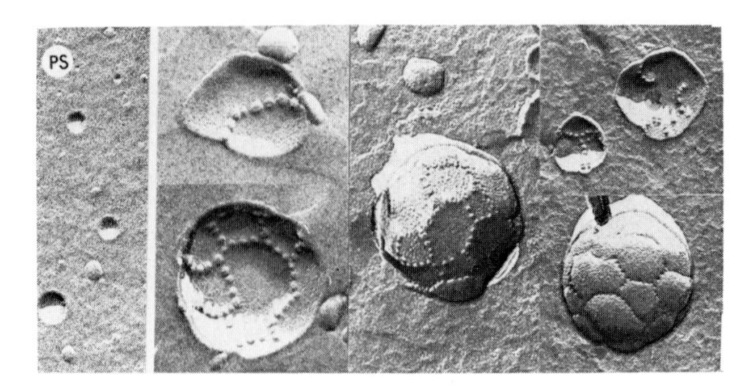

FIGURE 22. Freeze-fracture micrographs of small (sonicated) unilamellar vesicles composed of 80 mol% soya phosphatidylethanolamine (PE) and 20 mol% of egg phosphatidylserine (PS) before (left) and after (right) dialysis at 20°C against a pH 3 buffer. The length of dialysis time was 40 min. Samples were mixed with glycerol (25% vov/vol) and quenched from 20°C. It may be noted that large systems formed for the PE-PS systems after dialysis closely resemble the honeycomb type of structure (see Figure 26). (From Hope, M. J. et al., *Biochem. Biophys. Res. Commun.*, 110, 15, 1983. With permission.)

short lifetime such that they will escape detection by freeze-fracture.[57] Lipidic particles could well represent stable versions of such transitory intermediates. In any event, we take the general view that the presence of lipids which prefer H_{II} phase organization satisfies two fundamental requirements for fusion, irrespective of the detailed nature of the intermediate structure. These are, first, a condition of relatively low hydration allowing close apposition, and second, an ability to undergo the transitory departure from lamellar organization clearly required for fusion to proceed.

E. Fusion and the Mixing of Aqueous Contents of Vesicles

It has been argued that to demonstrate true fusion, mixing of two membrane-bound spaces must be demonstrated without appreciable leakage.[165] Several procedures have been designed to assay such mixing, all of which employ two populations of vesicles in which different agents are trapped which give rise to some response when mixed. The most popular protocol[165] relies on the formation of a fluorescent chelation complex between Terbium (Tb) and dipicolinic acid (DPA). In this procedure Tb^{3+}, weakly complexed to citric acid or nitrilotriacetic acid to prevent interaction with acidic phospholipids, is trapped in one vesicle population. A second population is prepared with DPA inside and when the internal contents mix, the fluorescent $Tb(DPA)_3^{3-}$ complex is formed. EDTA or Ca^{2+} inhibit formation of this complex in the external medium. The development of fluorescence is extremely rapid and therefore suitable for studying the kinetics of fusion events.[166] Employing PS vesicles, it has been demonstrated that Ca^{2+}-induced fusion can be detected just prior to collapse of the PS vesicles into cochleate structures, with the concomitant leakage of vesicle contents.[166] Consequently, mixing the vesicle contents is observed as a rapid burst of fluorescence which peaks and then decays as a result of inward leakage of Ca^{2+} or EDTA. These studies suggest that fusion of PS LUVs is a rather leaky process. It is interesting to note that fusion events occurring in CL-PC systems on addition of Ca^{2+}, which may be expected to involve non-bilayer intermediates, are much less leaky.[167]

These studies have been extended to a variety of lipid mixtures, including PA, PI, PS-PE, and PI-PE systems.[168,169] It was found that the presence of PC inhibits fusion of PS-PC systems, whereas the presence of PE facilitates fusion.[169] Further, whereas Ca^{2+} stimulates fusion of PA LUVs, fusion is not observed for vesicles composed of PI; however, PI-PE systems fuse readily.[169] The authors conclude that the results are not consistent with the previous hypothesis of Ca^{2+}-induced phase separation where fusion occurs at phase boundaries. It is suggested that fusion occurs as a result of the dehydration of the intermembrane space either due to the binding of Ca^{2+} or a combination of Ca^{2+} binding and the presence of the poorly hydrated PE head group which facilitates closer intermembrane contact. They suggest that nonbilayer intermediates are not involved in the fusion event as dimyristoyl PE (which does not adopt the H_{II} phase in isolation), capable of promoting fusion as are unsaturated (H_{II} phase preferring) PEs, and propose that close apposition promoted by PE and the acidic phospholipid-divalent cation complexes enhances fusion by "reducing the polarity" of the interbilayer space. Supposedly, then, this results in a reduced energetic barrier toward an intermixing of membrane components by some rather ill-defined mechanism.

With the possible exception of Ca^{2+}-stimulated fusion of PS vesicles, these results are broadly consistent with fusion proceeding via a transitory nonbilayer intermediate. The fact that both unsaturated and saturated varieties of PE can promote fusion in PE-PI systems may result from the ideal mixing properties of liquid crystalline lipid mixtures noted previously. This would be expected to produce an averaged hydrocarbon chain order common to all component lipids, and thus the phase properties of a mixture composed of a relatively saturated (bilayer phase) PE with relatively unsaturated PI may be similar to that of mixtures containing an unsaturated (H_{II} phase) PE.

F. Exocytosis

Exocytosis provides a useful model for biological fusion events, and there is now a considerable amount of information available on the secretion process in various systems.[170-172] In most stimulus-secretion coupled systems a transitory influx of extracellular Ca^{2+} (or a mobilization of intracellular pools of Ca^{2+}) triggers exocytosis. The chromaffin cell and granule provide one of the most readily studied exocytotic systems due mainly to the ease with which the secretory granules can be isolated in relatively large quantities and high purity. The influx of Ca^{2+} leading to secretion of chromaffin granule contents by adrenal medulla cells appears to lead to relatively specific granule-plasma membrane fusion rather than granule-granule fusion. This latter event is not commonly observed except perhaps at high (>10 m*M*) Ca^{2+} levels.[173] Studies of isolated chromaffin granules indicate that lower Ca^{2+} concentrations can induce aggregation, with the appearance of twinned granules.[174] Some of these observations have been interpreted as indicating fusion; however, the twinned systems exhibit a dumbell appearance, implying that each granule maintains its integrity. Larger, fully rounded fused systems are not observed.

Studies have been directed toward understanding the roles of the polymorphic properties of lipid in the exocytotic event.[175] Briefly, it was reasoned that the inner monolayer lipids of the plasma membrane of adrenal cells may play an active role. In particular, if this monolayer has a lipid composition approximately the same as that of the erythrocyte (primarily PE and PS), then the influx of Ca^{2+} will destabilize this monolayer, enhancing the preference of the lipids for a nonbilayer (H_{II}) organization. This instability could be relieved by formation of inverted micellar or inverted cylinder structure with the outer monolayer of closely apposed granules, thus initiating the fusion process.

The model system employed to test such speculation consists of chromaffin granules incubated with PE-PS SUVs.[175] The subsequent introduction of Ca^{2+} (2 m*M*) resulted in the immediate release of granule contents and massive granule-vesicle-granule fusion as indicated in the freeze-fracture micrographs of Figures 23 and 24. Such effects were not observed in the absence of PE-PS vesicles. These and other considerations (see Nayar et al.[175] for full details) led to the exocytotic model shown in Figure 25.

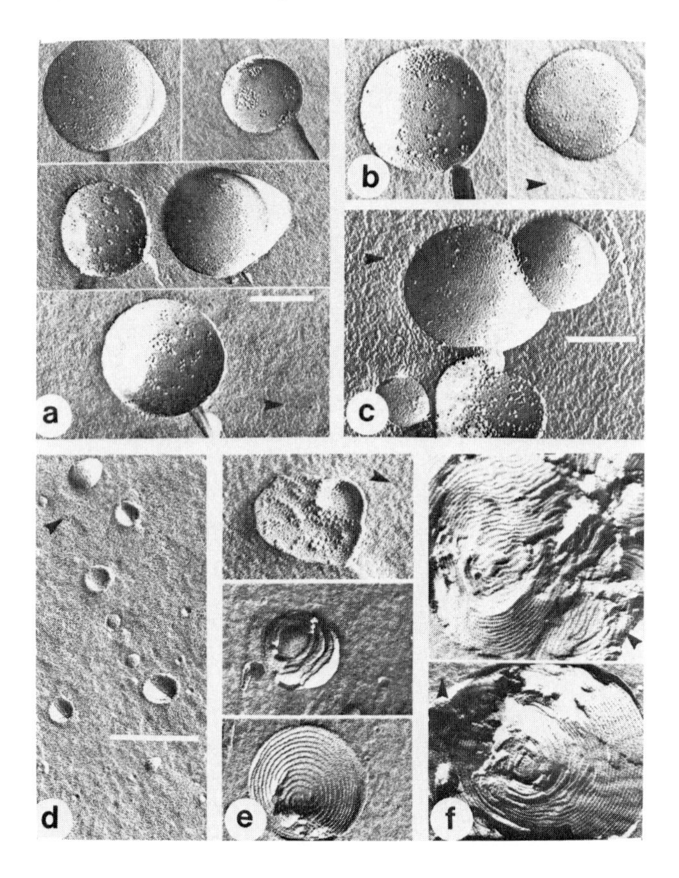

FIGURE 23. Freeze-fracture micrographs of the following: (a) chromaffin granules in the absence of Ca^{2+}, showing both convex (EF) and concave (PF) fracture faces; (b) chromaffin granules in the presence of 2 m*M* Ca^{2+}; (c) chromaffin granules in the presence of 5 m*M* Ca^{2+}; (d) sonicated PE-PS (3:1) vesicles in the absence of Ca^{2+}; (e) PE-PS (3:1) vesicles in the presence of 2 m*M* Ca^{2+} (the arrow indicates rows of lipidic particles); (f) PE-PS (3:1) vesicles in the presence of 5 m*M* Ca^{2+}. The white bars represent 200 nm, and the direction of shadowing is indicated by the arrowhead in each micrograph. (From Nayar, R. et al., *Biochemistry*, 21, 4583, 1982. With permission.)

G. Compartmentalization Within a Continuous Membrane Structure

MLVs composed of a mixture of bilayer and H_{II} lipids (e.g., unsaturated PE-PC-cholesterol systems) exhibit some interesting features. As indicated previously, the presence of 30 mol% or more of the bilayer (PC) species can stabilize a net bilayer organization. However, incubation at higher temperatures can result in the appearance of a narrow ^{31}P NMR peak, which can become the dominant spectral feature.[136] Remarkably, this behavior is often irreversible, for on cooling to the starting temperature the narrow resonance can remain the major, indeed often the only, ^{31}P NMR response. In conjunction with this behavior, freeze-fracture studies[164] reveal the presence of lipidic particles after the heating-cooling cycle in DOPC-DOPE-cholesterol systems which are not present in the original preparation. Lipid in these lipidic particles, interpreted as intrabilayer inverted micelles, would be expected to give rise to narrow "isotropic" ^{31}P NMR spectra. However, the amount of lipid in these intramembranous structures is always much less than the amount of lipid giving rise to isotropic ^{31}P NMR spectra, suggesting other sources of motional averaging. The ^{31}P NMR results are consistent with phospholipid in lamellar structures of 100 nm radius or less,

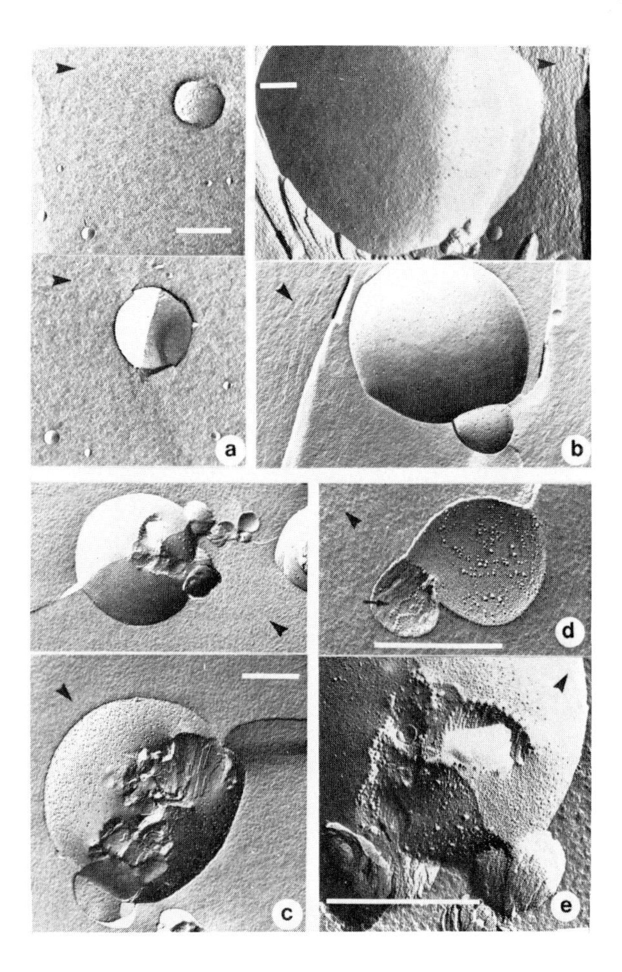

FIGURE 24. Freeze-fracture micrographs of: (a) chromaffin granules in the presence of sonicated PE-PS (3:1) vesicles where the ratio of chromaffin granule lipid to exogenous phospholipid is 1:4 and where the sample was prepared in the absence of Ca^{2+}; (b) the same preparation as (a) but incubated in the presence of 2 m*M* Ca^{2+}; (c) the same preparation as (a) but incubated in the presence of 5 m*M* Ca^{2+}; (d) a micrograph at higher magnification depicting the interaction between chromaffin granules and exogenous PE-PS (3:1) vesicular lipid after incubation in the presence of 2 m*M* Ca^{2+} (upper portion); (e) same as (d) but in the presence of 5 m*M* Ca^{2+}. Arrows indicate particles which exhibit characteristics of lipidic particles. The white bars represent 400 nm, and the direction of shadow is indicated by the arrowhead in each micrograph. (From Nayar, R. et al., *Biochemistry*, 21, 4583, 1982. With permission.)

where lateral diffusion processes alone can result in isotropic averaging.[176] These structures cannot be independent closed vesicles, as the corresponding lipid dispersions occur as large, visible globules of lipid, which can be isolated by low-speed centrifugation. This indicates that the small structures are integral subunits of a macroscopic entity.

The correlation of two events noted in freeze-fracture studies of these and related systems leads to a logical alternative. Briefly, lipidic particles are often arranged in organized rows, occurring at the fusion interface in fusing vesicular systems.[147,177] This suggests that the lipidic particles observed in the freeze-fracture of the multilamellar model systems actually correspond to regions of fusion between adjacent bilayers. This leads to the "honeycomb" model depicted in Figure 26, where heating the PE-PC-cholesterol LUVs to a temperature

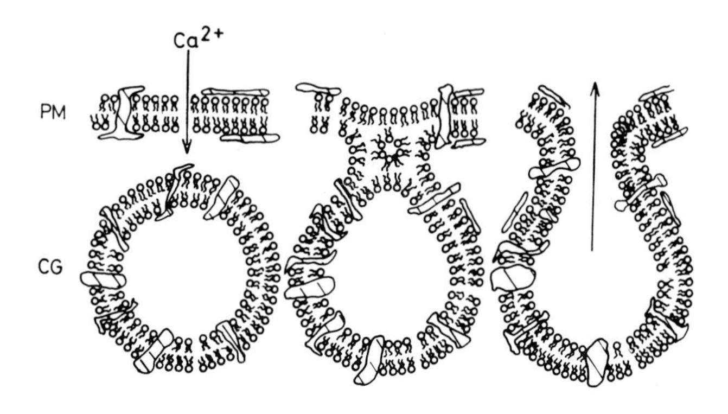

FIGURE 25. Proposed mechanism of release of chromaffin granule and other secretory granules in vivo. PM refers to the chromaffin cell plasma membrane, whereas CG denotes the chromaffin granule.

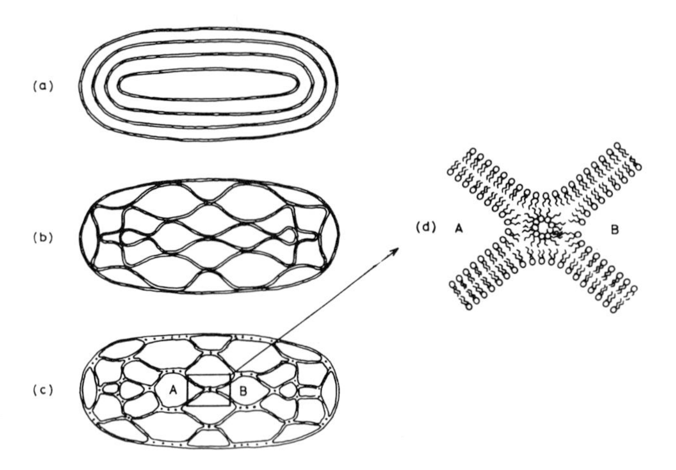

FIGURE 26. Mechanism of formation of a honeycomb structure from multilamellar PE-PC-cholesterol systems and other systems containing mixtures of bilayer and hexagonal (H_{II}) phase lipids. In (a) the usual depiction of multilamellar liposomes is given; however, it is clear that in general regions will exist where the layers are in close apposition. A stylized version of this is given in (b). In (c) it is postulated that these regions will undergo (partial) fusion with associated formation of lipidic particles at the interface. Compartmentalization in a continuous membrane structure then results, as indicated for compartments A and B in the expanded diagram of (d).

above the T_{BH} of the PE component promotes interbilayer fusion with associated lipidic particle formation. This results in the additional formation of interstitial regions where the lipids remain in a bilayer organization. If these interstitial regions are small enough (e.g., radius of 1000 Å or less) the component phospholipid will experience isotropic averaging and exhibit narrow ^{31}P NMR spectra as observed experimentally. This model is also fully consistent with the permeability behavior and trapped volumes observed for PE-PC-cholesterol systems.[177] A graphic example of honeycomb structure observed by fusing PE-PS-SUV systems via incubation at pH 4 is illustrated in Figure 27.

The ability of lipids to adopt structures such as those indicated in Figure 26 has potentially important implications for membrane morphology. In particular, such structures suggest the possibility that separate, isolated compartments may be connected by a continuous membrane

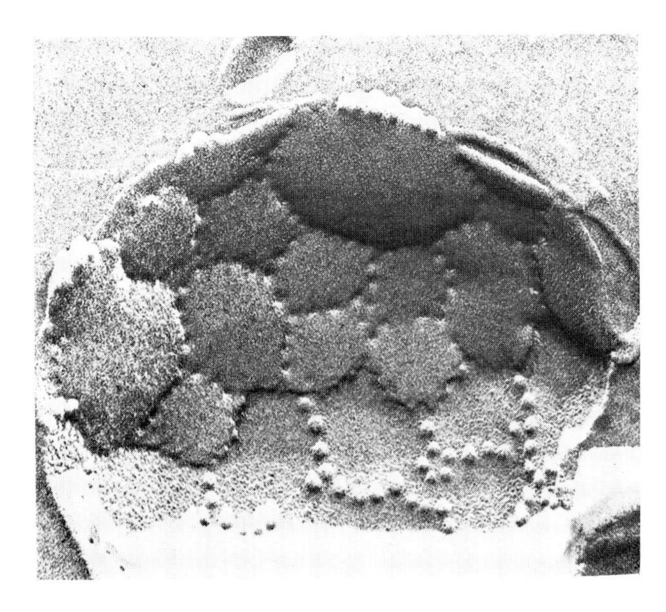

FIGURE 27. Freeze-fracture micrograph of a honeycomb structure produced by fusion of PE-PS structure produced by fusion of PE-PS-SUV systems. On dialysis against a pH 3 buffer (see Figure 22).

structure. This would allow unique opportunities for interorganelle communication and provides an understanding of how membrane-bound systems could maintain compartmentalization while being part of a continuous membrane structure. This may be related to continuities observed between the endoplasmic reticulum and the nuclear membrane, and the outer mitochondrial membrane and the Golgi apparatus.[178] Similar considerations may apply to the tight junction networks of epithelial cells (and the contact points between inner and outer membranes of mitochondria and bacteria) as discussed in Section VIII. All these could be considered as situations of ''arrested'' fusion.

H. Summary

The results presented in this section indicate very strongly, if not conclusively, a role for nonbilayer lipids and associated structures in membrane fusion phenomena. In unilamellar systems factors promoting inverted nonbilayer structures invariably promote fusion. Lipid-soluble fusogens which induce cell-cell fusion in vitro induce H_{II} organization for component lipid, indicating a similar mechanism. The circumstantial evidence for a role of nonbilayer lipid structure in exocytosis is strong. The lipid composition of biological membranes is compatible with such structural reorganization, and factors known to induce fusion in vivo can trigger nonbilayer organization in appropriate lipid systems. These general features contrast strongly with previous models of fusion events, and must be considered to establish fusion models based on the polymorphic capabilities of lipids as the most likely to reflect the in vivo event. However, several difficulties still remain. These include the relatively high levels of Ca^{2+} required to stimulate fusion between model systems and the exact details of the intermediate structure itself, which may occur on a time scale sufficiently rapid to escape detection by techniques such as freeze-fracture.[57]

VI. PERMEABILITY PROPERTIES OF LIPIDS

A. Introduction

To this stage we have summarized the structural preferences of liquid crystalline lipids as revealed in simple lipid-water dispersions and have indicated the strong evidence supporting a role of nonbilayer lipid structures in membrane fusion events. We now consider the properties of lipids as they relate to the primary function of the lipid bilayer, namely, to provide a selective permeability barrier between two aqueous environments.

There is a large amount of literature on the permeability properties of lipid bilayers to various solutes. Unfortunately, this does not yet provide a basis for understanding the roles of different lipid species in establishing permeability properties. It is instructive to examine the reasons for this. First, diverse model membrane systems have been employed to monitor bilayer permeability. Each system has associated limitations with resulting inconsistencies in permeability behavior. For example, the bilayer lipid membrane or black lipid membrane (BLM) is a popular system for monitoring bilayer permeability properties of various solutes. However, the BLM is usually formed by brushing a solution of a lipid dissolved in a nonpolar solvent (e.g., an *n*-alkane) across a hole in a partition between two aqueous compartments.[179] Excess lipid drains to the perimeter of the hole, leaving a bilayer leaflet in the center. Obviously such a system is most convenient for monitoring permeation from one compartment to the other; however, some hydrocarbon solvent remains in the bilayer leaflet and if a mixture of lipids is employed their proportions in the bilayer region are uncertain.[180]

The second group of model systems appropriate for permeability studies are the closed SUV, LUV, or MLV systems summarized in Section II. MLVs exhibit a broad size range and a large number of internal lamellae. Quantitative measures of diffusion rates are therefore difficult to achieve. The utility of small SUV systems for diffusion studies is limited by the small trapped volume as well as their highly curved nature which can influence permeability and other physical properties. Finally, most LUV preparations involve organic solvent or detergent, which is difficult to remove completely and may well influence permeability properties.

Irrespective of the presence of impurities in the model membrane system, other pitfalls exist. These include the presence of the ''unstirred layers'' adjacent to the membrane interface in which there is little or no convective mixing, resulting in a concentration gradient of a diffusing solute between the layer adjacent to the membrane and the bulk phase.[181] This can reduce the actual solute concentration gradient across the bilayer itself. Alternatively, in membrane-bound model systems, transbilayer diffusion of a charged species in the absence of an appropriate charge counterflow can result in a membrane potential which limits subsequent diffusion.

Given these and other difficulties it is not surprising that there is a lack of quantitative reliable data concerning the influence of lipid head group and acyl chain composition on bilayer permeability to ions and uncharged molecules. Here we briefly summarize those results which we consider to bear the most direct relation to biological systems; therefore, the discussion is restricted to liquid crystalline systems, omitting the large body of data showing enhanced permeability when gel and liquid crystalline domains coexist.[182] Given the problems inherent in the BLM model, the primary focus will concern closed lamellar systems. We first consider some basic theoretical aspects and subsequently the permeabilities of certain mono- and divalent ions and nonelectrolytes of biological interest.

B. Theory

The ability of a compound to diffuse across a membrane is usually defined in terms of the permeability coefficient P. This parameter arises from a version of Fick's law, which is that the diffusion rate of a given substance through a membrane (number of molecules per unit time, dn/dt) is directly proportional to the area (A) of the membrane and the concentration difference [$\Delta C(t)$] across the membrane. Thus $dn/dt \; \alpha \; A\Delta C(t))$ or $dn/dt = - PA \, \Delta C(t)$, where the permeability coefficient P has the units of length/time (e.g. cm/sec). If we consider the special case of a unilamellar vesicle of radius R containing an initial concentration of solute $C_I(o)$ and where the initial external concentration of this solute is zero, then it is straightforward to show that $\Delta C(t) = C_I(o) \exp(-3 Pt/R)$. Under conditions where the external volume is much greater than the internal trapped volume, then $\Delta C(t) \simeq C_I(t)$ (where $C_I[t]$ is the internal concentration), and thus $C_I(t) = C_I(o) \exp(-3 Pt/R)$. For a 100-nm diameter LUV it may be calculated that the time required for release of one half the entrapped material ($t_{1/2}$) is 0.1 sec for $P = 10^{-5}$ cm/sec, whereas for $P = 10^{-10}$ cm/sec, $t_{1/2} = 3.2$ hr.

As previously indicated, measures of the permeability of membranes to ions are complicated by the fact that for free permeation to proceed, a counterflow of other ions of equivalent charge is required; otherwise a membrane potential is established which is equal and opposite to the chemical potential of the diffusing species. As an example, consider the 100-nm diameter vesicle with a well-buffered interior pH of 4.0 and an exterior pH of 7.0 suspended in a Na^+ buffer. H^+ Ions appear to be much more permeable than Na^+ ions.[188] The H^+ ions can diffuse out, but Na^+ ions cannot move in. A membrane potential $\Delta\psi$ is therefore established according to the Nernst equation (interior negative) where $\Delta\psi = 59 \log([H^+]_i/[H^+]_o) = -177$ mV and the subsequent efflux of protons is coupled to the much slower influx of Na^+ ions. Assuming a membrane thickness of 4 nm and an interior dielectric constant of 2, the capacitance of the vesicle membrane can be calculated as $C \sim 0.5\ \mu F/cm^2$, and thus the number of protons that diffuse out to set up $\Delta\psi$ can be calculated via the relation $\Delta Q = C\Delta\psi$ to be $\sim$150. Subsequent H^+ efflux will only occur as Na^+ ions permeate in.

A further factor influencing permeability which bears directly on the properties of component lipids concerns the surface potential arising from the charge on phospholipid polar head groups. This gives rise to a "surface potential" (ϕ) according to the Gouy-Chapman relation.[183]

$$\phi = \frac{2kT}{ze} \sinh^{-1} \frac{\sigma_m}{(8kTn_o b\epsilon\epsilon_o)^{1/2}}$$

where k is Boltzmann's constant, T, the absolute temperature, z, the charge on the aqueous electrolyte, e, the electronic charge, n_o, the bulk (molecular) concentration of electrolyte, and σ_m, the surface charge density of the charged phospholipid species. For a monovalent electrolyte concentration of 150 m*M* at 20°C this reduces to $\phi = 0.052 \sinh^{-1}(\sigma/4.5)$ mV where σ is expressed as $\mu C/cm^2$ (C = coulombs). In order to illustrate the effects this surface potential can have, consider the erythrocyte membrane inner monolayer, which contains approximately 30% PS. Assuming an area per lipid molecule of 60 $Å^2$ this results in a surface charge density $\sigma = 8\ \mu C/cm^2$, giving rise to a negative surface potential $\phi = -69$ mV. This potential will repel anions from, and attract cations to, the lipid-water interface. The H^+ ion concentration at the inner monolayer interface, for example, would be increased by the Boltzmann factor $\exp(e\phi/kT) = 14.5$, giving rise to an (initial) rapid efflux of protons.

C. Bilayer Permeability of Water and Nonelectrolytes

This area has recently been reviewed by Fettiplace and Haydon.[184] Briefly, liquid crystalline lipid bilayers are remarkably permeable to water, which exhibits permeability coefficients in the range 10^{-2} to 10^{-4} cm/sec. Typically, these coefficients are obtained by monitoring the water flux through a membrane experiencing an osmotic gradient by measuring the change in volume of the compartments on one side of the membrane or the other. In the case of membrane-bound systems (e.g., MLVs) this is commonly accomplished by monitoring the absorbance at 450 nm which reflects their swelling or shrinking in nonisotonic media. Results employing BLM systems consisting of egg PC in an alkane solvent indicate permeability coefficients in the range of 3.3×10^{-3} cm/sec,[184] whereas for egg PC MLVs containing one to three bilayers water permeability coefficients of 4.5×10^{-3} cm/sec are indicated.[185] Activation energies in the range of 8 to 13 kcal/mol appear consistent with water diffusion across the hydrocarbon core rather than via transient pores in the membrane.[186]

The permeability properties of nonelectrolytes (uncharged polar molecules) across bilayer membranes have received less detailed attention. The study by Poznansky et al.[187] of the homologous series of amides (formamine to valeramide) is perhaps one of the most representative. Employing radiolabeled solutes and egg PC LUVs, permeability coefficients approximately two orders of magnitude slower than water were obtained. Similarly, permeability coefficients for urea of 4×10^{-6} cm/sec at 25°C were observed with an activation energy approximately the same as for water. The general observation that membrane permeability increases with increasing lipid solubility (observed as increased partitioning into the lipid phase) indicates that for most solutes the rate-limiting step involves partitioning into the bilayer.

D. Permeability of Ions

There is a great amount of literature on the permeability properties of ions of biological interest, including, H^+/OH^-, Na^+, K^+, Cl^-, and Ca^{2+}. In general, lipid bilayers are exceptionally impermeable to most ions, however, H^+ and Cl^- ions may constitute an exception.

1. Proton-Hydroxyl

This area has been reviewed by Deamer[188] and is an area of considerable controversy as there are as many as six orders of magnitude difference between net H^+/OH^- permeability reported in the literature. Nichols and Deamer,[189] employing egg PC in LUVs (ether injection) containing 2 mol% PA, demonstrated net H^+/OH^- permeabilities of 1×10^{-4} cm/sec. In contrast, work on BLM egg PC-*n*-decane systems[190] indicated no detectable H^+ flux except for HCl generated pH gradients. The HCl permeabilities obtained were remarkably high (~3 cm/sec), indicating that H^+ was transported as neutral HCl molecules. Evidence in support of this mechanism has been reported by Nozaki and Tanford[191] on EPC LUVs prepared by detergent dialysis.

These results emphasize the difficulties involved in obtaining unambiguous estimates of ion permeabilities. Deamer[188] has pointed out several problems with the studies indicating H^+ diffusion via HCl monomers. In particular, the BLM study,[190] very low pH values (pH ~1) were employed. The pK of the PC phosphate is ~2 and thus the membrane would exhibit a large positive surface potential. By the reasoning indicated in the previous section this would markedly decrease the H^+ concentration at the membrane interface, with a corresponding decrease in proton flux. Similarly, the results of Nozaki and Tanford can be questioned[188] on the basis that an initial rapid proton flux had set up a membrane potential, limiting subsequent flux to diffusion of HCl or counterions. In summary, the balance of reliable available evidence suggests H^+/OH^- permeability coefficients in the range of 10^{-4} cm/sec, at least for egg PC "etherosome" LUV systems.

2. Chloride

It has been known for some time that Cl^- ions exhibit permeability coefficients for lipid bilayers some[192-196] two to three orders of magnitude larger than for K^+ or Na^+.[6,192-196] The reasons for such enhanced permeability, which appears to reflect a relatively general property of enhanced permeability of anions as opposed to cations may be related to a positive interior potential of the lipid bilayer.[197] There is, however, growing evidence that Cl^- diffusion proceeds via an HCl complex, which may also involve an association with the phospholipid polar head group.[198]

3. Sodium and Potassium

Lipid bilayers are usually remarkably impermeable to K^+ and Na^+ ions. The permeability coefficients for Na^+ and K^+ generally fall into the range of 10^{-10} to 10^{-12} cm/sec for EPC LUV systems[189,191] or 10^{-14} cm/sec for sonicated SUV systems.[199] The reasons why SUV systems should exhibit reduced permeabilities are presently obscure, but could be related to effects of the unstirred layers.[181] These effects are expected to result in reduced observed permeabilities for smaller systems.

There are indications that K^+ ions are slightly more permeable than Na^+ ions in bilayer membranes.[200] This is supported by the observation that EPC LUV systems (produced by an extrusion techniques) with transmembrane K^+/Na^+ gradients can exhibit a membrane potential (interior negative) in the absence of a K^+ ionophore such as valinomycin.[16] This indicates that K^+ can leak out somewhat faster than Na^+ can permeate in.

E. Lipid Composition and Membrane Permeability

As should be clear from the preceding observations, the large majority of reliable quantitative measures of the permeability of various solutes through liquid crystalline bilayers have been performed on EPC systems. However, some qualitative indications of the effect of acyl chain and headgroup composition are available.

As a general observation, factors increasing the order in the hydrocarbon region decrease membrane permeability. Thus BLMs composed of monoglyceride films demonstrate decreased water permeabilities for increased acyl chain saturation and chain length.[201] Similar observations have been made for glycerol and erythritol permeabilities in MLV systems.[202] The ability of cholesterol to induce a significantly more ordered acyl chain region is also reflected by reduced permeability properties. Incorporation of cholesterol into EPC or monoglyceride films[201,203] reduces water permeability by a factor of 5 or more. Similar effects have been observed for HCl diffusion[190] and glycerol and erythritol permeability.[202] Similarly, increased temperatures which lead to increased acyl chain mobility (disorder) result in increased permeability. For liquid crystalline systems this is reflected by the high activation energies exhibited for permeability coefficients, which fall into the range of 8 to 20 kcal/mol for water,[185] protons, [204] and other ions.[197]

The influence of the phospholipid headgroup composition on membrane permeability is less clear. As indicated previously, the presence of acidic (negatively charged) phospholipids would be expected to increase membrane permeability for anions and decrease it for cations due to surface potential effects. This is not always true. For example, while PI and PA can decrease Cl^- flux and increase K^+ flux in MLV systems, PS apparently has little influence.[192] Incorporation of PA into saturated PC MLVs decreases cation permeability,[192] whereas in LUV EPC systems PA increases H^+ permeability.[188] At the moment there is therefore no simple predictive basis for deciding the influence of a charged head group on the bilayer permeability properties. This may be due, in part, to an increased area per molecule of the charged species due to inter-head group electrostatic repulsion, resulting in increased acyl chain disorder.

The influence of unsaturated PEs on membrane permeability has not received detailed attention. In isolation, the H_{II} phase is adopted which does not, of course, exhibit barrier properties. In mixtures with PC, however, a decrease in K^+ permeability has been observed.[192] This may be related to the increased acyl chain ordering observed in PE-PC bilayers in comparison to bilayers containing only PC.[275]

The permeability properties of Ca^{2+} and other divalent cations have also not been studied in detail. Systems such as unsaturated PE-PS LUVs are remarkably permeable to Mn^{2+} as indicated by the rapid "quenching" of the phospholipid ^{31}P NMR resonance in such systems.[276] This may be be attributed to fusion and H_{II} phase formation. The ability of PA to act as a Ca^{2+} ionophore has received somewhat more attention, which is of particular interest with respect to a possible role of PA to translate Ca^{2+} during events associated with the PI response:[101,205,206] however, the results obtained are equivocal. Serhan et al.[207,208] reported an ability of PA to increase Ca^{2+} influx into MLV systems and noted that PS, PE, and PG did not engender such effects. Similar results have been reported for EPC-PA LUV systems.[209] Other investigators have found PA to have no effect on Ca^{2+} permeability and attribute previous effects to the presence of oxidized fatty acids.[210]

F. Summary

Three main points stem from the preceding brief overview of membrane permeability of liquid crystalline lipid bilayers and the influence of lipid composition on permeability properties. First, our understanding of the specific roles of lipids in membrane permeability remains relatively primitive. This is mainly due to the diverse model systems employed and the contaminants present, as well as complications due to a requirement of counterion flux and the effects of unstirred layers. The need for a quantitative self-consistent determination of nonelectrolyte and ion permeabilities in well-defined LUV or BLM systems of varying lipid composition which do not contain residual solvent, is apparent. Second, as may be expected, there is strong evidence to support a close correlation between bilayer permeability and the order in the hydrocarbon. A quantitative theoretical and experimental evaluation of the connection between these two variables would be of particular interest. Finally, the little that is known allows intriguing possibilities for specific roles of lipids in determining and modulating membrane permeability properties. In particular, the possible influence of local modulation of lipid composition or transbilayer redistributions of lipid on the permeability of a given solute should be well defined before postulating direct roles of protein components.

VII. LIPID-PROTEIN INTERACTIONS

A. Introduction

Lipids clearly have important roles in determining membrane structure and permeability properties. In this section we consider additional roles of lipids as they relate to providing an appropriate environment for localization and function of membrane-associated protein. Again, the properties of liquid crystalline lipids will provide the primary focus, with particular emphasis given to the influence of protein on the structural (polymorphic) preferences of lipid systems. Before discussing these features a brief review of our current understanding of lipid-protein interactions is appropriate.

A large proportion of studies of lipid-protein interactions have been concerned with the possible regulatory roles of lipids on protein function. This has been envisaged to occur in two ways. The first involves modulation of the local order in the "annular" hydrocarbon adjacent to the hydrophobic lipid-protein interface. Alternatively, a particular lipid composition may be required in the annulus for protein activity. The only unambiguous influence of local lipid order on protein function has been observed for gel-state lipids, which can inhibit certain membrane-bound proteins.[211-214] The fact that eukaryotic membranes do not appear to contain gel state lipids suggests that such modulation is not likely to occur in vivo. In addition, evidence for a requirement for a specific lipid annulus for protein function is also not convincing. Many membrane enzymes, including cytochrome oxidase,[215-218] Na^+-K^+ ATPase,[219,220] and the sarcoplasmic reticulum ATPase[211,212,221] exhibit excellent activity in the presence of a variety of lipid species, including detergents.[222] Further, annular lipids appear to experience relatively rapid exchange with bulk bilayer lipids,[223,224] which further weakens the case for a specific annular lipid composition for function.

In Section IV it was emphasized that biological variables such as pH, ionic strength, and divalent cations can strongly influence the structural preferences of various lipid dispersions. Here we indicate that proteins can exhibit similar abilities, and discuss the possible roles of different lipid species to provide appropriate packing and sealing characteristics at the lipid-protein interface.

B. Influence of Proteins on Lipid Polymorphism

1. Extrinsic Proteins and Polypeptides

a. Poly-L-Lysine

Poly-L-lysine is a basic polypeptide which interacts strongly with acidic (negatively charged) phospholipids.[225] Studies have been directed towards determining whether this extrinsic protein analogue can influence lipid structure in a one-component system and can effect lateral segregation of negatively charged lipids in a two-component lipid system. It has been shown that polylysine can precipitate cardiolipin (CL) resulting in a multilamellar liquid crystalline structure.[226] Further, the ability of Ca^{2+} to induce H_{II} phase structure for CL is inhibited, indicating that polylysine effectively competes with Ca^{2+} for CL binding sites. In mixtures of CL and unsaturated PE (soya) which form bilayer structure, the addition of polylysine results in H_{II} phase structure.[226] This may originate from a polylysine-induced segregation of CL, thus allowing the PE to assume the structure it prefers in isolation.

b. Small Polypeptides

Many small, basic polypeptides interact strongly with membrane-containing acidic phospholipids, and appear to also penetrate the hydrophobic region. Examples include mellitin and cardiotoxins isolated from venoms.[227,228] For cardiotoxin IV, which is isolated from snake venom, it has been shown that in a lipid extract of complex composition, addition of the peptide results in formation of lipid particles and H_{II} phase structure.[229]

c. Cytochrome c and Apocytochrome c

Cytochrome *c* is a structured, nearly spherical basic polypeptide (8 positive charges per molecule at pH 7.0) that participates in the terminal electron transfer of the respiratory chain of mitochondria. The heme-free precursor, apocytochrome *c*, which is in a random coil configuration, is synthesized on free ribosomes in the cytosol and is subsequently imported to the outer surface of the inner mitochondrial membrane. Cytochrome *c* binds to and penetrates negatively charged lipid systems,[225,230,231] enhances cation permeability,[232] and can induce lipid phase separations.[231]

Addition of cytochrome *c* to beef heart CL results in (partial) H_{II} phase formation[233] and lipidic particle structure as indicated by freeze-fracture electron microscopy, whereas addition to PS or PE results in precipitation but no apparent nonbilayer structures. Apocytochrome *c* exhibits quite different behavior. Addition of this protein to CL and PS results in particulate structure as detected by freeze-fracture,[234] and recent experiments indicate that partial translocation of apocytochrome *c* can occur across PS bilayers.[235]

2. Intrinsic Proteins and Peptides

a. Gramicidin

Gramicidin, a hydrophobic channel-forming antibiotic, is a popular model for examining intrinsic protein-lipid interactions. In mixtures with PEs, gramicidin promotes H_{II} phase organization,[236] and is remarkably, able to induce H_{II} phase structure in PC systems.[237] This ability has been attributed to partial dehydration of the gramicidin-PC complexes and packing properties in the acyl chain regions.[238] A further contribution from the net "cone" shape of gramicidin may also be of importance as indicated by the observation that mixtures of lyso-PC and gramicidin can form lamellar structures.[239]

b. Glycophorin

Glycophorin is the major sialoglycoprotein of the erythrocyte membrane. This protein contains a hydrophobic sequence of some 23 amino acids which spans the bilayer in an α helical organization.[240] The c terminus is exposed to the cytoplasm and is highly negatively charged due to the presence of some 32 sialic acid residues. Glycophorin may be readily reconstituted into PC LUV systems, resulting in apparent packing defects which increase permeability and lipid flip-flop processes.[241,242]

In reconstituted systems with unsaturated PEs, glycophorin exhibits an ability to stabilize bilayer structure, resulting in LUV systems.[243] This bilayer stabilization capacity appears to involve the carbohydrate region, which may prevent the close intermembrane apposition required for H_{II} phase formation to proceed, as well as the hydrophobic sequences. This is illustrated by the fact that trypsin treatment of DOPE-glycophorin reconstitutes, which removes the carbohydrate moiety, still results in bilayer systems.[243]

C. Lipid-Protein Interactions and Membrane Permeability

As indicated previously, evidence in support of a role of lipid-protein interactions in regulating membrane protein function by modulating local fluidity or annular lipid composition is not strong. However, lipid-protein interactions must satisfy other demands related to sealing the protein into the bilayer matrix. This was originally proposed by Israelachvili,[244] who suggested that lipid with a variety of ''shapes'' may be required to match irregular contours at the protein-lipid hydrophobic interface. This is consistent with the observation that reconstitution of integral proteins with a single liquid crystalline lipid species leads to systems that are often highly permeable to charged and neutral solutes.[242,245] Alternatively, if a heterogeneous lipid mixture is employed, much lower permeabilities are observed. Glycophorin, for example, gives rise to systems permeable to molecules of 900 mol wt or less when reconstituted with $18{:}1_c/18{:}1_c$ PC (DOPC), which is not observed for glycophorin reconstituted with total erythrocyte lipids or DOPC-DOPE mixtures.[246]

These results support the proposal that integral proteins may require lipids of different shapes to ensure optimal packing and sealing at the lipid-protein interface. If this is the case, then it might be expected that membrane-bound enzymes such as cytochrome oxidase and the sarcoplasmic reticulum Ca^{2+} ATPase, which generate chemical gradients of H^+ and Ca^{2+}, respectively, may also display corresponding requirements. In particular, it may be expected that the coupling between oxidation of substrates (cytochrome *c*, ATP) and generation of ion gradients may depend on the presence of cone-shaped lipids such as PE, for example, to seal the protein in the membrane. Such a requirement is indeed exhibited by both cytochrome oxidase[216,217] and the sarcoplasmic reticulum ATPase.[221] However, these results should be treated with some caution, as the inclusion of increasing amounts of an unsaturated PE in PE-PC systems results in smaller reconstituted vesicle systems on detergent dialysis. For a given protein to lipid ratio this can influence the number of proteins present per vesicle, which can have surprising effects on the respiratory control exhibited by oxidase-containing systems.[217]

D. Summary

The results presented in this section clearly indicate the emergence of a new appreciation of lipid-protein interactions. In particular, such interactions may be envisioned as relatively dynamic processes, where the local concentration of a given lipid may be enhanced on a time-averaged basis due to electrostatic lipid-protein interactions or a local requirement for lipid with particular shape properties. This can result in local structural perturbations of the lipid bilayer or, alternatively, better sealing in the regions of proteins which penetrate the bilayer. A summary of these possibilities is indicated in Figure 28.

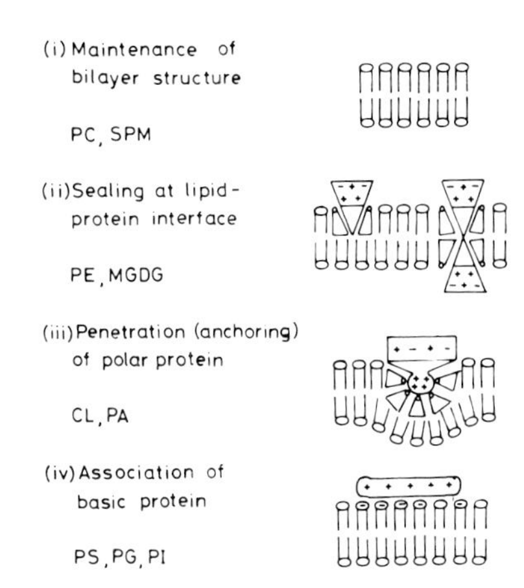

FIGURE 28. Potential roles of lipid as a result of shape and/or charge in maintaining the architecture of (bilayer) lipid-protein membrane systems: (a) maintenance of bilayer structure, (b) sealing at lipid-protein interface, (c) penetration (anchoring) of polar protein, (d) association of basic protein.

VIII. BIOLOGICAL MEMBRANES

The aim of model membrane studies is, of course, to gain a detailed understanding of the properties of and interactions between membrane components, enabling a detailed understanding of the relation between composition and function in biological membrane systems. Such understanding has not yet been achieved. Certain correlations do exist, however, between the observations made on model systems. Here we summarize those aspects we consider most instructive.

A. Erythrocyte Membrane

More than 95% of the phospholipid in the erythrocyte membrane is organized in a bilayer configuration.[247] With the exception of the induction of the H_{II} phase by chemical fusogens,[158,159] no other structural organization of the erythrocyte lipids has been observed even after such harsh treatments as extensive phospholipase degradation[248] or proteolytic digestion of the membrane proteins.[248] The stability of the bilayer structure might be related to the low metabolic activity of that membrane or to the large mechanical stresses the red cell experiences in the circulation.

The red cell maintains an asymmetric transbilayer distribution of lipids. The preferential localization of PS and PE in the inner monolayer suggests that the structural stability of that layer might be particularly sensitive to exogenous factors. As indicated previously, Ca^{2+} can trigger a bilayer H_{II} phase transition in dispersions with inner monolayer lipid composition,[249] which may be related to erythrocyte vesiculation in the presence of factors leading to increased Ca^{2+} concentrations in the cytosol.[38]

B. Endoplasmic Reticulum Membrane

The endoplasmic reticulum is a very metabolically active organelle involved in the biosynthesis of proteins, lipid, and carbohydrates. It is isolated in the form of "microsomes" (vesicles of 100 to 200 nm diameter) following cell rupture. The endogenous lipids are relatively unsaturated and the PE component (comprising ~30% of microsomal phospholipid) strongly prefers the H_{II} organization in isolation at 37°C[250] (see Figure 5). These characteristics, while consistent with overall lamellar organization, suggest a certain instability in that organization where relatively minor perturbations could result in local nonbilayer structure. This is consistent with the observation that dehydration of microsomal membranes leads to H_{II} phase formation[251] and could be correlated with the ability of the endoplasmic reticulum to vesiculate so readily on mechanical disruption. Similarly, the network morphology of the endoplasmic reticulum and the apparent interconnections between ER and the Golgi apparatus, the outer mitochondrial membrane, and the nuclear membrane[178] could be likened to a honeycomb compartmentalization as illustrated in Figure 27. The lipid composition is certainly compatible with these possibilities.

A system in which lipid bilayer stability is relatively easily perturbed would also be consistent with various difficulties encountered in characterizing microsomal preparations. For instance, efforts to characterize the transbilayer distributions of lipids across the microsomal have led to remarkably ambiguous results.[5] It may well be that the labeling or exchange techniques employed to ascertain such distributions perturbs bilayer integrity. Alternatively, such instability could be directly related to function. These include the possibilities or rapid vesiculation to form vesicles transporting membrane components to other organelles, or the processes of protein insertion during synthesis of protein on membrane-bound ribosomes.[238]

C. Inner Mitochondrial Membrane

The inner mitochondrial membrane is also a most metabolically active membrane, containing various components of the respiratory chain, among many other transport proteins. The H^+ permeability barrier presented by the lipid component is vital to the coupling between the proton gradient and production of ATP via the F_1F_0 complex. This agrees with ^{31}P NMR studies which indicate that over 95% of the endogenous lipids are in the bilayer organization in intact, functionally competent mitochondria.[252]

The lipid composition, while compatible with such an overall bilayer organization, again exhibits characteristics indicating that the bilayer structure may easily be disrupted. In particular, the phospholipids are highly unsaturated and are comprised of PC (39 mol%), PE (33 mol%), and CL (29 mol%).[253] The PE component adopts the H_{II} phase in isolation, as does CL in the presence of Ca^{2+} or Mn^{2+}. Thus, bilayer integrity could be expected to be particularly sensitive to the presence of divalent cations, which is supported by the observation that incubation with Mn^{2+} can give rise to small H_{II} phase regions in the inner mitochondrial membrane as detected by freeze-fracture.[254] It is intriguing that incubation with other cations (Ca^{2+}) induces contact points between the inner and outer mitochondrial membranes.[254] This supports the previously mentioned possibility that such contact points could correspond to arrested fusion events which rely on the ability of endogenous lipids to adopt nonbilayer structures.

D. Epithelial and Endothelial Membranes: Tight Junctions

Tight junction networks separate the "apical" and "basolateral" domains of "polarized" epithelial and endothelial cells, and are commonly visualized as intersecting thread-like networks on freeze-fracture electron micrographs. In addition to preventing diffusion of solutes from one side of the epithelial or endothelial barrier to the other, tight junctions also appear to act as a barrier to the diffusion of membrane protein and lipid components because the protein[255] and lipid[256] composition of apical and basolateral membranes are different. In particular, it has been demonstrated that lipids introduced into the outer monolayer of the apical membrane remain there, whereas lipids which can move to the inner monolayer redistribute over the entire cell membrane.[257]

Freeze-fracture and other studies[258,259] have resulted in the suggestion that the tight junction corresponds to formation of interbilayer inverted lipid cylinders (characteristic of the H_{II} phase) between polarized cells. This concept is illustrated in Figure 29 and would satisfy the diffusion barrier characteristics of tight junctions noted above. Evidence for similar "arrested fusion" phenomena between cells has been observed in rat adipose tissue.[260]

FIGURE 29. Diagram of a cross-section of the tight junction visualized as an inverted lipid cylinder combined with a freeze-fracture micrograph of a tight junction strand. (From Kachar, B. and Reese, T. S., *Nature (London)*, 296, 464, 1982. With permission.)

E. Retinal Disk Membranes

The retinal rod contains an ordered array of stacked disks associated with the transduction of light into nerve impulses. The principal disk membrane protein is rhodopsin. The lipid composition of this membrane is remarkably unsaturated, containing up to 60% docohexanoic (22:6) acid,[261] which when combined with the presence of 30 mol% PE strongly suggests a proclivity for H_{II} phase organization rather than bilayer structure. This is consistent with the observation that total lipid extracts of (bovine) retinal disks exhibit nonbilayer structure.[262] Phospholipids in intact disks are organized in a bilayer, indicating a strong bilayer stabilizing ability of rhodopsin.[262] The instability of this bilayer structure is indicated by freeze-fracture[263] and X-ray[264] studies, indicating the presence of H_{II} phase inclusions on limited dehydration.

F. Chloroplast Membranes

The plant chloroplast membranes contain the photosynthetic electron transfer system involved in the transduction of light to chemical energy via a chemiosmotic mechanism requiring an intact membrane permeability barrier. This occurs in the stacked thylakoid membrane, which exhibits a remarkably unsaturated lipid composition in which linolenic acid (18:3) is the primary acyl chain substituent.[265] The major lipid classes present are monogalactosyldiglyceride (MGalDG) and digalactosyldiglyceride (DalGDG), which constitue 45 and 25% of total polar lipid. As indicated previously, MGalDG adopts the H_{II} phase in isolation, whereas DGalDG prefers the bilayer organization, suggesting again that the net bilayer structure and thus the intact H^+ permeability barrier may be most easily disrupted. In this regard, it is interesting to note that MgalDG/DGalDG LUV systems can only be obtained for MGalDG/DGalDG ratios of ≤ 2,[266] similar to the ratios observed in vivo. Higher MGalDG contents lead to the formation of lipidic particles and other nonbilayer structures.

G. Prokaryotic Membranes

Certain prokaryotes have rather limited biosynthetic abilities for lipid biosynthesis, allowing appreciable manipulation of lipid composition and corresponding elucidation of factors regulating lipid composition. *Acholeplasma laidlawii* is perhaps the best example, as inhibition of endogenous fatty acid synthesis has led to the development of strains with essentially homogeneous fatty acid composition.[267] Studies on these systems have led to interesting observations regarding a balance between different molecular species of lipid, which can be interpreted in terms of a requirement for a balance between cone-shaped (H_{II}

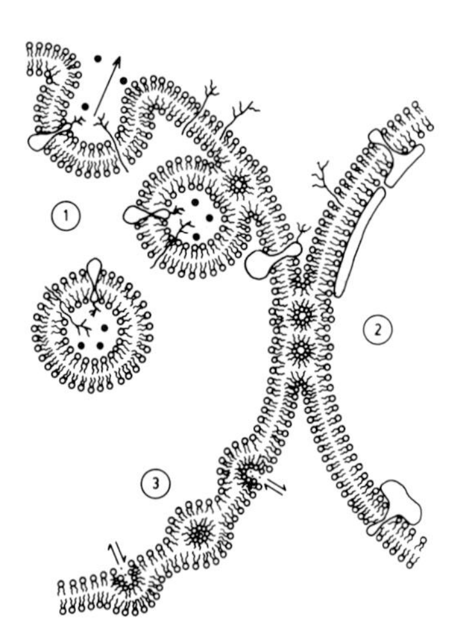

FIGURE 30. A metamorphic mosaic model of biological membranes illustrating various aspects of membrane morphology and function potentially involving nonbilayer lipid structure. In region (1) an exocytotic fusion event (see Figure 25) proceeding via an intermediate inverted micellar or inverted cylinder organization is shown, whereas in region (2) inverted cylinder structure allows a stable semi-fused interbilayer connection to exist, possibly corresponding to tight junctions (see Figure 29). In region (3) enhanced permeability to divalent cations is proposed to proceed via an inverted micellar intermediate, which may correspond to the ability of phosphatidic acid to act as a Ca^{2+} ionophore.

phase) lipids and cylindrical (bilayer) species. In particular, it is observed that as the acyl chain length or unsaturation is increased, the ratios of endogenous monoglucosyldiglyceride (MGluDG) to diglucosyldiglyceride (DGluDG) decrease dramatically.[267] As MGluDG is a cone-shaped (H_{II} phase) lipid and DGluDG a cylindrical (bilayer phase) lipid,[139] and as increases in chain length and unsaturation give rise to increased cone-shaped character, the changes in the MGluDG/DGluDG ratio are consistent with a need to conserve lipid shape distributions. Similarly, other workers have shown that higher levels of cholesterol (a cone-shaped molecule) result in lower MGluDG/DGluDG ratios,[268] whereas the inclusion of low levels of inverted cone (anesthetic) molecules in the growth medium has reverse effects.[269] These observations support the contention[268,270] that conservation of lipid shape properties may be more basic than maintenance of membrane lipid ''fluidity'' per se. As revolutionary as such a proposal may sound, it may be noted that changes in lipid composition noted for bacterial systems grown at different temperatures[271] (used to support conservation of ''fluidity'' arguments) is also fully consistent with this shape conservation hypothesis. This is because higher temperatures lead to greater acyl chain motion and increased cone shape character of lipids. Thus, it may be suggested that increased lipid saturation noted for *Escherichia coli* grown at higher temperatures,[271] for example, counters such effects and conserves lipid shape distributions. Further evidence in support of this shape conservation hypothesis has been summarized elsewhere.[270]

IX. CONCLUDING REMARKS

It should be clear from the material presented that our phenomenological understanding of the structural properties of lipids found in membranes, and factors which may regulate these phenomena is rapidly becoming a mature body of knowledge. Similarly, as reviewed in more detail elsewhere,[83] our understanding of the ''shape'' or packing properties of lipid which lead to various structural expressions of lipids is also developing into a mature science. These developments have two particularly important consequences. First, the fact that membrane lipids can adopt transitory or long-lived nonbilayer alternatives has direct application to membrane-mediated phenomena such as fusion, which require local departures from bilayer structure. In turn, this leads to new flexibility in modeling membrane morphology as summarized in the "metamorphic mosaic model" illustrated in Figure 30.

The second area concerns lipid-packing properties and their relation to basic membrane characteristics such as permeability. It is becoming clear that biological membranes appear to maintain a balance between lipids which prefer H_{II} structure in isolation (cone shaped) and lipids with cylindrical geometry preferring the bilayer phase. Indeed, as discussed in the previous section, this balance appears to be exceedingly delicate in more metabolically active membranes where relatively minor perturbations can lead to bilayer disruption and nonlamellar structure. It is unlikely that such a balance is maintained just to allow local departures from bilayer structure for fusion or related phenomena or to satisfy packing properties in the region of integral protein. It is, however, quite probable that this balance will strongly influence the permeability barrier characteristics of the membrane. As detailed in Section VI, our current understanding of the relation between lipid composition and related packing properties and membrane permeability is relatively unsophisticated. In our view, studies clarifying the relation between head group composition and acyl chain order and membrane permeability will constitute the next step toward obtaining an overall synthesis between lipid composition and membrane function.

ACKNOWLEDGMENTS

The research programs of Cullis, Hope, and Tilcock are supported by the Medical Research Council (MRC) of Canada, the British Columbia Heart Foundation, and the British Columbia Health Care Research Foundation. Dr. Cullis is an MRC Scientist. We thank Mr. M. Fettes for his expert assistance.

REFERENCES

1. **Silvius, J. R.,** Thermotropic phase transitions of pure lipids in model membranes and their modification by membrane proteins, in *Lipid-Protein Interactions*, Vol. 2, Jost, P. C. and Griffith, O. H., Eds., J. Wiley & Sons, 1982, chap. 7.
2. **Gorter, E. and Grendel, F.,** On bimolecular layers of lipids on the chromocytes of the blood, *J. Exp. Med.*, 41, 439, 1925.
3. **Kornberg, R. D. and McConnell, H. M.,** Lateral diffusion of phospholipids in a vesicle membrane, *Proc. Natl. Acad. Sci. U.S.A.*, 68, 2564, 1971.
4. **Singer, S. J. and Nicolson, G. L.,** The fluid mosaic model of the structure of cell membranes, *Science*, 175, 720, 1972.
5. **Op Den Kamp, J. A. F.,** Lipid asymmetry in membranes, *Annu. Rev. Biochem.*, 48, 47, 1979.
6. **Bangham, A. D., Standish, M. M., and Watkins, J. D.,** Diffusion of univalent ions across the lamellae of swollen phospholipids, *J. Mol. Biol.*, 13, 238, 1965.
7. **Huang, C.,** Studies on phosphatidylcholine vesicles. Formation and physical characteristics, *Biochemistry*, 8, 344, 1969.
8. **Cullis, P. R. and de Kruijff, B.,** ^{31}P NMR studies of unsonicated aqueous dispersions of neutral and acidic phospholipids: effects of phase transitions, p^{2}H and divalent cations on the motion in the phosphate region of the polar headgroup, *Biochim. Biophys. Acta*, 436, 523, 1976.
9. **De Kruijff, B., Cullis, P. R., and Radda, G. K.,** Outside-inside distributions and sizes of mixed phosphatitycholine-cholesterol vesicles, *Biochim. Biophys. Acta*, 436, 729, 1976.

10. **Schuh, J. R., Banerjee, U., Muller, L., and Chan, S. I.,** The phospholipid packing arrangement in small bilayer vesicles as revealed by proton magnetic resonance studies at 500 MHz, *Biochim. Biophys. Acta,* 687, 219, 1982.
11. **Ostro, M. J.,** *Liposomes,* Ostro, M. J., Ed., Marcel Dekker, New York, 1983, chap. 1.
12. **Mimms, L. T., Zampighi, G., Nozaki, Y., Tanford, C., and Reynolds, J. A.,** Formation of large unilamellar vesicles by detergent dialysis employing octylglucoside, *Biochemistry,* 20, 833, 1981.
13. **Batzri, S. and Korn, E. D.** Single bilayer liposomes prepared without sonication, *Biochim. Biophys. Acta,* 298, 1015, 1973.
14. **Deamer, D. and Bangham, A. D.,** Large volume liposomes by an ether vaporization method, *Biochim. Biophys. Acta,* 443, 629, 1976.
15. **Szoka, F. and Papahadjopoulos, D.,** Procedure for preparation of liposomes with large internal aqueous space and high capture by reverse phase evaporation, *Proc. Natl. Acad. Sci. U.S.A.,* 75, 4194, 1978.
16. **Hope, M. J., Bally, M. B., Webb, G., and Cullis, P. R.,** Production of large unilamellar vesicles by a rapid extrusion procedure. Characterization of size distribution, trapped volume and ability to maintain a membrane potential, *Biochim. Biophys. Acta,* 812, 55, 1985.
17. **Racker, E.,** Reconstitution and mechanism of action of ion pumps, in *A New Look at Mechanisms of Bioenergetics,* Academic Press, New York, 1976, 127.
18. **Safman, P. G. and Delbruck, M.,** Brownian motion in biological membranes, *Proc. Natl. Acad. Sci. U.S.A.,* 72, 3111, 1975.
19. **Cullis, P. R.,** Lateral diffusion rates of phosphatidylcholine in vesicle membranes. Effects of cholesterol and hydrocarbon phase transitions, *FEBS Lett.,* 70, 233, 1976.
20. **Lindblom, G., Johansson, L. B. A., and Arvidson, G.,** Effect of cholesterol in membranes, *Biochemistry,* 20, 2204, 1981.
21. **Ghosh, R. and Seelig, J.,** The interaction of cholesterol with bilayers of phosphatidylethanolamine, *Biochim. Biophys. Acta,* 691, 151, 1982.
22. **Seelig, J. and Seelig, A.,** Lipid conformation in model membranes and biological membranes, *Q. Rev. Biophys.,* 13, 19, 1980.
23. **Luzatti, V.,** X-ray diffraction studies of lipid-water systems, in *Biological Membranes,* Chapman, D., Ed., Academic Press, London, 1968, 71.
24. **Blaurock, A. E.,** Evidence of bilayer structure and of membrane interactions from X-ray diffraction analysis, *Biochim. Biophys. Acta,* 650, 167, 1982.
25. **Buldt, G., Gally, H. U., Seelig, J., and Zaccai, G.,** Neutron diffraction studies on phosphatidylcholine model membranes. I. Head group conformation, *J. Mol. Biol.,* 134, 673, 1979.
26. **Zaccai, G., Buldt, G., Seelig, A., and Seelig, J.,** Neutron diffraction studies on phosphatidylcholine model membranes. II. Chain conformation and segmental disorder, *J. Mol. Biol.,* 134, 693, 1979.
27. **Rand, R. P., Tinker, D. O., and Fast, P. G.,** Polymorphism of phosphatidylethanolamines from two natural sources, *Chem. Phys. Lipids,* 6, 333, 1971.
28. **Janiak, M. J., Small, D. M., and Shipley, G. G.,** Nature of the thermal pretransition of synthetic phospholipid: dimyristoyl and dipalmitoyllecithin, *Biochemistry,* 15, 4575, 1976.
29. **Lindblom, G., Larsson, K., Johansson, L., Fontell, K., and Forsen, S.,** The cubic phase of monoglyceride-water systems. Arguments for a structure based upon lamellar bilayer units, *J. Am. Chem. Soc.,* 101, 5465, 1979.
30. **Harlos, K. and Eibl, H.,** Influence of calcium on phosphatidylglycerol. Two separate lamellar structures, *Biochemistry,* 19, 895, 1980.
31. **Marsh, D. and Seddon, J. M.,** Gel-to-inverted hexagonal (L_β-H_{II}) phase transitions in phosphatidylethanolamines and fatty-acid phosphatidylcholine mixtures, demonstrated by ^{31}P NMR spectroscopy and X-ray diffraction, *Biochim. Biophys. Acta,* 690, 117, 1982.
32. **Luzatti, V., Tardieu, A., Gluik-Krzywicki, T., Rivas, E., and Reiss-Husson, F.,** Structure of the cubic phases of lipid-water systems, *Nature (London),* 220, 485, 1968.
33. **Larsson, K.,** On the structure of isotropic phases in lipid-water systems, *Chem. Phys. Lipids,* 9, 181, 1972.
34. **Boni, L. T. and Hui, S. W.,** Polymorphic phase behaviour of dilinoleylphosphatidylethanolamine and palmitoyloleylphosphatidylcholine mixtures. Structural changes between hexagonal, cubic and bilayer phase, *Biochim. Biophys. Acta,* 731, 177, 1983.
35. **Gruner, S. M., Milch, J. R., and Reynolds, G. T.,** Survey of two-dimensional electro-optical X-ray detectors, *Nucl. Instr. Methods,* 195, 287, 1982.
36. **Seelig, J.,** ^{31}P nuclear magnetic resonance and the head group structure of phospholipids in membranes, *Biochim. Biophys. Acta,* 515, 104, 1978.
37. **Kohler, S. J. and Klein, M. P.,** Orientation and dynamics of phospholipid head groups in bilayers and membranes determined from ^{31}P nuclear magnetic resonance chemical shielding tensors, *Biochemistry,* 16, 519, 1977.

38. **Cullis, P. R. and de Kruijff, B.,** Lipid polymorphism and the functional roles of lipids in biological membranes, *Biochim. Biophys. Acta,* 559, 399, 1979.
39. **Hemminga, M. A. and Cullis, P. R.,** ^{31}P NMR studies of oriented phospholipid multilayers, *J. Magnetic Resonance,* 47, 307, 1982.
40. **Taylor, M. G. and Smith, I. C. P.,** A comparison of spin-probe ESR, ^{2}H and ^{31}P nuclear magnetic resonance for the study of hexagonal phase lipids, *Chem. Phys. Lipids,* 28, 119, 1981.
41. **Thayer, A. M. and Kohler, S. J.,** Phosphorus-31 nuclear magnetic resonance spectra characteristic of hexagonal and isotropic phospholipid phases generated from phosphatidylethanolamine in the bilayer phase, *Biochemistry,* 20, 6831, 1981.
42. **Noggle, J. H., Maracek, J. F., Mandal, S. B., van Venetie, R., Rodgers, J., Jain, M. K., and Ramirez, F.,** Bilayers of phosphatidyldiacylglycerol and phosphatidylcholesterol give ^{31}P NMR spectra characteristic for hexagonal and isotropic phases, *Biochim. Biophys. Acta,* 691, 240, 1982.
43. **Hui, S. W., Stewart, T. P., Yeagle, P. L., and Albert, D.,** Bilayer to non-bilayer transitions in mixtures of phosphatidylethaonolamine and phosphatidylcholine. Implications for membrane properties, *Arch. Biochem. Biophys.,* 207, 227, 1981.
44. **Seddon, J. M., Kaye, R. D., and Marsh, D.,** Induction of the lamellar-inverted hexagonal phase transition in cardiolipin by protons and monovalent cations, *Biochim. Biophys. Acta,* 734, 347, 1983.
45. **Tilcock, C. P. S., Bally, M. B., Farren, S. B., Cullis, P. R., and Gruner, S. M.,** Cation-dependent segregation phenomena and phase behavior in model membrane systems containing phosphatidylserine. Influence of cholesterol and acyl chain composition, *Biochemistry,* 23, 2696, 1984.
46. **Vasilenko, I., de Kruijff, B., and Verkleij, A. J.,** The synthesis and use of thiophospholipids in ^{31}P NMR studies of lipid polymorphism, *Biochim. Biophys. Acta,* 685, 144, 1982.
47. **Tilcock, C. P. S. and Cullis, P. R.,** The polymorphic phase behavior of mixed phosphatidylserine-phosphatidylethanolamine model systems as detected by ^{31}P NMR. Effects of divalent cations and pH, *Biochim. Biophys. Acta,* 641, 189, 1981.
48. **Davis, J. H.,** The description of membrane lipid conformation, order and dynamics by ^{2}H NMR, *Biochim. Biophys. Acta,* 737, 117, 1983.
49. **Deamer, D. W., Leonard, R., Tardieu, A., and Branton, D.,** Lamellar and hexagonal lipid phases visualized by freeze-etching, *Biochim. Biophys. Acta,* 219, 47, 1970.
50. **Verkleij, A. J., van Echteld, C. J. A., Gerritsen, W. J., Cullis, P. R., and de Kruijff, B.,** The lipidic particle as an intermediate structure in membrane fusion processes and bilayer to hexagonal H_{II} transitions, *Biochim. Biophys. Acta,* 600, 620, 1980.
51. **Verkleij, A. J.,** Lipidic intramembranous particles, *Biochim. Biophys. Acta,* 779, 43, 1984.
52. **Sleytr, U. B. and Robard, A. W.,** Understanding the artefact problem in freeze-fracture replication, A review. *J. Microsc.,* 126, 101, 1982.
53. **Ververgaert, P. H. J. Th., Verkleij, A. J., Verhoeven, J. J., and Elbers, P. F.,** Spray freezing of liposomes, *Biochim. Biophys. Acta,* 311, 651, 1973.
54. **Zingsheim, H. P. and Plattner, H.,** Electron microscopic methods in membrane biology, in *Methods in Membrane Biology,* Vol. 7, Korn, E. D., Ed., Plenum Press, New York, 1976.
55. **Sen, A., Brain, A. P. R., Quinn, P. J., and Williams, W. P.,** Formation of inverted lipid micelles in aqueous dispersions of mixed sn-3-galactosyldiacylglycerols induced by heat and ethylene glycol, *Biochim. Biophys. Acta,* 686, 215, 1982.
56. **Tilcock, C. P. S. and Fisher, D.,** Interactions of glycerol monooleate and dimethylsulphoxide with phospholipids. A differential scanning calorimetry and ^{31} P NMR study, *Biochim. Biophys. Acta,* 686, 340, 1982.
57. **Siegel, D. P.,** Inverted micellar structures in bilayer membranes. Formation rates and half-lives, *Biophys. J.,* 45, 399, 1984.
58. **Wu, W., Huang, C., Conley, T. G., Martin, R. B., and Levin, I. W.,** Lamellar-micellar transition of 1-stearoyllysophosphatidylcholine assemblies in excess water, *Biochemistry,* 21, 5975, 1982.
59. **Cullis, P. R. and de Kruijff, B.,** The polymorphic phase behaviour of phosphatidylethanolamines of natural and synthetic origin. A ^{31}P NMR study, *Biochim. Biophys. Acta,* 513, 31, 1978.
60. **Mantsch, H. H., Martin, A., and Cameron, D. G.,** Characterization by infrared spectroscopy of the bilayer to nonbilayer phase transition of phosphatidylethanolamines, *Biochemistry,* 20, 3138, 1981.
61. **Fontell, K.,** Liquid crystalline behavior in lipid-water systems, *Prog. Chem. Fats Lipids,* 16, 145, 1978.
62. **Shipley, G. G.,** The liquid crystalline behavior of lipids, in *Fundamentals of Lipid Chemistry,* Burton, R. M. and Guerra, F. C., Eds., BI-Science, Mich., 1974.
63. **Farren, S. B., Sommerman, E., and Cullis, P. R.,** Production of specifically ^{2}H labelled unsaturated phospholipid in gram quantities: a convenient synthesis of [11,11-$^{2}H_2$] oleic acid, *Chem. Phys. Lipids,* 34, 279, 1984.
64. **Boroske, E. and Trahms, L.,** A ^{1}H and ^{13}C NMR study of motional changes of dipalmitoyllecithin associated with the pretransition, *Biophys. J.,* 42, 275, 1983.

65. **Hentschel, M., Hosemann, R., and Helfrich, W.**, Direct X-ray study of the molecular tilt in diplamitoyl lecithin bilayers, *Z. Naturforsch.*, 35, 643, 1980.
66. **Small, D.**, Phase equilibria and structure of dry and hydrated egg lecithin, *J. Lipid Res.*, 8, 551, 1967.
67. **Reiss-Husson, F.**, Structure des phases liquide-cristallines de differents phospholipides, monoglycerides, sphingolipides, anhydres ou en presence d'eau, *J. Mol. Biol.*, 25, 363, 1967.
68. **Luzatti, V., Gulik-Krzywicki, T., and Tardieu, A.**, Polymorphism of lecithins, *Nature (London)*, 218, 1031, 1968.
69. **Mantsch, H. H., Cameron, D. G., Tremblay, P. A., and Kates, M.**, Phosphatidylsulfocholine bilayers. An infrared spectroscopic characterization of the polymorphic phase behaviour, *Biochim. Biophys. Acta*, 689, 63, 1982.
70. **Jarrell, H. C., Byrd, R. A., Deslauriers, R., Ekiel, I., and Smith, I. C. P.**, Characterization of the phase behavior of phosphonolipids in model and biological membranes by ^{31}P NMR, *Biochim. Biophys. Acta*, 648, 80, 1981.
71. **Deslauriers, R., Ekiel, I., Byrd, R. A., Jarrell, H., and Smith, I. C. P.**, A ^{31}P NMR study of structural and functional aspects of phosphate and phosphonate distribution in Tetrahymena, *Biochim. Biophys. Acta*, 720, 329, 1982.
72. **Shipley, G. G., Avecilla, L. S., and Small, D. M.**, Phase behaviour and structure of aqueous dispersions of sphingomyelin, *J. Lipid Res.*, 15, 124, 1974.
73. **Untracht, S. H. and Shipley, G. G.**, Molecular interactions between lecithin and sphingomyelin, temperature and composition-dependent phase separation, *J. Biol. Chem.*, 252, 4449, 1977.
74. **Calhoun, W. I. and Shipley, G. G.**, Sphingomyelin-lecithin bilayers and their interaction with cholesterol, *Biochemistry*, 18, 1717, 1979.
75. **Hui, S. W., Stewart, T. P., and Yeagle, P. L.**, Temperature-dependent morphological and phase behaviour of sphingomyelin, *Biochim. Biophys. Acta*, 601, 271, 1980.
76. **Cullis, P. R. and Hope, M. J.**, The bilayer stabilizing role of sphingomyelin in the presence of cholesterol. A ^{31}P NMR study, *Biochim. Biophys. Acta*, 597, 533, 1980.
77. **Harlos, K. and Eibl, H.**, Hexagonal phases in phospholipids with saturated chains, *Biochemistry*, 20, 2888, 1981.
78. **Tilcock, C. P. S. and Cullis, P. R.**, The polymorphic phase behaviour and miscibility properties of synthetic phosphatidylethanolamines, *Biochim. Biophys. Acta*, 684, 212, 1982.
79. **Seddon, J. M., Cevc, G., and Marsh, D.**, Calorimetric studies of the gel-fluid (Lb-La) and lamellar-inverted hexagonal (La-H_{II}) phase transition in dialkyl and diacyl phosphatidylethanolamines, *Biochemistry*, 22, 1280, 1983.
80. **Hardman, P. D.**, Spin-label characterization of lamellar-to-hexagonal (H_{II}) phase transition in egg phosphatidylethanolamine aqueous dispersions, *Eur. J. Biochem.*, 124, 95, 1982.
81. **Boggs, J. M., Stamp, D. W., and Deber, C. M.**, Influence of ether linkage on the lamellar to hexagonal phase transition of ethanolamine phospholipids, *Biochemistry*, 20, 5728, 1981.
82. **Rilfors, L., Khan, A., Brentel, I., Wieslander, A., and Lindblom, G.**, Cubic liquid-cristalline phase with phosphatidylethanolamine from *Bacillus megaterium* containing branched acyl chains, *FEBS Lett.*, 149, 293, 1982.
83. **Israelachvili, J. N., Marcelja, S., and Horn, R. G.**, Physical principles of membrane organization, *Q. Rev. Biophys.*, 13, 121, 1980.
84. **Gally, H. N., Pluschke, G., Overath, P., and Seelig, J.**, Structure of *E. coli* membranes. Fatty acyl chain order parameters of inner and outer membranes and derived liposomes, *Biochemistry*, 19, 1638, 1980.
85. **Tilcock, C. P. S., Bally, M. B., Farren, S. B., and Cullis, P. R.**, Influence of cholesterol on the structural preferences of dioleoylphosphatidylethanolamine-dioleoylphosphatidylcholine systems: a phosphorus-31 and deuterium magnetic resonance study, *Biochemistry*, 21, 4596, 1982.
86. **Casal, H. L. and Mantsch, H. H.**, The thermotropic phase behavior of N-methylated dipalmitoyl phosphatidylethanolamines, *Biochim. Biophys. Acta*, 735, 387, 1983.
87. **Kirk, G., Gruner, S. M., Tilcock, C. P. S., and Cullis, P. R.**, manuscript in preparation.
88. **Jacobson, K. and Papahadjopoulos, D.**, Phase transitions and phase separations in phospholipid membranes induced by changes in temperature, pH and concentration of divalent cations, *Biochemistry*, 14, 152, 1975.
89. **van Dijck, P. W. M., Ververgaert, P. H. J. Th., Verkleij, A. J., van Deenen, L. L. M., and de Gier, J.**, Comparative studies on the effects of pH and Ca^{2+} on bilayers of various negatively charged phospholipids and their mixtures with phosphatidylcholine, *Biochim. Biophys. Acta*, 512, 84, 1978.
90. **Portis, A., Newton, C., Pangborn, W., and Papahadjopoulos, D.**, Studies on the mechanism of membrane fusion: evidence for an intermembrane Ca^{2+}-phospholipid complex, synergism with Mg^{2+} and inhibition by spectrin, *Biochemistry*, 18, 780, 1979.
91. **Hope, M. J. and Cullis, P. R.**, Effects of divalent cations and pH on phosphatidylserine model membranes: a ^{31}P NMR study, *Biochem. Biophys. Res. Commun.*, 92, 846, 1980.

92. **Cevc, G., Watts, A., and Marsh, D.,** Titration of the phase transition of phosphatidylserine bilayer membranes, *Biochemistry,* 20, 4955, 1981.
93. **Papahadjopoulos, D., Vail, W. J., Jacobson, K., and Poste, G.,** Chochleate lipid cylinders: formation by fusion of unilamellar lipid vesicles, *Biochim. Biophys. Acta,* 394, 483, 1975.
94. **Hark, S. K. and Ho, J. T.,** Raman study of calcium-induced fusion and molecular segregation of phosphatidylserine dimyristoylphosphatidylcholine membranes, *Biochim. Biophys. Acta,* 601, 54, 1980.
95. **van Dijck, P. W. M., Ververgaert, P. H. J. Th., Verkleij, A. J., van Deenen, L. L. M. and de Gier, J.,** Influence of Ca^{2+} and Mg^{2+} on the thermotropic behavior and permeability properties of liposomes prepared from dimyristoylphosphatidylglycerol and mixtures of dimyristoylphosphatidylglycerol and dimyristoylphosphatidylcholine, *Biochim. Biophys. Acta,* 406, 465, 1975.
96. **Findlay, E. J. and Barton, P. G.,** Phase behaviour of synthetic phosphatidylglycerols and binary mixtures with phosphatidylcholines in the presence and absence of calcium ions, *Biochemistry,* 17, 2400, 1978.
97. **Sacre, M. M., Hoffman, W., Turner, M., Tacanne, J. F., and Chapman, D.,** Differential scanning calorimetry studies of some phosphatidylglycerol lipid-water systems, *Chem. Phys. Lipids,* 25, 69, 1979.
98. **Harlos, K. and Eibl, H.,** Influence of calcium on phosphatidylglycerol. Two separate lamellar structures, *Biochemistry,* 19, 895, 1980.
99. **Ranck, J. L., Keira, T., and Luzatti, V.,** A novel packing of the hydrocarbon chain in lipids. The low temperature phases of dipalmitoyl phosphatidylglycerol, *Biochim. Biophys. Acta,* 488, 432, 1976.
100. **Farren, S. B. and Cullis, P. R.,** Polymorphism of phosphatidylglycerol-phosphatidylethanolamine model membrane systems. A ^{31}P NMR study, *Biochem. Biophys. Res. Commun.,* 97, 182, 1980.
101. **Michell, R. H.,** Inositol phospholipids and cell surface receptor function, *Biochim. Biophys. Acta,* 415, 81, 1975.
102. **Eibl, H. and Blume, A.,** The influence of charge on phosphatidic acid bilayer membranes, *Biochim. Biophys. Acta,* 553, 476, 1979.
103. **Jahnig, I., Harlos, K., Vogel, H., and Eibl, H.,** Electrostatic interactions at lipid membranes. Electrostatically induced tilt, *Biochemistry,* 18, 1459, 1979.
104. **Harlos, K., Stumple, J., and Eibl, H.,** Influence of pH on phosphatidic acid multilayers. A rippled structure at high pH values, *Biochim. Biophys. Acta,* 555, 409, 1979.
105. **Blume, A. and Eibl, H.,** The influence of charge on bilayer membranes. Calorimetric investigations of phosphatidic acid bilayers. *Biochim. Biophys. Acta,* 558, 13, 1979.
106. **Papahadjopoulos, D., Vail, W. J., Pangborn, W. A., and Poste, G.,** Studies on membrane fusion. II. Induction of fusion in pure phospholipid membranes by calcium ions and other divalent metals, *Biochim. Biophys. Acta,* 448, 265, 1976.
107. **Koter, M., de Kruijff, B., and van Deenen, L. L. M.,** Calcium-induced aggregation and fusion of mixed phosphatidylcholine-phosphatidic acid vesicles as studied by ^{31}P NMR, *Biochim. Biophys. Acta,* 514, 255, 1978.
108. **Verkleij, A. J., de Maagd, R., Leunissen-Bijvelt, J., and de Kruijff, B.,** Divalent cations and chlorpromazine can induce non-bilayer structures in phosphatidic acid-containing model membranes, *Biochim. Biophys. Acta,* 684, 255, 1982.
109. **Farren, S. B., Hope, M. J., and Cullis, P. R.,** Polymorphic phase preferences of phosphatidic acid. A ^{31}P and ^{2}H NMR study, *Biochem. Biophys. Res. Commun.,* 111, 675, 1983.
110. **Rainier, S., Jain, M. K., Ramirez, F., Ioannou, P. V., Maracek, J. F., and Wagner, R.,** Phase transition characteristics of diphosphatidylglycerol (cardiolipin) and stereoisomeric phosphatidyldiacylglycerol bilayers. Mono- and divalent metal ion effects, *Biochim. Biophys. Acta,* 558, 187, 1979.
111. **Rand, R. P. and Sengupta, S.** Cardiolipin forms hexagonal structures with cations, *Biochim. Biophys. Acta,* 255, 484, 1972.
112. **Cullis, P. R., Verkleij, A. J., and Ververgaert, P. H. J. Th.,** Polymorphic phase behaviour of cardiolipin as detected by ^{31}P NMR and freeze-fracture techniques. Effects of calcium, dibucaine and chlorpromazine, *Biochim. Biophys. Acta,* 513, 11, 1978.
113. **Vasilenko, I., de Kruijff, B., and Verkleij, A. J.,** Polymorphic phase behaviour of cardiolipin from bovine heart and from *Bacillus subtilis* as detected by ^{31}P NMR and freeze-fracture techniques. Effects of Ca^{2+}, Mg^{2+} and temperature, *Biochim. Biophys. Acta,* 684, 282, 1982.
114. **de Kruijff, B., Verkleij, A. J., Leunissen-Bijvelt, J., van Echteld, C. J. A., Hille, J., and Rijnbout, H.,** Further aspects of the Ca^{2+}-dependent polymorphism of bovine heart cardiolipin, *Biochim. Biophys. Acta,* 693, 1, 1982.
115. **Seddon, J. M., Kaye, R. D., and Marsh, D.,** Induction of the lamellar-inverted hexagonal phase transition in cardiolipin by protons and monovalent cations, *Biochem. Biophys. Acta,* 734, 347, 1983.
116. **Nayar, R., Schmid, S. L., Hope, M. J., and Cullis, P. R.,** Structural preferences of phosphatidylinositol and phosphatidylinositol-phosphatidylethanolamine model membranes. Influence of Ca^{2+} and pH, *Biochim. Biophys. Acta,* 688, 169, 1982.

117. **Kanfer, J. N. and Hakamori, S.,** *Sphingolipid Biochemistry, Handbook of Lipid Research,* Vol. 3, Plenum Press, New York, 1983.
118. **Pascher, I. and Sundell, S.,** Molecular arrangements in sphingolipids. The crystal structure of cerebroside, *Chem. Phys. Lipids,* 20, 175, 1977.
119. **Abrahamsson, S., Pascher, I., Larsson, K., and Karlsson, K.-A.,** Molecular arrangements in glycolipids, *Chem. Phys. Lipids,* 8, 152, 1972.
120. **Cullis, P. R., de Kruijff, B., Hope, M. J., Nayar, R., and Rietveld, A.,** Structural properties of phospholipids in the rat liver mitochondrial membranes: ^{31} NMR study, *Biochim. Biophys. Acta,* 600, 625, 1980.
121. **Ruocco, M. J., Atkinson, D., Small, D. M., Skarjune, R. P., Oldfield, E., and Shipley, G. G.,** X-ray diffraction and calorimetric study of anhydrous and hydrated N-palmitoylgalactosylsphingosine (cerebroside), *Biochemistry,* 20, 5975, 1981.
122. **Bunow, M. R. and Levin, I. W.,** Molecular conformation of cerebrosides in bilayers determined by Raman spectroscopy, *Biophys. J.,* 32, 1007, 1981.
123. **Curatolo, W., Small, D. M., and Shipley, G. G.,** Phase behaviour and structural characteristics of hydrated bovine brain gangliosides, *Biochim. Biophys. Acta,* 468, 11, 1977.
124. **Klopfenstein, W. E., de Kruijff, B., Verkleij, A. J., Demel, R. A., and van Deenen, L. L. M.,** Differential scanning calorimetry on mixtures of lecithin, lysolecithin and cholesterol, *Chem. Phys. Lipids,* 13, 215, 1974.
125. **van Echteld, C. J. A., de Kruijff, B., and de Gier, J.** Differential miscibility properties of various phosphatidylcholine-lysophosphatidylcholine mixtures, *Biochim. Biophys. Acta,* 595, 71, 1980.
126. **Jain, M. K. and de Haas, G. H.,** Structure of 1-acyl lysophosphatidylcholine and fatty acid complex in bilayers, *Biochim. Biophys. Acta,* 642, 203, 1981.
127. **Wu, W., Huang, C., Conley, T. G., Martin, R. B., and Levin, I. W.,** Lamellar-micellar transition of 1-stearoyllysophosphatidylcholine assemblies in excess water, *Biochemistry,* 21, 5957, 1982.
128. **Hauser, H., Guyer, W., Spiess, M., Pascher, I., and Sundell, S.,** The polar group conformation of a lysophosphatidylcholine analogue in solution. A high-resolution nuclear magnetic resonance study, *J. Mol. Biol.,* 137, 265, 1980.
129. **Quinn, P. J. and Williams, W. P.,** The structural role of lipids in photosynthetic membranes, *Biochim. Biophys. Acta,* 737, 223, 1983.
130. **Sen, A., Mannock, D. A., Collins, D. J., Quinn, P. J., and Williams, W. P.,** Thermotropic phase properties and structure of 1,2-distearoylgalactosyl glycerols in aqueous systems, *Proc. R. Soc. London Ser. B:* 218, 349, 1983.
131. **Rivas, E. and Luzatti, V.,** Polymorphisme des lipides polaires et des galactolipides de chloroplastes de mais, en presence d'eau, *J. Mol. Biol.,* 41, 261, 1969.
132. **Shipley, G. G., Green, J. P., and Nichols, B. W.,** The phase behaviour of monoglactosyl, digalactosyl and sluphoquinovosyl diglycerides, *Biochim. Biophys. Acta,* 311, 531, 1973.
133. **Wieslander, A., Ulmius, J., Lindblom, G., and Fontell, K.,** Water binding and phase structure for different *Acholeplasma laidlawii* membrane lipids studied by deuteron magnetic resonance and X-ray diffraction, *Biochim. Biophys. Acta,* 512, 241, 1978.
134. **Wieslander, A., Christiansson, A., Rilfors, L., and Lindblom, G.,** Lipid bilayer stability in membranes. Regulation of lipid composition in *Acholeplasma laidlawii* as governed by molecular shape, *Biochemistry,* 19, 3650, 1980.
135. **Wieslander, A., Rilfors, L., Johansson, L. B., and Lindblom, G.,** Reversed cubic phase with membrane glucolipids from *Acholeplasma laidlawii.* ^{1}H, ^{2}H and diffusion nuclear magnetic resonance measurements, *Biochemistry,* 20, 730, 1981.
136. **Cullis, P. R., van Dijck, P. W. M., de Kruijff, B., and de Gier, J.,** Effect of cholesterol on the properties of equimolar mixtures of synthetic phosphatidylethanolamine and phosphatidylcholine. A ^{31}P NMR and differential scanning calorimetry study, *Biochim. Biophys. Acta,* 513, 21, 1978.
137. **Bally, M. B., Tilcock, C. P. S., Hope, M. J., and Cullis, P. R.,** Polymorphism of phosphatidylethanolamine-phosphatidylserine model systems. Influence of cholesterol and Mg^{2+} on Ca^{2+} triggered bilayer to hexagonal (H_{II}) transitions, *Can. J. Biochem. Cell Biol.,* 61, 346, 1983.
138. **Demel, R. A. and de Kruijff, B.,** The function of sterols in membranes, *Biochim. Biophys. Acta,* 457, 109, 1976.
139. **Cullis, P. R. and de Kruijff, B.,** Polymorphic phase behaviour of lipid mixtures as detected by ^{31}P NMR. Evidence that cholesterol may destabilize bilayer structure in membrane systems containing phosphatidylethanolamine, *Biochim. Biophys. Acta,* 507, 207, 1978.
140. **Gally, J. and de Kruijff, B.,** Correlation between molecular shape and hexagonal H_{II} phase promoting ability of sterols, *FEBS Lett.,* 143, 133, 1982.
141. **Seeman, P.,** The molecular mechanism of anesthesia, *Pharmacol. Rev.,* 24, 583, 1972.

142. **Verkleij, A. J., de Maagd, R., Leunissen-Bijvelt, J., and de Kruijff, B.,** Divalent cations and chlorpromazine can induce non-bilayer structures in phosphatidic acid containing model membranes, *Biochim. Biophys. Acta,* 684, 225, 1982.
143. **Cullis, P. R. and Verkleij, A. J.,** Modulation of membrane structure by Ca^{2+} and dibucaine, *Biochim. Biophys. Acta,* 552, 546, 1979.
144. **Cullis, P. R., Hornby, A. P., and Hope, M. J.,** Effects of anaesthetics on lipid polymorphism, in *Molecular Mechanisms of Anesthesia: Progress in Anesthesiology,* Vol. 2, Fink, B. R., Ed., Raven Press, New York, 1980, 397.
145. **Hornby, A. P. and Cullis, P. R.,** Influence of local and neutral anaesthetics on the polymorphic phase preferences of egg yolk phosphatidylethanolamine, *Biochim. Biophys. Acta,* 647, 285, 1981.
146. **Tilcock, C. P. S., Hope, M. J., and Cullis, P. R.,** Influence of cholesterol esters of varying unsaturation on the polymorphic phase preferences of egg phosphatidylethanolamine, *Chem. Phys. Lipids,* 35, 363, 1984.
147. **de Kruijff, B., Verkleij, A. J., van Echteld, C. J. A., Gerritsen, W. J., Mombers, C., Noordam, P. C., and de Gier, J.,** The observation of lipid particles in lipid bilayers as seen by ^{31}P NMR and freeze-fracture electron microscopy, *Biochim. Biophys. Acta,* 555, 200, 1979.
148. **Hauser, H., Pascher, L., Pearson, R. H., and Sunbell, S.,** Preferred conformation and molecular packing of phosphatidylethanolamine and phosphatidylcholine, *Biochim. Biophys. Acta,* 650, 21, 1981.
149. **Madden, T. D. and Cullis, P. R.,** Stabilization of bilayer structure for unsaturated phosphatidylethanolamines by detergents, *Biochim. Biophys. Acta,* 684, 149, 1982.
150. **Rand, R. P., Pangborn, W. A., Purdon, A. D., and Tinker, D. O.,** Lysolecithin and cholesterol interact stoichiometrically forming bimolecular lamellar structures in the presence of excess water, or lysolecithin, or cholesterol, *Can. J. Biochem.,* 53, 189, 1975.
151. **Jain, M. K., van Echteld, C. J. A., Ramirez, F., de Gier, J., de Haas, G. H. and van Deenen, L. L. M.,** Association of lysophosphatidylcholine with fatty acids in aqueous phase to form bilayers, *Nature (London),* 284, 486, 1980.
152. **de Kruijff, B., Cullis, P. R., Verkleij, A. J., Hope, M. J., van Echteld, C. J. A., and Taraschi, T. F.,** Lipid polymorphism and membrane function, in *Enzymes of Biological Membranes,* Martinosi, A., Ed., Plenum Press, New York, 1984.
153. **Cullis, P. R., de Kruijff, B., Hope, M. J., Nayar, R., and Schmid, S. L.,** Phospholipids and membrane transport, *Can. J. Biochem.,* 58, 1091, 1980.
154. **Ahkong, Q. F., Fisher, D., Tampion, W., and Lucy, J. A.,** The fusion of erythrocytes by fatty acids, esters, retinol and α-tocophenol, *Biochem. J.,* 136, 147, 1973.
155. **Kennedy, A. and Rice-Evans, C.,** A spectrofluorometric study of the interaction of glycerol monooleate with human erythrocyte ghosts, *FEBS Lett.,* 69, 45, 1976.
156. **Ahkong, Q. F., Fisher, D., Tampion, W. and Lucy, J. A.,** Mechanism of membrane fusion, *Nature (London),* 253, 194, 1975.
157. **Chandler, D. E. and Heuser, J. E.,** Arrest of membrane fusion events in mast cells by quick freezing, *J. Cell Biol.,* 86, 666, 1980.
158. **Cullis, P. R. and Hope, M. J.,** Effects of fusogenic agents on the membrane structure of erythrocyte ghosts and the mechanism of membrane fusion, *Nature (London),* 271, 672, 1978.
159. **Hope, M. J. and Cullis, P. R.,** The role of non-bilayer lipid structure on the fusion of human erythrocytes induced by lipid fusogens, *Biochim. Biophys. Acta,* 640, 82, 1981.
160. **Papahadojopoulos, D., Portis, A., and Pangborn, W.,** Calcium induced lipid phase transitions and membrane fusion, *Ann. N.Y. Acad. Sci.,* 308, 50, 1978.
161. **Papahadjopoulos, D., Poste, G., and Vail, W. J.,** Studies on membrane fusion with natural and model membranes, in *Methods in Membrane Biology,* Vol. 10, Korn, E. D., Ed., Plenum Press, New York, 1978, 1.
162. **Papahadjopoulos, D., Poste, G., Schaeffer, B. E., and Vail, W. J.,** Membrane fusion and molecular segregation in phospholipid vesicles, *Biochim. Biophys. Acta,* 352, 10, 1974.
163. **Hope, M. J., Walker, D. C., and Cullis, P. R.,** Ca^{2+} and pH induced fusion of small unilamellar vesicles consisting of phosphatidylethanolamine and negatively charged phospholipids: a freeze-fracture study, *Biochem. Biophys. Res. Commun.,* 110, 15, 1983.
164. **Verkleij, A. J. van Venetie, R., Leunissen-Bijvelt, J., de Kruijff, B., Hope, M. J., and Cullis, P. R.,** Membrane fusion and lipid polymorphism in *Physical Methods on Biological Membranes and Their Models,* Conti, F., Ed., Plenum Press, New York in press.
165. **Wilschut, J. and Papahadjopoulos, D.,** Ca^{2+} induced fusion of phospholipid vesicles monitored by mixing of aqueous contents, *Nature (London),* 281, 690, 1979.
166. **Wilschut, J., Düzgünes, N., Fraley, R., and Papahadjopoulos, D.,** Studies on the mechanism of membrane fusion: kinetics of calcium ion induced fusion of phosphatidylserine vesicles followed by a new assay for mixing of aqueous vesicle contents, *Biochemistry,* 19, 6011, 1980.

167. **Wilschut, J., Holsappel, M., and Jansen, R.,** Ca^{2+} induced fusion of cardiolipin-phosphatidylcholine vesicles monitored by mixing of aqueous contents, *Biochim. Biophys. Acta,* 690, 297, 1982.
168. **Sundler, R. and Papahadjopoulos, D.,** Control of membrane fusion by phospholipid head groups. I. Phosphatidate/phosphatidylinositol specificity, *Biochim. Biophys. Acta,* 649, 743, 1981.
169. **Sundler, R., Düzgünes, N., and Papahadjopoulos, D.,** Control of membrane fusion by phospholipid head groups. II. The role of phosphatidylethanolamine in mixtures with phosphatidic acid and phosphatidylinositol, *Biochim. Biophys. Acta,* 649, 751, 1981.
170. **Smith, A. D. and Winkler, H.,** Fundamental mechanisms in the release of catecholamines, in *Handbook of Experimental Pharmacology,* Blashko, H. and Musholl, E., Eds., Springer-Verlag, New York, 1972, 538.
171. **Trifaro, J. M.,** Common mechanisms of hormone secretion, *Ann. Rev. Pharmacol. Toxicol.,* 17, 27, 1977.
172. **Meldolisi, J., Bargese, N., de Camilli, P., and Ceccarelli, B.,** Cytoplasmic membranes and the secretory process, *Cell. Surface Rev.,* 5, 509, 1978.
173. **Edwards, W., Phillips, J. H., and Morris, S. J.,** Structural changes in chromaffin granules induced by divalent cations, *Biochim. Biophys. Acta,* 356, 164, 1974.
174. **Ekerdt, R., Dahl, G., and Gratzl, M.,** Membrane fusion of secretory vesicles and liposomes. Two different types of fusion, *Biochim. Biophys. Acta,* 646, 10, 1981.
175. **Nayar, R., Hope, M. J., and Cullis, P. R.,** Phospholipid as adjuncts for Ca^{2+} stimulated release of chromaffin granule contents. Implications for mechanisms of exocytotic release, *Biochemistry,* 21, 4583, 1982.
176. **Burnell, E. E., Cullis, P. R., and de Kruijff, B,** Effects of tumbling and lateral diffusion on phosphatidylcholine model membrane ^{31}P NMR lineshapes, *Biochim. Biophys. Acta,* 603, 63, 1980.
177. **Noordam, P. C., van Echteld, C. J. A., de Kruijff, B., Verkleij, A. J., and de Gier, J.,** Barrier characteristics of membrane model systems containing unsaturated phosphatidylethanolamines, *Chem. Phys. Lipids,* 27, 221, 1980.
178. **Morré, D. J., Kartenbeck, J., and Franke, W. W.,** Membrane flow and interconversion among endomembranes, *Biochim. Biophys, Acta,* 559, 71, 1979.
179. **Fettiplace, R., Gordon, I. G. H., Hladky, S. B., Requens, J., Zingshen, H. B., and Haydon, D. A.,** Techniques in the formation and examination of black lipid bilayer membranes, in *Methods in Membrane Biology,* Vol. 4, Korn, E. D., Ed., Plenum Press, New York, 1974, 1.
180. **Pagano, R. E., Ruysschaert, J. M., and Miller, I. R.,** The molecular composition of lipid bilayer membrane in aqueous solution, *J. Membrane Biol.* 10, 11, 1972.
181. **Hanai, T. and Haydon, D. A.,** The permeability to water of bimolecular lipid membranes, *J. Theor. Biol.,* 11, 370, 1966.
182. **Blok, M. C., van Deenen, L. L. M., and de Gier, J.,** Effect of the gel to liquid crystalline phase transition on the osmotic behaviour of phosphatidylcholine liposomes, *Biochim. Biophys. Acta,* 433, 1, 1976.
183. **Bard, A. J. and Faulkner, L. R.,** *Electrochemical Methods: Fundamentals and Applications,* Wiley, New York, 1980.
184. **Fettiplace, R. and Haydon, D. A.,** Water permeability of lipid membrane, *Physiol. Rev.,* 60, 510, 1980.
185. **Reeves, J. P. and Douben, R. M.,** Water permeability of phospholipid vesicles, *J. Membrane Biol.,* 3, 123, 1970.
186. **Trauble, H.** The movement of molecules across lipid membranes. A molecular theory, *J. Membrane Biol.,* 4, 193, 1971.
187. **Poznansky, M., Tang, S., White, P. C., Milgram, J. M., and Selenen, M.,** Non-electrolyte diffusion across lipid bilayer systems, *J. Gen. Physiol.,* 67, 45, 1976.
188. **Deamer, D. W.,** Proton permeability in biological and model membranes, in *Intracellular pH: Its Measurement, Regulation and Utilization in Cellular Functions,* Alan R. Liss, New York, 1982, 173.
189. **Nichols, J. W. and Deamer, D. W.,** Net proton-hydroxyl permeability of large unilamellar liposomes measured by an acid-base titration technique, *Proc. Natl. Acad. Sci. U.S.A.,* 77, 2038, 1980.
190. **Gutknecht, J. and Walter, A.,** Transport of protons and hydrochloric acid through lipid bilayer membranes, *Biochim. Biophys. Acta,* 641, 183, 1981.
191. **Nozaki, Y. and Tanford, C.,** Proton and hydroxide ion permeability of phospholipid vesicles, *Proc. Natl. Acad. Sci. U.S.A.,* 78, 4324, 1981.
192. **Papahadjopoulos, D. and Watkins, J. C.,** Phospholipid model membranes. II. Permeability properties of hydrated liquid crystals, *Biochim. Biophys. Acta,* 135, 639, 1967.
193. **Bangham, A. D.,** Lipid bilayers and biomembranes, *Ann. Rev. Biochem.,* 41, 753, 1972.
194. **Singer, M. A.,** Transfer of anions across phospholipid membranes, *Can. J. Physiol. Pharmacol.,* 51, 523, 1973.
195. **Nicholls, P. and Miller, N.,** Chloride diffusion from liposomes, *Biochim. Biophys. Acta,* 356, 184, 1974.
196. **Toyoshima, Y. and Thompson, T. E.,** Chloride flux in bilayer membranes. Chloride permeability in aqueous dispersions of single walled bilayer vesicles, *Biochemistry,* 14, 1525, 1975.

197. **Haydon, D. A. and Hladky, S. B.,** Ion transport across thin lipid membranes. A critical discussion of mechanisms in selected systems, *Q. Rev. Biophys.,* 5, 187, 1972.
198. **Robertson, R. N. and Thompson, T. E.,** The function of phospholipid polar headgroups in membranes, *FEBS Lett.,* 76, 16, 1977.
199. **Hauser, H., Phillips, M. C., and Stubbs, M.,** Ion permeability of phospholipid bilayers, *Nature (London),* 239, 342, 1972.
200. **Jain, M. K. and Wagner, R. C.,** *Introduction to Biological Membranes,* Wiley, New York, 1980, 134.
201. **Fettiplace, R.,** The influence of the lipid on the water permeability of artificial membranes, *Biochim. Biophys. Acta,* 513, 1, 1978.
202. **de Gier, J., Mandersloot, J. S., and van Deenen, L. L. M.,** Lipid composition and permeability of liposomes, *Biochim. Biophys. Acta,* 150, 666, 1968.
203. **Finkelstein, A. and Cass, A.,** Effect of cholesterol on the water permeability of thin lipid membranes, *Nature (London),* 216, 717, 1967.
204. **Rossignol, M., Thomas, P., and Grignon, C.,** Proton permeability of liposomes from natural phospholipid mixtures, *Biochim. Biophys. Acta,* 684, 195, 1982.
205. **Putney, J. W., Weiss, S. J., van der Waal, C. M., and Haddas, R. A.,** Is phosphatidic acid a calcium ionophore under neurohumoral control? *Nature (London),* 284, 345, 1980.
206. **Salmon, D. M. and Honeyman, T. W.,** Proposed mechanism of cholinergic action in smooth muscle, *Nature (London),* 284, 344, 1980.
207. **Serhan, C. N., Anderson, P., Goodman, E., Dunham, P., and Wiessmann, G.,** Phosphatidate and oxidized fatty acids are calcium ionophores, *J. Biol. Chem.* 256, 2736, 1981.
208. **Serhan, C. N., Fridovitch, J., Goetzl, E. J., Dunham, P., and Wiessmann, G.,** Leukotriene B_4 and phosphatidic acid are calcium ionophores, *J. Biol. Chem.,* 257, 4746, 1982.
209. **Nayar, R., Mayer, L. D., Hope, M. J., and Cullis, P. R.,** Phosphatidic acid as a calcium ionophore in large unilamellar vesicle systems, *Biochim. Biophys. Acta,* 777, 343, 1984.
210. **Holmes, R. P. and Yoss, N. L.,** Failure of phosphatidic acid to translocate Ca^{2+} across phopsphatidylcholine membranes, *Nature (London),* 305, 637, 1983.
211. **Warren, G. B., Toon, P. A., Birdsall, N. J. M., Lee, A. G., and Metcalfe, J. C.,** Reversible lipid titrations of the activity of pure adenosine triphosphatase-lipid complexes, *Biochemistry,* 13, 5501, 1974.
212. **Warren, G. B., Houslay, M. D., Metcalfe, J. C., and Birdsall, N. J. M.,** Cholesterol is excluded from the phospholipid annulus surrounding an active calcium transport protein, *Nature (London),* 255, 684, 1975.
214. **Kimelberg, H. K.,** *Dynamic Aspects of Cell Surface Organization,* Poste, G. and Nicolson, G. L., Eds., North-Holland, Amsterdam, 1977, 205.
215. **Vik, S. B. and Capaldi, R. A.,** Lipid requirements for cytochrome *c* oxidase activity, *Biochemistry,* 16, 5755, 1977.
216. **Madden, T. D., Hope, M. J., and Cullis, P. R.,** Lipid requirements for coupled cytochrome oxidase vesicles, *Biochemistry,* 22, 1970, 1983.
217. **Madden, T. D., Hope, M. J., and Cullis, P. R.,** Influence of vesicle size and oxidase content on respiratory control in reconstituted cytochrome oxidase vesicles, *Biochemistry,* 23, 1413, 1984.
218. **Racker, E.,** Reconstitution of cytochrome oxidase vesicles and conferral of sensitivity to energy transfer inhibitors, *J. Membrane Biol.,* 10, 221, 1973.
219. **de Pont, J., van Prooijen, A., and Bonting, S. L.,** Role of negatively charged phospholipids in Na^+, K^+ ATPase from rabbit kidney, *Biochim. Biophys. Acta,* 508, 464, 1978.
220. **Hilden, S. and Hokin, L.,** Coupled Na^+-K^+ transport in vesicles containing purified Na^+, K^+ ATPase and only phosphatidylcholine, *Biochem. Biophys. Res. Commun.,* 69, 521, 1976.
221. **Navarro, J., Toivio-Kinnucan, M., and Racker, E.,** Effect of lipid composition on the Ca^{2+}/ATP coupling ratio of the Ca^{2+}-ATPase of sarcoplasmic reticulum, *Biochemistry,* 23, 130, 1984.
222. **Dean, W. L. and Tanford, C.,** Reactivation of a lipid depleted Ca^{2+} ATPase by a non-ionic detergent, *J. Biol. Chem.,* 252, 3551, 1977.
223. **Oldfield, E. Gilmore, R., Glaser, M., Gutowsky, H. S., Hshung, J. C., Kang, S. Y., King, J. E., Meadows, M., and Rice, D.,** Deuterium nuclear magnetic resonance investigation of the effects of proteins and polypeptides on hydrocarbon chain order in model membrane systems, *Proc. Natl. Acad. Sci. U.S.A.,* 75, 4657, 1978.
224. **Rice, D. M., Hsung, J. C., King, T. E. and Oldfield, E.,** Protein-lipid interactions. High field ^2H and ^{31}P NMR investigation of the cytochrome *c* oxidase-phospholipid interaction and the effects of cholate, *Biochemistry,* 18, 5885, 1979.
225. **Papahadjopoulos, D., Moscarello, M., Eylan, E. H., and Isac, T.,** Effects of proteins on thermotropic phase transitions of phospholipid membranes, *Biochim. Biophys. Acta,* 401, 317, 1975.
226. **de Kruijff, B. and Cullis, P. R.,** The influence of poly-L-lysine on phospholipid polymorphism. Evidence that electrostatic polypeptide-phospholipid interactions can modulate bilayer-nonbilayer transitions, *Biochim. Biophys. Acta,* 601, 235, 1980.

227. **Habermann, E.,** Mellitin structure and activity, in *Natural Toxins*, Eaher, A. and Wadstrom, B., Eds., Permagon Press, New York, 1980, 173.
228. **Duforq, C. J., Faucon, J. F., Bernard, E., Perolet, M., Tessier, M., Bougis, P., van Rietschoten, J., Delori, P., and Rochat, H.,** Structure-function relationships for cardiotoxins interacting with phospholipids, *Toxicon*, 20, 165, 1982.
229. **Gulik-Krzywicki, T., Balerna, M., Vincent, J. P., and Lazdunski, M.,** Freeze-fracture study of cardiotoxin action on axonal membrane and axonal membrane lipid vesicles, *Biochim. Biophys. Acta*, 643, 101, 1981.
230. **Nicholls, P.,** Cytochrome *c* binding to enzymes and membranes, *Biochim. Biophys. Acta*, 346, 261, 1974.
231. **Brown, L. R. and Wuthrich, K.,** NMR and ESR studies of the interactions of cytochrome *c* with mixed cardiolipin-phosphatidylcholine vesicles, *Biochim. Biophys. Acta*, 468, 389, 1977.
232. **Kimbelberg, H. K. and Papahadjopoulos, D.,** Interactions of basic proteins with phospholipid membranes. Binding and changes in the sodium permeability of phosphatidylserine vesicles, *J. Biol. Chem.*, 246, 1142, 1971.
233. **de Kruijff, B. and Cullis, P. R.,** Cytochrome *c* specifically induces non-bilayer structures in cardiolipin containing model membranes, *Biochim. Biophys. Acta*, 602, 477, 1980.
234. **Rietveld, A., Syens, P., Verkleij, A. J., and de Kruijff, B.,** Interaction of cytochrome *c* and its precursor apocytochrome *c* with various lipids, *EMBO J.*, 2, 907, 1983.
235. **Rietveld, A. and de Kruijff, B.,** Is the mitochondrial precursor protein apocytochrome c able to pass a lipid bilayer?, *J. Biol. Chem.*, 259, 6704, 1985.
236. **van Echteld, C. J. A., van Stigt, R., de Kruijff, B., Leunissen-Bijvelt, J., Verkleij, A. J., and de Gier, J.,** Gramicidin promotes formation of the hexagonal H_{II} phase in aqueous dispersions of phosphatidylethanolamine and phosphatidylcholine, *Biochim. Biophys. Acta*, 648, 287, 1981.
237. **van Echteld, C. J. A., de Kruijff, B., Verkleij, A. J., Leunissen-Bijvelt, J., and de Gier, J.,** Gramicidin induces the formation of non-bilayer structures in phosphatidylcholine dispersions in a fatty acid chain length-dependent way, *Biochim. Biophys. Acta*, 692, 126, 1982.
238. **de Kruijff, B., Cullis, P. R., Verkleij, A. J., Hope, M. J., van Echteld, C. J. A., Taraschi, T. F., van Hoogevest, P., Killian, J. A., and van der Steen, A. T. M.,** Modulation of lipid polymorphism by lipid-protein interactions, in *Progress in Protein-Lipid Interactions*, Vol. 1, in press.
239. **Killian, J. A., de Kruijff, B., van Echteld, C. J. A., Verkleij, A. J., Leunissen-Bijvelt, J., and de Gier, J.,** Mixtures of gramicidin and lysophosphatidylcholine form lamellar structures, *Biochim. Biophys. Acta*, 728, 141, 1983.
240. **Tomita, M. and Marchesi, V. T.,** Amino acid sequence and oligosaccharide attachment sites of human erythrocyte glycophorin, *Proc. Natl. Acad. Sci. U.S.A.*, 72, 2964, 1975.
241. **de Kruijff, B., van Zoelen, E. J. J., and van Deenen, L. L. M.,** Glycophorin facilitates the transbilayer movement of phosphatidylcholine in vesicles, *Biochim. Biophys. Acta*, 509, 537, 1978.
242. **van Hoogevest, P., du Maine, A. P. M., and de Kruijff, B.,** Characterization of the permeability increase induced by the incorporation of glycophorin in phosphatidylcholine vesicles. Determination of the size of the non-specific permeation pathway, *FEBS Lett.*, 157, 41, 1983.
243. **Taraschi, T. F., de Kruijff, B., Verkleij, A. J., and van Echteld, C. J. A.,** Effect of glycophorin on lipid polymorphism. A ^{31}P NMR study, *Biochim. Biophys. Acta*, 685, 153, 1982.
244. **Israelachvili, J. N.,** Refinement of the fluid mosaic model of membrane structure, *Biochim. Biophys. Acta*, 469, 221, 1977.
245. **van der Steen, A. J. M., de Kruijff, B., and de Gier, J.,** Glycophorin incorporation increases the bilayer permeability of large unilamellar vesicles in a lipid dependent manner, *Biochim. Biophys. Acta*, 691, 13, 1982.
246. **van Hoogevest, P., du Maine, A. P. M., de Kruijff, B., and de Gier, J.,** The influence of lipid composition on glycophorin induced bilayer permeability, *Biochim. Biophys. Acta*, 771, 119, 1984.
247. **Cullis, P. R. and Grathwohl, Ch.,** Hydrocarbon phase transitions and lipid-protein interactions in the erythrocyte membrane, *Biochim. Biophys. Acta*, 471, 213, 1977.
248. **van Meer, G., de Kruijff, B., op den Kamp, J. A. F., and van Deenen, L. L. M.,** Preservation of bilayer structure in human erythrocytes and erythrocyte ghosts after phospholipase treatment, *Biochim. Biophys. Acta*, 600, 1, 1980.
249. **Hope, M. J. and Cullis, P. R.,** The bilayer stability of inner monolayer lipids from the human erythrocyte, *FEBS Lett.*, 107, 323, 1979.
250. **de Kruijff, B., Rietveld, A., and Cullis, P. R.,** ^{31}P NMR studies on membrane phospholipids in microsomes, rat liver slices and intact perfused rat liver, *Biochim. Biophys. Acta*, 600, 343, 1980.
251. **Crowe, L. M. and Crowe, J. H.,** Hydration dependent hexagonal phase lipid in a biological membrane, *Arch. Biochem. Biophys.*, 217, 582, 1982.
252. **de Kruijff, B., Nayar, R., and Cullis, P. R.,** ^{31}P NMR studies on phospholipid structure in membranes of intact, functionally active, rat liver mitochondria, *Biochim. Biophys. Acta*, 684, 47, 1982.

253. **Krebs, J. J. R., Hauser, H., and Carafoli, E.,** Asymmetric distribution of phospholipids in the inner membrane of beef heart mitochondria, *J. Biol. Chem.*, 254, 5308, 1979.
254. **van Venetie, R. and Verkleij, A. J.,** Possible role of non-bilayer lipids in the structure of mitochondria, *Biochim. Biophys. Acta*, 692, 397, 1982.
255. **Boulan, E. and Sabatini, D. D.,** Asymmetric budding of viruses in epithelial monolayers. A model system for study of epithelia polarity, *Proc. Natl. Acad. Sci. U.S.A.*, 75, 5071, 1978.
256. **van Meer, G. and Simons, K.,** Viruses budding from either the apical or basolateral domain of MDCK cells have unique phospholipid compositions, *EMBO J.*, 1, 847, 1982.
257. **Dragsten, P. R., Blumenthol, R., and Handler, J. S.,** Membrane asymmetry in epithelia. Is the tight junction a barrier to diffusion in the plasma membrane? *Nature (London)*, 294, 1981.
258. **Kachar, B. and Reese, T. S.** Evidence for the lipidic nature of tight junction strands, *Nature (London)*, 296, 464, 1982.
259. **Pinto de Silva, P. and Kachar, B.,** On tight junction structure, *Cell*, 28, 441, 1982.
260. **Blanchette-Mackie, E. J. and Scow, R. O.,** Membrane continuities within cells and intercellular contacts in white adipose tissue of young rats, *J. Ultrastruct. Res.*, 77, 277, 1981.
261. **Miljanich, E. P., Nemes, P. P., White, D. L., and Dratz, E. A.,** The asymmetric transmembrane distribution of phosphatidylethanolamine, phosphatidylserine and fatty acids of the bovine retinal rod outer segment disk membrane, *J. Membrane Biol.*, 60, 249, 1981.
262. **de Grip, W. J., Drenthe, E. H. S., van Echteld, C. J. A., de Kruijff, B., and Verkleij, A. J.,** A possible role of rhodopsin in maintaining bilayer structure in the photoreceptor membrane, *Biochim. Biophys. Acta*, 558, 330, 1979.
263. **Corless, J. M. and Costello, M. J.,** Paracrystalline inclusions associated with disk membranes of frog retinal rod outer segments, *Exp. Eye Res.*, 32, 217, 1981.
264. **Gruner, S. M., Rothschild, K. J., and Clark, N. A.,** X-ray diffraction and electron microscope study of phase separation in rod outer segment photoreceptor membrane multilayers, *Biophys. J.*, 39, 241, 1982.
265. **Williams, J. P., Simpson, E. E., and Chapman, D. J.,** *Plant Physiol.*, 63, 669, 1979.
266. **Sprague, S. G. and Staehelin, L. A.,** A rapid reverse phase evaporation method for the reconstitution of uncharged thylakoid membrane lipids that resist hydration, *Plant Physiol.*, 75, 502, 1984.
267. **Silvius, J. R., Mak, N., and McElhaney, R. N.,** Lipid and protein composition and thermotropic lipid phase transitions in fatty acid homogeneous membranes of *Acholeplasma laidlawii*, B., *Biochim. Biophys. Acta*, 597, 199, 1980.
268. **Wieslander, A., Christiansson, A., Rilfors, L., and Lindblom, G.,** Lipid bilayer stability in membranes, *Biochemistry*, 19, 3650, 1980.
269. **Christiansson, A., Gutman, H., Wieslander, A., and Lindblom, G.,** Effects of anesthetics on water permeability and lipid metabolism in *Acholeplasma laidlawii* membranes, *Biochim. Biophys. Acta*, 645, 24, 1981.
270. **Rilfors, L., Wieslander, A., Lindblom, G., and Christiansson, A.,** Lipid bilayer stability in biological membranes, in *Biomembranes*, Vol. 12, *Membrane Fluidity*, Manson, L. A. and Kates, M., Eds., Plenum Press, New York, in press.
271. **Cronan, J. E. and Gelmann, E. P.,** *Bact. Rev.*, 39, 232, 1975.
272. **van Dijck, P. W. M., de Kruijff, B., van Deenen, L. L. M., de Gier, J., and Demel, R. A.,** The preference of cholesterol for phosphatidylcholine in mixed phosphatidylcholine-phosphatidylethanolamine bilayers, *Biochim. Biophys. Acta*, 455, 576, 1976.
273. **Dekker, C. J., Geurts van Kessel, W. S. M., Klomp, J. P. G., Pieters, J., and de Kruijff, B.,** Synthesis and polymorphic phase behaviour of polyunsaturated phosphatidylcholines and phosphatidylethanolamines, *Chem. Phys. Lipids*, in press.
274. **Cullis, P. R., de Kruijff, B., Hope, M. J., Verkleij, A. J., Nayar, R., Farren, S. B., Tilcock, S. P. S., Madden, T. D., and Bally, M. B.,** Structural properties of lipids and their functional roles in biological membranes, in *Membrane Fluidity in Biology*, Vol. 1, Aloia, R. C., Ed., Academic Press, New York, 1983, 39.
275. **Tilcock, C. P. S. and Cullis, P. R.,** Unpublished data.
276. **Cullis, P. R.,** Unpublished data.
277. **Tilcock, C. P. S.,** Unpublished data.

Chapter 2

TECHNIQUES FOR STUDYING PHOSPHOLIPID MEMBRANES

Alfred H. Merrill, Jr. and J. Wylie Nichols

TABLE OF CONTENTS

I. INTRODUCTION

The major organizing structures of cells are membranes and cytoskeletal elements. Membranes determine which molecules enter and exit the cell and how the components within the cell are distributed. Cytoskeletal elements affect cell shape, movement of structures within cells, and as is becoming more apparent, the organization of some cytoplasmic components.[1] To understand many aspects of cellular regulation, therefore, it is necessary to determine the molecules directly involved in the event of interest as well as evaluate their relationship with these organizational structures.

As background to this series on phospholipids and cellular regulation, this chapter briefly describes the structures of phospholipids, some of their characteristics in membranes, and the techniques used to acquire such information. Because there have been many excellent articles, reviews, and even books on this subject, the primary objective of this review is to give those whose major interest and background has been in some other aspect of regulation a starting point from which to approach this literature. Accordingly, the references do not necessarily acknowledge the authors who deserve credit for a particular idea or technique, but rather, illustrate or review the topic under discussion. The reader is also referred to the chapters by Cullis[2] and Lambeth[3] in this volume.

II. LIPID COMPONENTS OF MEMBRANES

The membranes of mammalian cells typically contain over 1000 different compounds[4] including different lipid classes (i.e., glycerolipids, sphingolipids, sterols, and others), variants within lipid classes (e.g., different head groups and fatty acids), and a host of proteins with structural, transport, receptor, and catalytic functions.[5] Any attempt to elucidate the properties of such complex mixtures, to fully characterize interactions between a single component and the rest of the biological membrane, or to understand even simple membrane models begins with a consideration of the molecular features of the components. One source of confusion for the newcomer to lipids is the variety of names for a single compound (e.g., lecithin, phosphatidylcholine, 1,2-diacylphosphatidylcholine), which persists despite the existence of a systematic nomenclature.[6] This review generally uses the preferred names for these compounds, except where the recommendations have not been applied generally, which could add to the confusion.

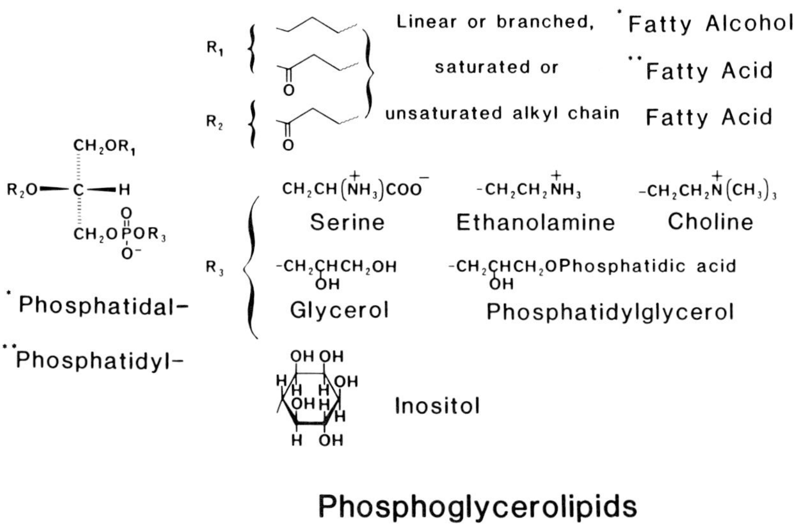

R_1 Linear or branched, saturated or unsaturated alkyl chain — *Fatty Alcohol, **Fatty Acid
R_2 Fatty Acid

CH_2OR_1
R_2O—C—H
CH_2OPOR_3

*Phosphatidal–
**Phosphatidyl–

R_3:
$CH_2CH(\overset{+}{N}H_3)COO^-$ Serine
$-CH_2CH_2\overset{+}{N}H_3$ Ethanolamine
$-CH_2CH_2\overset{+}{N}(CH_3)_3$ Choline
$-CH_2CH(OH)CH_2OH$ Glycerol
$-CH_2CH(OH)CH_2O$Phosphatidic acid — Phosphatidylglycerol
Inositol

Phosphoglycerolipids

FIGURE 1. Major constituents of phosphoglycerolipids. The common names for the phosphoglycerolipids are phosphatidyl (or phosphatidal) head group, except when the head group is phosphatidylglycerol, which yields the compound diphosphatidylglycerol or cardiolipin. Phosphatidylinositols also include the 4-phosphate and 4,5-*bis* phosphates.

A. Glycerolipids

Acylglycerols, phosphoglycerolipids, and alkyl ether lipids have *sn*-glycerol as a common structural and biosynthetic (via glycerol 3-phosphate) backbone to which different substituents have been added (Figure 1).[7] The simplest phospholipid is phosphatidic acid (1,2-diacylglycerol 3-phosphate), which exists in membranes primarily as an intermediate in the metabolism of more complex phosphoglycerolipids (although it has been suggested that a phosphatidic acid/diacylglycerol cycle might function in vesicle formation and fusion).[8] This functional group is referred to as a "phosphatidyl" moiety and more complex structures with a head group attached via phosphodiester linkage at the 3 position are named phosphatidyl (head group). The major phosphoglycerolipids in mammalian systems have for the head group amino alcohols (phosphatidylcholine, phosphatidylethanolamine, and phosphatidylserine), glycerol (phosphatidylglycerol and diphosphatidylglycerol, also named cardiolipin), and inositol and its phosphates (which have been implicated in receptor action)[9] (Figure 1). At physiological pH, phosphatidylglycerol, phosphatidylinositol, and cardiolipin are negatively charged due to ionization of the phosphate groups (with pKa near 3) and have a net charge of -1, -2 for cardiolipin, and more for the polyphosphoinositides. Phosphatidylserine has two types of negatively charged groups, phosphate and carboxyl (pKa $\sim$4 to 5), and one positively charged amino (pKa $\sim$10) which has a net charge of -1. Phosphatidylcholine and phosphatidylethanolamine (pKa $\sim$8 to 10) are zwitterionic and have no net charge at physiological pH. The assignments of pKa for all of these groups are only approximate because they are affected by the membrane environment and presence of ions, especially divalent cations.[9a]

There is considerable variation in the fatty acids of the phosphoglycerolipids.[10] In most mammalian membranes, a saturated fatty acid (palmitate or stearate) is usually found at position 1, whereas unsaturated fatty acids (oleate, linoleate, linolenate, and arachidonate) predominate at position 2. Notable exceptions to this generalization are the disaturated phosphatidylcholine and phosphatidylglycerol of pulmonary surfactant. Diet, as well as numerous disease states, can affect the lipid compositions of membranes.[11]

A structurally and biosynthetically related group of glycerolipids are the alkyl-ether lip-

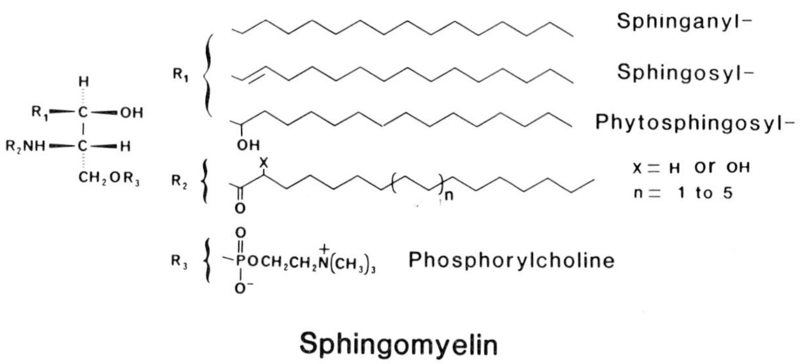

Sphinganyl-
Sphingosyl-
Phytosphingosyl-
X = H or OH
n = 1 to 5
Phosphorylcholine

Sphingomyelin

FIGURE 2. Major constituents of sphingomyelin. The terms sphinganyl-, sphingosyl-, and phytosphingosyl- denote the 18-carbon long-chain (sphingoid) bases which have the alkyl side chains designated R1 and 3-carbon backbone shown on the left.

ids.[12,13] These contain a fatty alcohol in ether linkage to the 1 position of the glycerol backbone, and in the case of the plasmalogens, the fatty alcohol is also α,β-unsaturated. The backbone for alkyl ether lipids is sometimes referred to as a "phosphatidal moiety". The most common alkyl-ether lipids of mammalian membranes are plasmalogens with ethanolamine as the head group.

Lysophospholipids lack an esterified fatty acid either at position 1 or 2. Although these compounds are present in membranes in small amounts and are usually considered as intermediates in phospholipid synthesis or degradation, their levels bound to circulating serum albumin are considerable.[14]

B. Sphingolipids

Sphingolipids[15-17] are elaborations of aliphatic amino alcohols called "long-chain" or sphingoid "bases". The most prevalent long-chain bases are sphingosine (*trans*-4-sphingenine, or [2S,3R,4R]-2-amino-4-octadecene-1,3-diol), sphinganine (dihydrosphingosine, or [2S,3R-2-amino-1,3-octadecane), and phytosphingosine (4- D-hydroxysphinganine or [2S, 3S, 4R] -2 amino -1,3,4-octadecanetriol), plus homologs of these compounds with different alkyl-chain lengths (Figure 2). Ceramides are derived from these compounds by the addition of long-chain fatty acids in amide linkage to the 2 position. Some sphingolipids are composed of α-hydroxy fatty acids or very long chain (i.e., with 22 to 26 carbon atoms) fatty acids.

Sphingomyelin (ceramidylphosphorylcholine) is essentially the only phosphosphingolipid found in mammals (some other organisms have additional head groups, including phosphonates).[18] A greater variety of sphingolipids is manifested in the glycosphingolipids, which have sugar moieties at the 1 position of the ceramide backbone (i.e., at R3 in Figure 2).[17] Glycolipids have been grouped into ganglio-, globo-, lacto-, and mucoglycosyl-ceramide series, as well as, groupings depending on whether or not they contain sulfate (i.e., sulfatides). These molecules contribute to the structure of the outer leaflet of the plasma membrane and function as cell surface markers and antigens, and in less understood ways, are related to some aspects of cell-cell interaction, differentiation, and cell growth. They also affect receptors and the interaction between membrane proteins and bacterial toxins and hormones, *inter alia*.[17]

C. Other Lipids

The other major lipids are the sterols, of which cholesterol and cholesterol esters are the most prevalent. Membranes also contain varying amounts of dolichols, free fatty acids (and in some cases fatty acyl-CoA), quinones (e.g., ubiquinone), retinoids and β-carotenes,

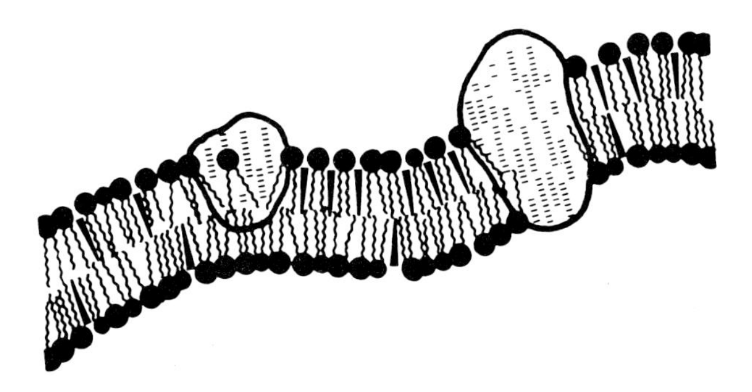

FIGURE 3. Model of bilayer membrane containing asymmetrically distributed lipids and proteins. The phospholipids have different head group sizes and acyl chains, and are neighbored by varying amounts of cholesterol, which is represented by the wedge-shaped structure.

waxes, α-tocopherol, and other hydrophobic compounds. In addition, numerous drugs (especially anesthetics) partition into bilayers and can contribute significantly to the properties of membranes.[19]

III. PROPERTIES OF PHOSPHOLIPIDS IN MEMBRANES

The prevailing model (the fluid-mosaic model)[20] for membrane structure depicts the membrane as a phospholipid bilayer into which proteins are partially to fully inserted (Figure 3), although this has been amended to include regions with nonbilayer structures.[2,21] In addition to variation in the types and amounts of lipids and proteins in different membranes, the components are often asymmetrically distributed on the two sides of the bilayer.[22] Bilayers form spontaneously from phospholipids dissolved in a dilute aqueous solution. The favorable thermodynamic free energy for bilayer formation exists because the hydrocarbon regions of the dissolved phospholipid monomers force the surrounding water molecules into a more ordered structure, decreasing their freedom of motion. Thus, the entropy of the system is increased and the free energy is reduced when the number of phospholipid monomer-water interactions are reduced to a minimum by the aggregation of the monomers into bilayers, such that only the polar head groups face the water phase. In addition to these physical-chemical forces, the precise structure of biological membranes is determined by their lipid composition and the order of assembly, and is maintained in vivo by metabolic and translocation processes.

Membranes are dynamic structures of lipids and proteins which both undergo rapid rotation, rocking, and lateral movement. Because there is order to the arrangement of the lipids, such as the approximately parallel alignment of the alkyl chains, even the most mobile (fluid) state of the membranes is referred to as "liquid crystalline". Lowering the temperature decreases the mobility of the alkyl chains, yielding a more ordered gel state. Pure phospholipids undergo this phase transition over a narrow temperature range, the midpoint of which is called the "transition temperature" (Tm). The Tm for a given membrane is a function of the type of phospholipid head group(s), degree of hydration, acyl chain composition, and the presence of other molecules such as proteins, cholesterol, or divalent cations.[23-26] Ether lipids often have slightly higher phase transition temperatures, but generally have properties similar to the diacyl analogs.[13] Examples of phase transition temperatures for different types of phospholipids are given in Table 1. In general, decreasing the alkyl chain length or introduction of a *cis*-double bond lowers the Tm. The melting behavior of a mixture of phospholipids can be simple (i.e., a single transition that is a function of the Tm for the individual phospholipids) or complex (i.e., a broad transition or multiple tran-

Table 1
MAJOR PHASE TRANSITION TEMPERATURES FOR PHOSPHOLIPIDS

Phospholipid	Tm (°C)	Ref.
Phosphatidylcholines		
1,2-Lauroyl	−1.8	220
1,2-Myristoyl	23.9	220
1,2-Palmitoyl	41.4	220
1,2-Stearoyl	54.9	220
1,2-Oleoyl	−22	24
1-Palmitoyl, 2-myristoyl	27.2	221
1-Myristoyl, 2-palmitoyl	35.3	221
1-Palmitoyl, 2-stearoyl	47.4	221
1-Stearoyl, 2-palmitoyl	44.0	221
Phosphatidylethanolamines		
1,2-Lauroyl	30.5	29
1,2-Myristoyl	49.5	220
1,2-Palmitoyl	63.1	29
Phosphatidylserines		
1,2-Myristoyl	36	222
1,2-Palmitoyl	51—53	222
Sphingomyelins		
N-Palmitoylsphinganylphosphorylcholine	47.8	223
N-Palmitoylsphingosylphosphorylcholine	41.3	223
N-Stearoylsphingosylphosphorylcholine	52.8	223
N-Lignoceroylsphingosylphosphorylcholine	48.6	223
N-Nervonoylsphingosylphosphorylcholine	25—30	224

sitions — sometimes due to separation of the lipids into different phases in the bilayer). This multifaceted behavior of phospholipids has stimulated many theoretical analyses;[27-32] however, many properties of phospholipids and proteins in biological membranes are not yet understood. Some of the properties of these molecules are described below.

A. Intermolecular Interactions

1. Geometric Considerations[33-38]

Since the early Greeks, it has been appreciated that nature abhors a vacuum. This is equally applicable to membranes, which are composed of molecules that pack in a way that not only are van der Waals interactions maximized, but also spaces between atoms will be minimized. The surface areas for representative fatty acids, phospholipids, and cholesterol (as well as phospholipid and cholesterol mixtures) are given in Table 2. It is apparent that the areas vary considerably among different fatty acids, phospholipids containing different fatty acids, and phospholipids with different head groups but identical acyl chains. It has been noted that differences in the area of the head groups vs. the acyl chains could affect the structures formed from the phospholipids.[34] Lysophosphatidylcholine, for example, has an unusually large head group size compared to acyl chain diameter; therefore, it is not surprising that it tends to form spheres and cylinders with the head groups packed hexaginally facing the outside (Figure 4), which are referred to as "HI hexagonal structures". Phosphatidylethanolamine (especially 1,2-dioleoylphosphatidylethanolamine), which has a greater acyl chain to head group diameter tends to form spheres and cylinders hexagonally packed with the head groups inside, which are termed "HII hexagonal structures". Factors that change the effective size of the head groups and acyl chains (such as increasing the mobility

Table 2
AREAS FOR MEMBRANE LIPIDS[a]

Lipid	Area (Å²/molecule)	Surface pressure (dyn/cm)
Palmitic acid	24	5
Palmitic acid	20	24
Oleic acid	48	5
Dimyristoylphosphatidylcholine	72	12
Dipalmitoylphosphatidylcholine[b]	58	12
Distearoylphosphatidylcholine	46	12
Dioleoylphosphatidylcholine	83	12
Dilinoleoylphosphatidylcholine	99	12
1-Stearoyl,2-oleoyl-phosphatidylcholine	75	12
Egg Lecithin (mainly 1-palmitoyl,2-oleoyl)	96	33
Dimyristoylphosphatidylethanolamine[b]	50	12
Distearoylphosphatidylethanolamine[b]	41	12
Cholesterol[c]	37—38	—
Mixtures (1:1) with cholesterol		
Distearoylphosphatidylcholine	46	12
1-Stearoyl,2-oleoyl-phosphatidylcholine	64	12
Egg lecithin	56	33
Dioleoylphosphatidylcholine	77	12

[a] Data from Reference 225 except where noted.
[b] Data from Reference 226.
[c] Based on approximate dimensions of 7.2 × 5 × 20 Å.

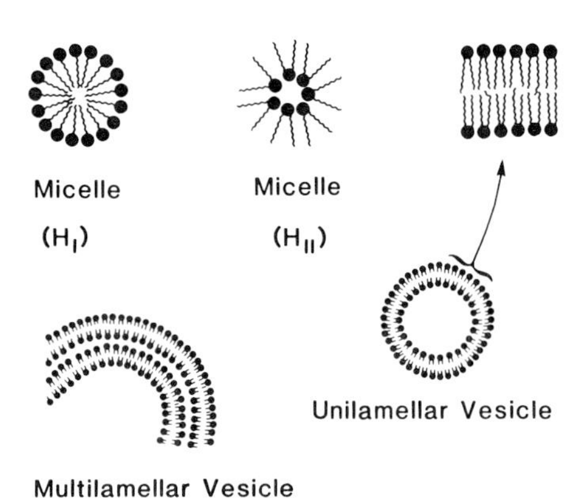

FIGURE 4. Structures of phospholipids in aqueous solutions.

of the sidechains by raising the temperature or decreasing electrostatic repulsions of anionic phospholipids by lowering the pH) alters their tendency to form bilayers vs. hexagonal phases. The area occupied by phospholipids can also be affected by their membrane environment. The head group and acyl chain areas of phosphatidylcholine on the outer leaflet of highly curved membranes have been reported to be 72 and 45 Å², respectively, while the same compound on the inner leaflet occupies 60 and 90 Å².[35] Cholesterol affects the packing of phospholipids (Table 2) by adding a hydrophobic "spacer" between neighboring molecules at the same time that it presents a rigid, planar surface that decreases the local mobility of neighboring acyl chains.[25,39] Cholesterol can "fill in" gaps in the hydrophobic region of membranes, as exemplified by the ability of mixtures of lysophospholipids plus cholesterol to form bilayers, although neither do so alone.

Interestingly, increasing the alkyl chain length affects both the area occupied per molecule (which increases approximately 15 nm^2 for phosphatidylcholines of fatty acids from 12 to 18 carbons in length) and the bilayer thickness (which increases approximately 0.5 nm over the same range).[40] *Cis*-unsaturated, branched chain, and cyclopropane (the latter two being mostly found in organisms other than mammals) fatty acids disrupt a simple alignment of all-*trans* saturated fatty acids in the hydrophobic core of the bilayer, and therefore decrease the Tm for transition from the gel to liquid crystalline states. Fatty acids with *trans*-double bonds also perturb the structure of the bilayer somewhat.[41]

There has been considerable interest in the relationships between the structures of phosphatidylcholine and sphingomyelin, the two choline-containing phospholipids. In general, sphingomyelin tends to condense the bilayer, perhaps due to hydrogen bonding between the amide bonds or via the free hydroxyl group; however, it has been noted that their effects on membrane acyl chain order are comparable.[42] Comparisons of models of sphingomyelin, phosphatidylcholine, and cholesterol, supported by additional physical measurements,[43] suggest that these molecules can achieve optimal packing geometries and, therefore, may be associated in membranes.

2. Interactions Between Phospholipids

The hydrocarbon and polar portions of phospholipid molecules are aligned in such a way that they interact primarily with each other. The hydrocarbon chains are attracted by van der Waals interactions, whereas the polar groups undergo dipolar, hydrogen bonding, and electrostatic interactions.

Van der Waals forces (dispersive forces) are attractive forces that occur between very closely opposed molecules. The attraction arises from fluctuations in the electron density of a molecule which result in instantaneous charge dipoles. The dipole in one molecule induces an oppositely oriented dipole in its neighbor, resulting in a net attraction of the two. The magnitude of the attraction energy is highly dependent on the separation distance (d) between the two molecules (inversely proportional to d^6). The van der Waals attraction energy has been theoretically calculated to be in the range of 27 kcal/mol for a phospholipid having two saturated 18-carbon acyl chains assuming a separation distance of 5 Å.[44] Although this enthalpic energy of attraction is large, it is not considered to be significant in determining the formation of phospholipid bilayers from monomers in aqueous solution since the van der Waals energy between soluble phospholipid monomers and surrounding water molecules is also large such that the net change in enthalpy between the two states is small. However, since the length, degree of unsaturation, and molecular packing of the acyl chains are all involved in determining the magnitude of the van der Waals interactions, the degree to which they favor bilayer formation, and conformations within bilayers, depends highly on the composition of the membrane.

The electrostatic interactions between the polar head groups are primarily hydrogen bonds or salt bridges. These interactions have been studied in detail using X-ray diffraction of single crystal structures of phosphatidylcholine and phosphatidylethanolamine.[38] The only difference in these two molecules is in the amine group — phosphatidylcholine has a quaternary amine while phosphatidylethanolamine has a primary amine. However, the more diffuse spreading of the positive charge in the quaternary amine reduces its ability to form strong salt bridges between adjacent negatively charged phosphate oxygens. As a result, the phosphate-nitrogen dipole of the phosphatidylethanolamine head group lies in the plane of the bilayer surface. The ammonium groups interact with the unesterfied phosphate oxygens by very short bonds having the character of both hydrogen bonds and salt bridges, which results in a close packed, rigid head group network at the bilayer surface. The weaker interactions and the slight tilt of the phosphate-nitrogen dipole of phosphatidylcholine give rise to a thicker polar region with a relatively low molecular packing area.[38]

In general, studies on fully hydrated phospholipid bilayers using nuclear magnetic resonance (NMR) confirm that the structures deduced from X-ray diffraction of crystals are similar to the time-averaged interactions of phospholipids in aqueous dispersions.[45]

The backbone atoms are also able to undergo dipolar or hydrogen-bonding interactions; however, these are considered to be a significant mainly for sphingolipids, which have one or more free hydroxyls and an amide bond.[15]

B. Movement of Membrane Lipids

1. Motion Within Each Molecule

Phospholipids undergo three major types of internal motion: rotation about the carbon-carbon bonds of the fatty acid side chains, motion about the atoms of the backbone, and rotation around the bonds of the head groups. Rotation about the saturated carbon-carbon bonds of aliphatic side chains requires little energy and is rapid at room temperature. The conformation produced by rotation of a single bond from the *"trans"* conformation (the preferred term is "anti") to a *gauche* conformation is shown in Figure 5. The effect of this single rotation would be to send the remaining carbons perpendicular to the original chain and the rest of the bilayer. This conformation is therefore not likely to exist except around "shallow" components of the membrane or at the ends of long aliphatic chains. A second rotation of this same molecule, however, returns the alkyl chain to a more linear conformation. This produces a "pocket" in the acyl chain that is large enough to contain several water molecules, and therefore has been proposed to explain the ease of transport of water molecules through membranes.[46] Rotation about double bonds is not energetically feasible and *cis*-double bonds introduce a "kink" into the alkyl chain, as shown in Figure 5. This disrupts the bilayer in its vicinity, but this effect can be minimized by rotation of a nearby carbon-carbon bond (as described above) to return the remainder of the chain toward a more linear path.

A second type of rotation can occur between the carbons of the glycerol moiety or the first three carbons of the long-chain bases. This motion is severely restricted to a rocking motion by the bilayer structure, which will tend to hold the polar head group and fatty acyl chains in the configuration typically shown for phospholipids (i.e., the chains together and the head group in the opposite direction).

Third, head groups can assume multiple conformations by rotation about the phosphodiester bond (Figure 5). Two primary examples of these are the species with the head group lying over the glycerol backbone of the same phospholipid vs. extended over adjacent regions of the membrane. Different phospholipids exist predominately in one or the other of these structures, as exemplified by the tendency of phosphatidylcholine to exist in the structure on the left and phosphatidylethanolamine to undergo more intermolecular hydrogen bonding by forming the alternative structure.

2. Motion Within Bilayers

Phospholipids undergo rotational and rocking motions *in situ*, lateral movement, and translation across the bilayer (which is usually referred to as "flip-flop") (Figure 6). The rates at which these occur are determined by the type of phospholipid and its environment and the temperature. The phospholipid environment includes neighboring lipids and proteins, which may cause the behavior of a subclass of the phospholipids to be different from the rest of the bilayer by promoting the formation of patches of lipid with different properties or by specific binding and relative "immobilization" of the phospholipid. These factors

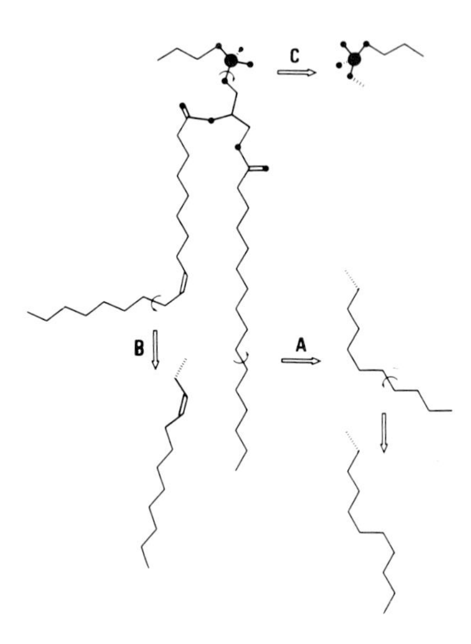

FIGURE 5. Structures resulting from rotational motions of phospholipids in membranes. Shown is 1-palmitoyl, 2-oleoyl-phosphatidylethanolamine, and examples of the conformations resulting from the following types of rotation. (A) Rotation about a saturated carbon-carbon bond to introduce a *gauche* conformation (first arrow), followed by a similar rotation about a nearby bond (second arrow); (B) rotation about a saturated carbon-carbon bond to introduce a *gauche* conformation near a *trans* double bond; (C) rotation about the phosphorus-oxygen bond of the head group. Similar rotations can occur around the carbon-carbon, carbon-oxygen, and other oxygen-phosphorus bonds of the head group. Not shown is the partial rotation around the carbon-carbon atoms of the glycerol backbone, which causes a "scissoring" effect.

contribute to the viscosity of the core of the membrane bilayer, which is on the order of 1 poise, or about 100 times more viscous than water.[25]

Rotational motion exists along the long axis of the phospholipid and as a rocking motion perpendicular to this axis (Figure 6). Both are very rapid in liquid crystalline phases. The rate of lateral diffusion is also fast (diffusion coefficients are on the order of 1 to 3×10^{-8} cm^2/sec) in fluid membranes (both for models and biological membranes, although the movement of proteins is usually 10 to 10,000 times slower than that of the lipids) and slows ($<5 \times 10^{-11}$ cm^2/sec) when the membrane is cooled below the Tm.[25,47] The higher diffusion constant corresponds to the exchange of a neighboring phospholipid 10^7 times per second, or the translocation of a given phospholipid by 1 μm/sec in the plane of the bilayer.[26]

Because it involves transfer of the polar head group through the hydrophobic core, trans-bilayer movement is generally very slow (half-times of days to months). The exceptions to this may be some subcellular membranes (e.g., microsomes) which do not appear to have asymmetric phospholipid distributions[48] and may contain proteins (often referred to as "flipases") that facilitate rapid equilibration across the membrane. Lipids lacking substantial polar character, such as cholesterol, diacylglycerols, and ceramides (which have a single hydroxyl group), undergo relatively rapid transbilayer movement. The half-time for trans-bilayer movement of cholesterol is on the order of seconds to minutes.[49]

3. Motion Between Bilayers

Due to their extremely low monomer concentrations in aqueous solution (critical micellar concentrations for phospholipids are typically $<10^{-9}$ *M*) and their slow rates of dissociation

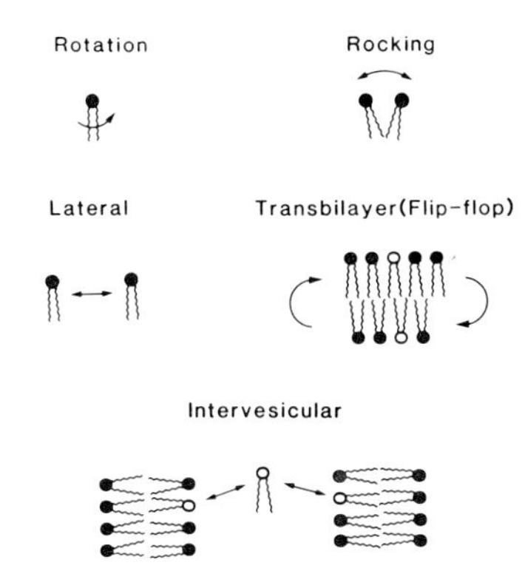

FIGURE 6. Types of motion of whole phospholipid molecules in membranes.

from the bilayer surface, phospholipid movement between bilayers is very long (half-times of days at 37°C, for example).[50,51] In vitro, the predominant mechanism of spontaneous transfer between bilayers is by the diffusion of soluble monomers through the aqueous phase, rather than vesicle collision or fusion.[50,52] Spontaneous transfer can be accelerated considerably by making the phospholipid more aqueous soluble (as in phospholipids with shorter acyl chains or with polar groups attached to the fatty acid moiety).[50] In addition, cytosolic proteins have been isolated that have the capacity to transfer phospholipids between membranes. There is a range of these proteins with varying specificity for the type of lipid transferred.[53-55] The mechanisms for transport involve the actual binding and carrying of the phospholipid for some,[55] while others seem to facilitate the transfer by increasing the rate of monomer diffusion between bilayers.[56]

Sphingomyelin also undergoes spontaneous[57,58] and protein-mediated transfer between bilayers and apolipoprotein-phospholipid recombinants.[59] Based upon the temperature dependence of the spontaneous transfer rate, it has been suggested that sphingomyelin can exist in a gel-like phase in a membrane composed of other lipids in liquid-crystalline states.[57] The slow rate of spontaneous exchange of sphingomyelin has been interpreted as evidence that this process contributes little to the transfer of this molecule in vivo,[57] although different methods have led to other conclusions.[58]

IV. ISOLATION OF BIOLOGICAL MEMBRANES

Many investigations of biological membranes[60-64] begin with the isolation of cellular subfractions enriched in the membranes of interest. If one needs a relatively pure preparation of a prototypic membrane (e.g., microsomes or mitochondria) then a tissue should be selected that is in ready abundance, contains almost entirely one cell type, reproducibly yields well-defined membrane fractions, and has been well characterized. Rat liver is probably most widely used because it meets these criteria, even though some of its membranes are complex (as exemplified by the three or more types of plasma membranes: sinusoidal, bile cannicular, and intermediate).[65] If instead, an overriding objective is to characterize a membranous process in another organism or organelle, care must be taken to establish the composition of the membrane fractions that are obtained by the methods employed. This is usually performed by microscopy and analyses of characteristic marker enzymes, complex carbohydrates, or antigens.

The quality of the membranes depends on the methods used for disruption of the tissues or cells in culture. Glass-glass or glass-Teflon® mortar and pestle homogenizers are employed frequently (e.g., Potter-Elvehjem or TenBroeck-type homogenizers) and can be purchased or modified to have different clearances, thereby causing various degrees of disruption of larger and smaller organelles. Mechanical devices that shear the samples, such as Waring Blenders, Omnimixers, and Polytron homogenizers *inter alia,* are often more effective on large samples and organs with considerable amounts of connective tissue, but can be more disruptive to subcellular organelles. For some materials, especially microorganisms, rapid pressure changes using a French pressure cell are effective. Other methods of cell disruption which are particularly useful with cultured cells include sonication, rapid shifts in ionic strength, enzyme treatments, and addition of detergents.

Membranes are usually separated by centrifugation of various types, including differential rate, density gradients, isopycnic density gradients, density perturbation, and affinity density perturbation. In some instances, the sedimentation of certain membranes can be altered by pretreating with the appropriate agents (e.g., lysosomes by the addition of digitonin). Other methods that have been successful in specific instances are the isolation of plasma membranes with DEAE-Sephadex® beads[66] and affinity chromatography using immobilized lectins or immunoglobulins.[67] Serum lipoproteins, which possess a phospholipid monolayer as part of their structure, have been isolated by centrifugation,[68] gel filtration,[69] and precipitation.[70]

Whatever approach is taken, the investigator should be aware of potential artifacts that may arise during isolation of the membranes. Although these will vary with the particular source of the membranes, general examples of artifacts during isolation are

1. The disruption or aggregation of the membranes. The (fortuitous!) formation of microsomal particles from the endoplasmic reticulum and the less desirable loss of respiratory control with mitochondria not handled carefully exemplify this phenomena.
2. The loss of components. Loosely bound proteins (e.g., extrinsic proteins) can be removed from the membrane surface during cell disruption and centrifugation. If certain protein or lipid species are particularly susceptible to chemical (e.g., oxidation of polyunsaturated fatty acids) or enzymatic (e.g., hydrolysis by esterases) degradation, they may undergo modification.
3. The addition of extraneous molecules. Some species, such as nucleic acids, can bind to membranes and accompany them through the isolation procedures.

V. DISRUPTION OF MEMBRANE COMPONENTS

A. Phospholipid Extraction

Phospholipids are efficiently recovered from most biological membranes using the extraction methods published by Folch et al.[71] and Bligh and Dyer,[72] as modified and applied by subsequent investigators (see References 73 to 76 for examples). These procedures involve the mixing of organic and aqueous solvents (e.g., chloroform/methanol/water) in ratios to achieve a single phase for maximal penetration of the sample, followed by addition of additional amounts of the immiscible components (e.g., chloroform and water) to yield two phases. Because these solvents are relatively toxic, an alternative extraction mixture using hexane and isopropanol (3:2, v/v) has been developed.[77] If small amounts of radiolabeled phospholipids are to be recovered by these methods, it is advisable to add unlabeled carrier to minimize losses at the interface and on glass surfaces. Many investigators include antioxidants, such as butylated hydroxytoluene (BHA), butylated hydroxyanisole (BHA), or α-tocopherol and metal chelaters (EDTA) during the extraction to protect polyunsaturated fatty acids. The extracts are stored at low temperature under nitrogen or argon. A useful index for lipid peroxidation during storage is the absorbance at 215 and 230 nm which represents the conjugated dienes.[78]

Individual molecular species can be resolved according to head group or fatty acyl chain differences by one- and two-dimensional thin-layer chromatography (TLC)[79-80] or by high performance liquid chromatography (HPLC).[81-82] Phospholipids can be quantitated by a number of methods,[79-85] examples of recently cited methods being phosphate assays[83,84] and charring.[85]

B. Membrane Solubilization

For the more gentle removal of lipids from membrane components (also for reconstitution of purified proteins into membranes) the delivery of lipids to membranes, and studies of proteins and lipids in a nonbiological background, the use of various detergents has been developed and characterized.[86-92] These compounds vary in structure from neutral compounds such as octyl-glucopyranoside, digitonin, and the polyoxyethylene series (Triton®, Tween®, and Brij®, which have varying numbers of ethylene groups — Triton® X-100, for example, has an average of 9.5 per molecule) to highly ionized molecules (cholate, deoxycholate, cetyltrimethylammonium bromide, and zwitterionic surfactants *inter alia*). All possess both polar and nonpolar moieties and this amphipathic (or alternatively amphiphilic) nature enables them to exist stably in aqueous solution while also interacting with highly hydrophobic lipids or proteins. The most effective detergents usually exist both free in solution and in aggregates called micelles, which form above the so-called "critical micelle concentration". When added to membranes, the monomers partition into the bilayer and eventually disrupt its structure to create mixed micelles containing detergent, lipids, and proteins. Conversely, mixed micelles can be used to solubilize otherwise aqueous immiscible compounds and facilitate their delivery to enzymes, membranes, or cells. In one version of this use, the detergent is diluted far below its CMC and the "carried" substance is forced to either enter the membrane or aggregate; careful optimization of the conditions is necessary to obtain the desired result without adding sufficient detergent to perturb the other components of the mixture.

Although selection of the best detergent for a particular application is often determined empirically, a number of parameters are worth considering. The detergent of choice should have characteristics compatible with subsequent analyses. For example, Triton® X-100 absorbs UV light and therefore interferes with spectroscopic studies,* ionic detergents can preclude purification steps that involve ion-exchange chromatography, and detergents that form aggregates with molecular weights comparable to the proteins of interest diminish the already poor resolution that can be obtained by gel filtration chromatography. The detergent should solubilize the desired components without irreversibly denaturing them (loss of activity may not be undesirable, however, if it can be recovered upon reconstitution of the proteins with lipids). Generally, nonionic detergents are less denaturing; however, they can be some of the more difficult to remove later (Bio-beads SM-2 selectively and rapidly removes Triton® X-100 and is used extensively for this purpose).[93] Polyoxyethylene-type detergents are susceptible to peroxidation of the ether moiety and may need to be treated before use and handled carefully if the sample has sensitive sulfhydryl or fatty acid groups.[94] In some cases, extraction of the membranes with sequentially increasing concentrations of a single detergent, several detergents, or even detergent mixtures may both release the desired protein and separate it from some other species. Cost will also be a consideration if large volumes of buffers containing the more expensive detergents will be needed for chromatographic procedures.

The concentration of the detergent to use is determined by its CMC and by the amount of lipid in the sample. The objective of the initial solubilization step is to divide the membrane into as many separate species as possible, yielding ideally one polypeptide per micelle. The detergent to phospholipid ratio necessary to form a stable micelle and the aggregation number

* A reduced form of Triton® X-100 has recently been reported to have similar properties but lacks UV absorbance.

(i.e., number of monomers per micelle) affect the concentration of detergent needed.[86-88] For subsequent steps, it is mainly necessary for the detergent to be above the CMC (it is worthwhile to determine if detergents are required after the initial solubilization; some proteins have limited solubility once the lipids have been removed). Replacement of one detergent with another can be accomplished by dialysis (if the detergent is dialyzable; Triton® X-100, for example, is not) or by binding the protein to a support (e.g., hydroxylapatite) washing with buffer containing the second detergent, and displacing the protein in the new micelle.

Membrane proteins have also been solubilized using phospholipases or proteases. While the intent of the former is to generate lysophospholipids, which are good detergents, the commercial preparations often contain proteases that can "solubilize" the proteins by clipping between the hydrophobic and hydrophilic domains. If, however, this does not matter, proteases can be added directly to obtain the same objective.

VI. PREPARATION OF ARTIFICIAL MEMBRANES

Because biological membranes are so complex, investigations of membrane-associated processes eventually attempt to reduce the number of species to only the proteins or lipids (or both) which are absolutely required to reproduce the phenomena. The process of preparing defined membranes from phospholipids, however, is itself somewhat complex because they form many different structures with properties that depend on the method of preparation. Furthermore, the type of experiment often dictates the properties that the membrane model should have. For example, physical studies require membranes of homogeneous size whenever possible and the smaller vesicles best meet this criteria, whereas transport experiments are easier when the internal volume of the vesicle (which is called the encapsulation capacity and expressed as ℓ/mol lipid) is large. Enzyme studies are often conducted using whatever works.

The success of any model depends on the type and quality of the phospholipids. Phosphatidylcholine from different sources (especially eggs, which is predominately 1-palmitoyl, 2-oleoyl-phosphatidycholine) forms stable vesicles readily. Acidic phospholipids work best when mixed with other lipids, probably because of repulsion by the head groups. In most cases, formation of lipid structures is best conducted above the Tm for the lipids. Outlined below are some of the major models that are used for characterization of phospholipid membranes.[95-101]

A. Models for Membranes (Figure 5)

1. Micelles

Lysophospholipids and mixtures of detergents and phospholipids in high ratios of detergent to phospholipid form relatively stable micelles and can be used, for example, to measure protein-lipid interactions. References to the properties and use of detergent micelles were presented in Section V. Even lysophospholipids and other single chain amphiphiles can form bilayers if cholesterol is added.[102,103] Lysophosphatidylcholine has been used to solubilize microsomal membranes and, in a rather novel approach to purifying and reconstituting lysophosphatidylcholine acyltransferase, to form phosphatidylcholine bilayers synthetically upon addition of fatty acyl-CoA.[104]

2. Multilamellar Vesicles

When phospholipids are deposited on the inner surface of a container (usually by adding the lipid in a volatile organic solvent and evaporating the solvent) and water is added, they tend to become hydrated in different forms of unstable tubular micelles. These myelin-like figures have interesting properties, such as the tendency to orient in a magnetic field.[105]

Mild agitation of hydrated phospholipids causes their spontaneous conversion to concentric spheres of lamellar phospholipids which are termed "multilamellar vesicles" (MLVs) or "Banghamsomes", after their developer.[106] Although MLVs were employed in early studies of membrane structure, transport, the use of liposomes to transport drugs, and have the advantage of being very simple to prepare,[97] the fact that only the outermost layer is available to interact with external agents limits their utility as a model for biological membranes. These structures are usually large (1000 nm) and have encapsulation volumes ranging from 1 to 4 ℓ/mol.

3. Small and Large Unilamellar Vesicles

When MLVs are sonicated extensively, the preparations become clear due to the formation of small unilamellar veiscles (SUVs), which are generally defined as vesicles less than 100 nm in diameter). SUVs prepared this way typically have diameters of 25 to 50 nm and only a small fraction of the total have vesicles within other vesicles. Other methods for forming SUV include ethanol injection,[107] sonication followed by high-speed centrifugation,[108] and passage of MLVs through a French press.[109] SUVs have the advantage of being fairly uniform and easy to produce; however, the geometric constraints imposed by their high degree of curvature yield an inhomogeneous distribution of phospholipids on the two sides of the bilayer (approximately 60 to 70% of the phospholipid is on the outer leaflet). Furthermore, their small size severely limits the amount and size of molecules that can be incorporated inside the vesicle (molecules greater than 40,000 have limited encapsulation). Encapsulation capacities range from 0.2 to 1.5 ℓ/mol.

Large unilamellar vesicles (LUV) have typical dimensions from 100 to 1000 nm and larger and can entrap relatively large volumes (2 to 17 ℓ/mol). LUVs are prepared by (1) injecting the phospholipids dissolved in an organic solvent (ether or ethanol work well, and the latter can yield vesicles of controlled size ranges)[107] into buffer (often referred to as the ether or ethanol-injection method),[110] (2) sonication in a mixture of water and organic solvent (called reversed phase evaporation),[97] (3) removal of detergent from mixtures of detergent and phospholipids, dialysis or gel filtration (called detergent dilution),[97,111,112] (4) freezing and thawing of sonicated phospholipids,[113] (5) by injection of lipid suspensions through polycarbonate filters,[114] and (6) fusion of small vesicles.[114a] Different methods and the use of a single method with different lipids yield vesicles with some degree of heterogeneity around an average vesicle size. If a more homogeneous population is desired, this can be obtained by gel filtration chromatography or passage through millipore filters.[96,115]

5. Black Lipid Membranes

For many studies of conductance and transport across membranes, the most useful model is a single bilayer that can be formed across a small hole (typically 1 to 100 mm^2), dividing two aqueous chambers. These are called black lipid membranes because they were first formed by painting a concentrated lipid solution across the orifice and watching it thin until it becomes unilamellar and no longer reflects light.[116-119] These membranes have been prepared with a variety of phospholipids and mixtures and are particularly useful for electrochemical analyses of membranes. The presence of solvent (generally decane) in the bilayer is a major obstacle for making comparisons with biological membranes. However, solvent-free black lipid membranes (Montal films)[120] can be made by forming phospholipid monolayers on the surfaces of opposing chambers and gradually raising the surface. As the surface monolayer crosses the orifice, the monolayers meet and form a bilayer. This technique has the additional advantage that asymmetric bilayers can be formed.

B. Reconstitution of Proteins into Membrane Vesicles

Proteins have usually been added to phospholipids by two approaches, either to add the protein, which is often associated with detergent, to already formed membrane vesicles or to mix detergent-solubilized phospholipids and protein and remove the detergent and form the vesicles in a combined step. The details of both have been described in excellent reviews of this topic.[3,121] An additional method, where feasible, is to mix the protein and lipids in apolar solvents, evaporate the organic solvent, rehydrate the sample, and isolate the vesicles by centrifugation.[122] Another way to reconstitute proteins with phospholipids is to use synthetic compounds with short chain fatty acids (such as 1,2-dioctoyl-phosphatidylcholine), since these have greater aqueous solubility and can be used to titrate specific phospholipid-binding sites.[123]

C. Artifacts of Artificial Membrane Preparation

The most difficult to control variable in reconstituting membrane processes is the exact composition of the final preparation. In the preparation of the phospholipid vesicles, traces of residual organic solvents, detergents, metals (i.e., divalent cations and pieces of the metal sonicator probe), other lipids (e.g., fatty acids or lysophospholipids), and peroxide degradation products, may be present and affect the properties of the bilayers. Upon adding protein solutions, detergent and phospholipids tightly bound by the protein will be incorporated into the vesicle. The above uncertainties, as well as the heterogeneity in the vesicle population with respect to the size, composition, and the possible asymmetric distribution of some of the species across the bilayer should be considered when interpreting data for reconstituted membrane preparations. It is perhaps because membranes are naturally heterogeneous and mobile that these problems tend not to prevent sophisticated analyses of proteins in reconstituted systems.

VII. TECHNIQUES APPLIED TO PHOSPHOLIPID MEMBRANES

A. Electron Microscopy

1. Thin-Section Electron Microscopy

Electron microscopy of thin sections of tissue provided the earliest pictures of cellular membranes. In such preparations,[124-126] specimens are treated with a fixative (e.g., glutaraldehyde alone or in conjunction with osmium tetroxide) to preserve cellular substructure and with a heavy metal (such as osmium tetroxide) to enhance the electron opacity of the membranes. The fixed specimens are dehydrated and embedded in wax or epoxy resin to provide a stronger supporting medium for sectioning, a process in which they are cut into 3- to 8- μm thin sections with a microtome.

Alternatively, the specimen can be deposited on coated grids and negatively stained by adding a solution of a heavy metal salt to the surface and allowing it to dry. The salt will fill the crevices between components such as proteins yielding electron density on the surface of the membrane bilayer. Biological membranes prepared with negative staining usually appear as two dark lines (due to reaction of the heavy metal with the proteins and polar head groups at the surface of the bilayer) sandwiching a relatively clear center (the hydrophobic core).

These methods are only of limited utility for detailed studies of membrane phospholipids because the treatments can extract some of the lipids and/or cause considerable deformation of the membranes. Thin sectioning is, however, particularly effective for investigations of proteins or enzymes that can be localized immunochemically or using a relatively specific histochemical stain.

2. Freeze-Fracture Electron Microscopy

The problem of distortion of the membranes during sample preparation has largely been solved by the freeze-fracture technique.[125-130] This method involves rapid freezing of the

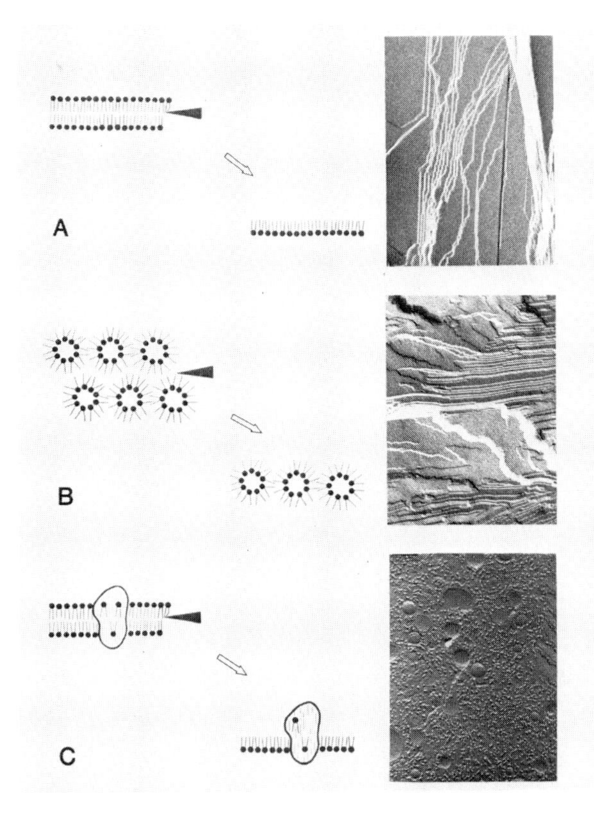

FIGURE 7. Appearance of different lipid structures seen by freeze-etching electron microscopy. (A) Smooth faces of lamellar structures; (B) long, parallel lines of tubular hexagonal-II phases; (C) irregular surface of membranes containing proteins. These electron micrographs were contributed by Dr. David W. Deamer.

specimen to trap the membranes in their native configuration.[130] Rapid freezing by the spray freezing[131,132] or copper-plate[133] methods are especially effective. Next, the specimen is fractured under vacuum, which splits at least some of the membranes through the hydrophobic core and gives two fracture faces (Figure 7), each constituting a side of the membrane bilayer.[128] In some instances, the samples are then etched by vacuum sublimation to remove a portion of the surrounding water, which enhances the image. Lastly, the specimen is replicated by shadowing with a more electron-dense material, such as platinum/carbon. In this process the electron-dense material is added at an angle to create dark regions on the raised edges and light areas where the surface is indented (see Figure 7). When platinum/carbon is used, the resolution will be about 30 Å due to the size of the platinum grain.

Using freeze-fracture electron microscopy, phase transitions of membrane phospholipids have been directly visualized. Studies of liposomes prepared from synthetic phosphatidylcholines, for example, have shown that the membranes have a smooth appearance in the gel and liquid crystalline states, but are rippled between the pretransition and the main transition.[127] More complex lipid mixtures may exhibit multiple transitions, depending at least in part on the lateral separation of the species. Freeze-fracture electron microscopy has also been used to demonstrate that membranes composed of some lipids are homogeneous (i.e., the patterns are similar to those for membranes with single components) while others are segregated into separate domains (i.e., multiple patterns are visible).

As was noted in Section III, some lipids tend to form spherical and tubular micelles which have been termed ''hexagonal phases''. For hexagonal I phases, the hydrophobic acyl chains are located inside and the head groups are on the surface of the micelle in contact with water. For hexagonal II phases, the polar head groups are inside the micelle. As is shown in Figure 7, fracturing of these structures will yield "bumps" and "pits" if the micelles are spherical or lines if the micelles are cylindrical (sometimes parallel lines if they exist in multiples). Some have been observed by freeze-fracture electron microscopy[127,130] and can have diameters from 20 to 120 Å. These structures resemble proteins dissolved in a planar membrane bilayer and their identification in biological membranes has usually involved the combination of microscopy and ^{31}P NMR spectroscopy.[127,134]

To more specifically observe individual lipids or proteins in membranes it is necessary to use cytochemical probes[135] or autoradiography.[136] Artifacts have been noted in the use of autoradiography to localize phospholipids.[137] Proteins can be visualized using ferritin-conjugated antibodies and lectins, or other markers such as hemocyanin, silica spheres, latex particles, horseradish peroxidase, or colloidal gold.[124,135] Whereas these commonly involve treatment of the membranes before freezing, newer approaches use critical point drying to facilitate cytochemical labeling of the fracture faces.[138,139]

Cholesterol can be localized by treating membranes with agents such as filipin, digitonin, and tomatin, and observing the regions that become deformed as a result of interactions between these compounds and 3β-hydroxysteroids.[135] Filipin treatment causes small (approximately 250 Å) circular bumps or pits in the membrane; digitonin and tomatin induce long tubular ridges or furrows (400 to 600 Å wide) that may curve and interweave. It appears that under the proper conditions these agents can be used to conclusively demonstrate the presence of cholesterol in membranes and provide strong, but less compelling evidence for the absence of this molecule. An example of the latter is the discovery that coated pits appears to be relatively free of cholesterol.[140]

Cytochemical agents have also been used to visualize phospholipids in membranes.[135] Polymyxin B has a strong affinity for negatively charged lipids and will deform membranes upon complex formation. The sensitivity to polymixin differs with the particular phospholipid; however, membranes composed of phospholipid mixtures will respond even though the major species does not bind polymixin tightly. The limitation of this approach, so far, is that reagents are not available for more specific interaction with particular phospholipid classes. Since antibodies have been prepared against sphingomyelin analogs,[141] it might be possible to use these to visualize this phospholipid in membranes.

B. Langmuir Trough

Oil placed on water spreads until it forms a thin film of approximately 1 atom thick (a point that was even noted by Benjamin Franklin). Due to this property, the molecular areas and volumes of lipids, as well as surface tension and potential at the air-water interface can be determined using a Langmuir trough[28,29,142,143] apparatus. This consists of a pan containing the aqueous subphase, a small boat to rest against and apply lateral pressure on the lipid monolayer, and a very sensitive monitor of the surface pressure (such as a Langmuir or Wilhelmy surface balance) (Figure 8). As the movable barrier (boat) is moved through the aqueous subphase, the lipid molecules become compressed until they are maximally packed as monolayers. The resistance to additional packing by the boat is recorded and from the amount of lipid added and the geometry of the device, the area per molecule is calculated. This method is not only informative about dimensions of individual phospholipids, but can also establish the extent to which mixtures represent a sum of the areas of the component species, or exhibit tighter or less dense packing due to lipid-lipid interactions. Similar studies can be conducted with mixtures of phospholipids and other agents, such as anesthetics. Since the area per molecule depends on the surface pressure (Table 2 and Figure 8), comparisons of different molecules should be made at similar pressures, or by determining the

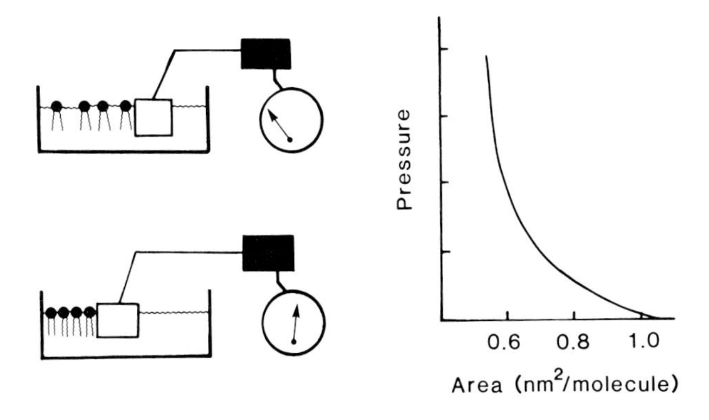

FIGURE 8. A representation of a Langmuir-trough-type apparatus used for measuring the area of phospholipids as monolayer films, and a typical graph of the relationship between the applied pressure and the area of the phospholipid.

limiting area, which is estimated from the pressure range over which essentially no further compression occurs.

C. X-Ray and Neutron Diffraction[144,147]

X-Ray diffraction patterns arise from bombardment of lipid specimens (such as lipids dried onto small glass supports or dispersed in buffer) with a beam of relatively parallel and monochromatic radiation (usually with a wavelength of 1.54 Å) and detecting the X-rays diffracted by collision with atomic electrons on film or position-sensitive electronic counters. Undiffracted radiation is absorbed by a small metal beam stop around which will appear arc rings of diffracted X-rays. Oriented samples give arcs instead of rings, which shortens the time required for data collection and allows diffraction resulting from structure components parallel and perpendicular to the bilayer plane to be detected. Different procedures are used for oriented and random samples; these and other details of methodology are discussed in the referenced reviews.

Information that can be initially derived from diffraction patterns is whether the sample consists of single- or multi-walled vesicles. The former are characterized by broad, circularly symmetric bands, whereas the latter give sharper, evenly spaced bands. The bilayer thickness can be estimated from the position of the bands[144] or the repeated distance between the bands (called Bragg reflections). Care should be taken in interpreting these distances as the thickness of the bilayer, however, because they actually represent repeats of bilayers and can be influenced by the amount of water that is present between the multilayers. Additional information is obtained if the bands are diffuse. There are two types of diffuse bands with lipids. The first are equatorial bands that arise at about 4.6 Å and occur when lipids are in a fluid state. The other type are found roughly equally spaced in a direction perpendicular to the bilayer plane and arise from the transbilayer density distribution. These can, in principle, be seen with lipids above and below the phase transition.

A much more detailed description of the membranes can be derived from analyses of the intensity of the diffraction bands. These analyses are usually conducted by Fourier transform or autocorrelation methods, with supplemental experimentation to solve the phases. Once the electron densities are calculated, it is possible to represent the results of an X-ray density profile as is shown in Figure 9 for a mixture of phosphatidylcholine and cholesterol. Since electron densities decrease in the order: polar groups > hydrocarbon chains > methyl groups (coherent scattering amplitudes of the individual atoms are S > P > O > N > C > H/D),

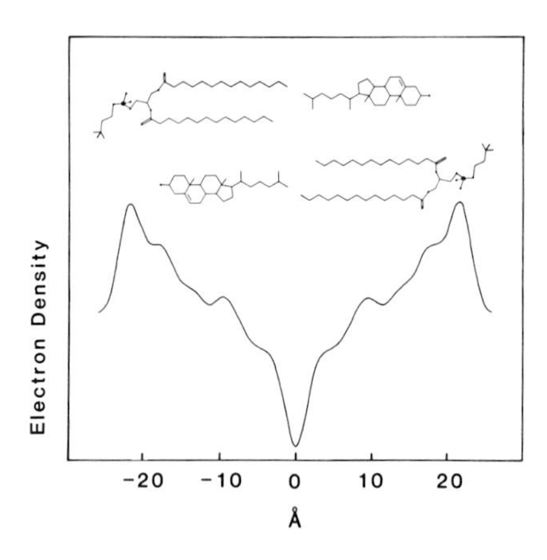

FIGURE 9. A typical electron density profile for a membrane bilayer and a model of the approximate positions of the atoms contributing to the density. The electron density profile comes from the data for a 1:1 mixture of dimyristoylphosphatidylcholine and cholesterol presented in the review by Franks and Lieb.[144]

the phosphate groups appear as peaks while the terminal methyl groups of the fatty acids in the middle of the hydrocarbon core are represented by the trough in the middle.

Neutron diffraction utilizes scattering from atomic nuclei to derive additional information about lipids in bilayers.[144] A major difference between neutron and X-ray diffraction is that the coherent scattering differs in amplitude and order (e.g., N > O > P > S). An important advantage of this is that the amplitude for deuterium is much greater than that for hydrogen and hydrogen/deuterium exchange can be used to localize certain portions of the structure without major modification of the molecules. If a sample can be analyzed by both X-ray and neutron diffraction, assignments can be made with greater certainty due to the differences in relative amplitudes for the major groups of phospholipids.

D. Optical Spectroscopy

1. Absorbance Spectroscopy

Absorption spectra are sensitive to the electronic states of the chromophore and therefore can provide information about its redox state, ionization, and environment. Absorbance spectroscopy[148-153] has been applied mostly to study protein chromophores (aromatic amino acids, flavins, and hemes, among others). (For a more detailed discussion of this use, see Volume II, Chapter 7 in this volume.) If the chromophore is optically active or becomes so upon insertion into an asymmetric site, it is possible to get information about its conformation and environment by other optical techniques. These include circular dichroism (CD), which measures the difference in absorbance of left and right circularly polarized light and optical rotatory dispersion (ORD), which measures the optical rotation as a function of wavelength. With the exception of the sphingolipids, which possess optical activity due to their D-*erythro* configuration, the application of CD and ORD to phospholipids has been less fruitful than it has been for proteins.[151] Sphingomyelin exhibits a strong ellipticity (often referred to as a Cotton effect) around 200 nm, the position and magnitude of which is affected by the type of lipid structure assumed by this lipid.[154,155] It has been noted that sphingomyelin contributes 20% of the CD signal below 200 nm for human erythrocyte ghosts and could cause misinterpretation of the CD spectra of membrane proteins.[15]

2. *Fluorescence Spectroscopy*[156-159]

Some excited molecules transfer a portion of the absorbed energy to the environment and decay to a lower energy singlet state before returning to the ground state with the emission of light of longer wavelengths than was absorbed. This process is called "fluorescence". The efficiency with which a quanta of absorbed light is reemitted as fluorescence is termed the "quantum efficiency" and is very sensitive to the environment of the fluorophor. Since molecules remain excited for 10^{-9} to 10^{-7} sec (compared to 10^{-15} sec for the initial absorption of energy), this process takes place over a time frame similar to the mobility of phospholipids and can, therefore, be employed for measurements of lipid mobility.

Fluorescence is generally more sensitive than absorbance and has found wide utility in studies of membrane lipids and proteins. It has the advantage that not all molecules fluoresce; therefore, a single fluorescent species can be measured in a membrane containing numerous nonfluorescent chromaphores. It has the disadvantage that many studies require the use of an artificial fluorophor, which might perturb the properties of the membrane. Nonetheless, the wide variety of molecules used have generally provided similar answers, many of which have been confirmed by other physical techniques. Among the various probes that have been attached to phospholipids or added alone as reporters of membrane viscosity are pyrenes, 1-anilinonaphthalene-8-sulfonate (ANS), various anthroyloxy-fatty acids, β-parinaric acid, fluoresceine and rhodamine conjugates, and 7-nitro-2,1,3-benzoxadiazol-4-yl acyl (NBD) labeled fatty acids and head groups.

The quantum yield, the shape of the fluorescence emission spectrum, and the fluorescence lifetime and polarization are the most frequently measured parameters. Some lipids, such as β-parinaric acid, are much more fluorescent when in a nonpolar environment, and can be used to measure the relative "hydrophobicity" of its environment when protein bound or it is transferred from an aqueous solution into a hydrophobic bilayer.[160] Many compounds decrease the quantum efficiency of a fluorophor (which is generally referred to as "quenching") by absorbing light of the wavelengths emitted by the fluorophore. Quenching is also caused by collision of molecules like O_2 and I^- with the excited singlet. Collisional quenching allows the determination of the fraction of a particular fluorophore on the outer leaflet of a bilayer since only that portion will interact with the quencher. This has been used to measure the location of fluorescent probes within a bilayer by looking at the effect on the fluorescence of a fluorophore (such as an anthroyloxyfatty acid) of quenchers at the surface of the membrane (Cu(II) and N,N-dimethylaniline) vs. ones in the hydrocarbon bilayer (16-nitroxidestearic acid).[161] Quenching can occur when fluorophores (called donors) are near other molecules (called the acceptors) that can receive energy directly from the excited state of the donor and themselves become excited (with or without subsequent fluorescence). This process is called fluorescence-, Forster-type-, or singlet-singlet energy transfer and is extremely sensitive to the distance between the two species (decreasing with $1/R^6$ where R is the distance between the chromophores). This process is usually thought of in the context of transer between two chromophoric compounds, but can also occur between paramagnetic species such as nitroxide spin labels.[162] Fluorescence energy transfer has been used to measure surface density[163] and intravesicular transfer of phospholipids.[56]

The apparent viscosity of membranes can be analyzed by determining the fluorescence anisotropy,[25,26,157] which involves exciting the sample with polarized light and measuring the fluorescence parallel and perpendicular to the incident light. This can be used to calculate an anisotropy parameter, r, which is related to the apparent viscosity of the solvent in the immediate vicinity of the probe and the effective rotational volume of the probe. One method to determine the relative viscosity of a sample is to measure the anisotropy of a probe in solutions of known viscosities and compare the results to that with membranes. For the most widely used probes, such as 1,6-diphenyl hexatriene, the relationship between the effective rotational volume of the probe and r has been determined.[164] Membrane order can also be assessed by measuring the lifetime for fluorescence emission.[165] A more sophisticated approach to analyzing membrane dynamics using fluorescence is to measure the differential lifetime of vertically and horizontally polarized components.[166]

Lateral movement of lipids and proteins can also be measured by fluorescence correlation spectroscopy and fluorescence photobleaching and recovery.[25,157] Fluorescence correlation spectroscopy[167] is conducted by illuminating a given region of a membrane containing the fluorophore with a laser beam and analyzing the fluctuations in fluorescence about the average as molecules randomly enter and exit. These fluctuations are related to the lateral diffusion of the fluorescent probe. In fluorescence photobleaching and recovery,[168] a flash of laser light is used to bleach the fluorophore and the reappearance of fluorescence as new molecules diffuse into the bleached spot is measured. This has been broadly applied to the analysis of lateral diffusion of surface molecules in animal cells and tissues.[168,169]

3. *Infrared*[170-172] *and Raman*[171-174] *Spectroscopy*

Irradiation of samples with light of longer wavelengths (i.e., beginning around 1000 nm) provides only enough energy for vibrational transitions, which are of course themselves interesting because they reveal information about subtle motions and interactions between molecules. Absorptions in the standard (i.e., 4000 to 200 cm^{-1}) and far (200 to 10 cm^{-1}) infrared (IR) have been examined, although most studies use the former. With the proper instrumentation, IR dichroicism can be measured. For examples of information gained from IR spectroscopic analyses of phospholipids, in phosphatidylcholine preparations the methylene vibrations appear around 1461 to 1469 and 2848 to 2959 cm^{-1} and the latter shifts upon phase transition. The vibration about the phosphate ester (i.e., P = O bond) differs for phosphatidylcholine (1254 to 1260 cm^{-1}) and phosphatidylethanolamine (1215 to 1260 cm^{-1}), which is thought to be due to the greater tendency of the latter to form a hydrogen bond between the NH and O atoms in adjacent phosphatidylethanolamine atoms.

In Raman spectroscopy, a sample is illuminated with a high intensity light source (such as a laser tuned to 488 nm), which induces an oscillating dipole in the molecule. If the polarizability of the oscillation is affected by its nuclear motions, the reemitted radiation can appear at different frequencies, one of which is higher than the original exciting light (called the anti-Stokes band). This is measured and since it is usually related to the IR bands of the molecule, the radiation is interpreted with respect to the vibrational motions of the group. An advantage of Raman spectroscopy vs. IR is that water has a fairly weak Raman spectrum, therefore is less complicating. Furthermore, Raman spectra can be complimentary to IR spectra since some groups that give weak IR signals yield more intense Raman spectra. Raman signals can be enhanced by employing Raman active probes, such as conjugated polyenes or deuterated phospholipids.

Raman spectra of phospholipids exhibit bands due to the stretching of carbon-hydrogen bonds between 2700 and 3100 cm^{-1} and bending bands between 1400 and 1500 cm^{-1}. Also informative are the C-C stretching band between 1050 and 1150 cm^{-1} (which has been interpreted as reflecting the *trans* and *gauche* chain conformations) and the acoustical and lattice vibrations (which reflect the accordion-like motions of the alkyl chains) that occur below 400 cm^{-1}. For example, upon heating a preparation of dipalmitoylphosphatidylcholine from below to above its main transition temperature, the loss of chain-chain interactions is observed with a sharp increase in *gauche* bonds. IR and Raman spectra for dipalmitoylphosphatidylcholine bilayers show different bands for the carbonyl groups at position 1 (1740 cm^{-1}) and 2 (1721 cm^{-1}), which has been interpreted as reflecting the greater hydrophobicity of the environment of the first carbonyl. This is perpendicular to the bilayer compared to the more parallel and bent conformation of the second carbonyl.[175] Cholesterol shifts the absorbance of the 2 carbonyl to the same position as the first.

E. Magnetic Resonance Spectroscopy[176]

Unpaired electrons and some atomic nuclei have spin angular momenta and will assume distinct orientations with different energies when placed in an external magnetic field. Irradiation of these samples at the same frequency (energy) as the quantized spacing of the energy levels will stimulate transitions from the lower to higher energy states. When the atoms resume the equilibrium distribution, the reemitted energy is collected to generate a record of the transition. These basic principles underlie nuclear magnetic resonance (NMR) and electron paramagnetic resonance (EPR) spectroscopy.

1. Nuclear Magnetic Resonance (NMR) Spectroscopy

a. General Information

NMR spectroscopy[25,177-181] provides information about molecular structure, the environment of a particular nucleus, and its mobility. Electrons exert a "screening" effect on a nucleus placed in a magnetic field; hence, those surrounded by greater electron density will absorb radiation at higher magnetic fields. This phenomenon results in a "shift" in the signal compared to a standard. The position of the signal relative to the standard, which is called the "chemical shift", can be used to identify the compound or when already known, can be used to characterize its environment. Chemical shifts are affected by nearby electron withdrawing or donating substituents, ring currents, and hydrogen-bonding and ionization. Since the magnetic field around a given nuclei is also affected by the magnetic moments of neighboring nuclei, the signal can be split into multiple components. The degree of splitting of a signal will usually depend on the number of nonequivalent nuclei in the vicinity. If the shielding of the nucleus is anisotropic, which is particularly the case with ^{31}P, the local magnetic field will fluctuate as the molecule tumbles and the signal will exhibit a shoulder (see Figure 10). This behavior is referred to as the chemical shift anisotropy (CSA) and provides information about the conformation and dynamics of that group. Interactions with paramagnetic groups in the solution (such as lanthenide ions) shift and/or broaden the signal and can be used to resolve otherwise overlapping signals, measure distances between groups, or establish whether or not a given species is in contact with the solution containing the reagent.

Additional characteristics of the NMR signal are the spin-lattice or longitudinal relaxation time (T_1) and the spin-spin or transverse relaxation time (T_2). T_1 and T_2 are the rates of return to equilibrium of the magnetic moments of the nuclei in the magnetic field. They are usually measured by (1) sequentially pulsing the sample with another magnetic field to cause the nucleus to precess about the main axis and (2) measuring the rate of return to equilibrium for each pulse. This relaxation involves energy transfer from the nuclei to the environment; hence, the spin-lattice relaxation time is affected by the surrounding medium. When molecular motion is severely restricted, as in solids or viscous liquids, T_1 will be very large and the sample will not return to equilibrium in the NMR time scale (NMR is insensitive to motions with a frequency of less than 10^5/sec). Because the signal in an NMR experiment is related to the number of nuclei that can be stimulated to undergo transitions from the lower to the higher energy states, the system will be "saturated" and the line will broaden until it disappears (Figure 10B). In systems with less restriction of molecular motion, the value of T_1 can be used as an index of the environment of the nuclei. T_2 is also affected by the environment of the nucleus, and together T_1 and T_2 are determinants of the line width of a particular signal.

For some nuclei (primarily ^{31}P), structural information can be gained by analyzing the chemical shift anisotropy (CSA), which is a function of the orientational and motional properties of the group, or the nuclear Overhauser effect (NOE), which can detect interactions between two different nuclei (such as ^1H in the head group methylenes and ^{31}P in the

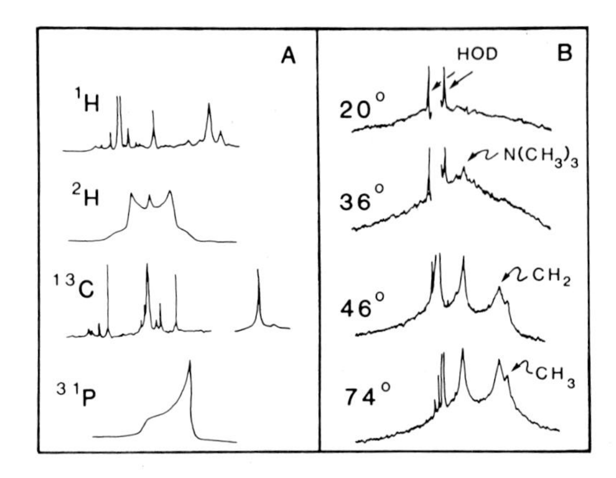

FIGURE 10. Examples of the NMR spectra of different nuclei of phospholipids. (A) ^1H spectrum of phosphatidylcholine (for assignments, see B), ^2H spectrum of phosphatidylcholine with deuterium in position 1 of the choline headgroup, ^{13}C spectrum of phosphatidylcholine with natural abundance ^{13}C (left) compared to that with ^{13}C enriched in the fatty acyl side chain (right), and ^{31}P spectrum of phosphatidylserine, demonstrating the asymmetry of the signal termed the chemical shift anisotropy.[177] (B) Assignments for different nuclei in the ^1H NMR spectra of phosphatidylcholine in D_2O and demonstration of the effect of temperature (i.e., lipid mobility) on the signals.[24]

phosphate of phosphatidylcholine). NOE occurs when the magnetic moments of two dissimilar nuclei are able to interact (both due to their spatial nearness and the fortuitous coupling of the nuclei by dipolar interactions) and irradiation of one of the atoms results in a cancellation or enhancement of the NMR signal of the other.

b. Application to Phospholipids

The major atomic nuclei used in studies of phospholipids are ^1H, ^2H, ^{13}C, ^{15}N, ^{19}F, and ^{31}P (Figure 10A). With the exception of ^{19}F, introduction of these into phospholipids causes little or no perturbation of the structure of the phospholipid. Each nucleus has advantages and limitations that have been discussed in detail.[177] Phosphorus, ^{13}C, and proton NMR spectra can be performed with native phospholipids and membranes; however, even the other nuclei can be incorporated into phospholipids synthetically or by feeding them to the appropriate microorganisms.

Phosphorus NMR has been very useful in determining the head group structure of phospholipids in membranes. Randomly oriented phosphate groups give symmetric signals, whereas phospholipids in lamellar and hexagonal-II phase structures exhibit shoulders at high or low fields, respectively. This, combined with electron microscopy, has identified hexagonal-II phases as natural features of vesicles prepared from some phospholipids (such as unsaturated phosphatidylethanolamines) and biological membranes.[182] Molecular information concerning the relatively rigid nature of the head group of phosphatidylserine compared to phosphatidylcholine and phosphatidylethanolamine has been obtained by examining the ^{31}P spin relaxation times.[177] Interactions between phospholipid head groups have been examined by the ^{31}P [^1H] nuclear Overhauser effect (NOE), although recent studies indicate that interpretation of these spectra can be very complicated.[183] ^{31}P NMR analyses of phospholipids can be conducted in detergents[184] with better resolution of the resonances of individual phospholipids.

The line shapes and relaxation times of various nuclei (e.g., ^2H, ^{13}C, and ^{19}F) have been used to describe the orientation and motion of different atoms in a given phospholipid.[177,185,186] These are reported as an order parameter or correlation time. An order parameter of 1.0 corresponds to a highly ordered system, 0 represents a disoriented chain. In general, the order parameter begins at around 0.2 to 0.4 and remains relatively constant along the acyl chains between carbon atoms 2 to 8 (numbering from the carboxyl group) until dropping off to 0.1 and below near the terminal methylene.[177] There appears to be little difference between the order parameters for equivalent regions of fatty acids attached to the *sn*-1 and *sn*-2 position of the glycerol backbone. Cholesterol has been found to increase the order parameter to as much as 0.85, indicating a high degree of ordering, despite the fact that the bilayers were still in the liquid crystalline state.[177] In the region of *cis*-unsaturated fatty acids, there is a sharp drop in the order parameter.[187] It has been noted, however, that membrane motions of larger amplitude, such as wave-like motions in the bilayer, could be part of the membrane dynamics that yield these variations in relaxation times.[188]

To simplify analyses of biological membranes with minimal perturbation of their native structure, lipids enriched in the nuclei of interest have been prepared and inserted into membranes biosynthetically by feeding the precursors to microorganisms. For examples of the information thus derived, *Escherichia coli* grown on deuterated glycerol or fatty acids[189] exhibited some effect of proteins on the head group structure, but not on the fatty acid or glycerol moieties. In general, the glycerol backbone, with or without proteins in the vicinity, is relatively immobile. Most other NMR studies of the effects of proteins on the hydrocarbon region of membranes have not seen large effects; however, recent use of the ^{13}C cross-polarization technique found that proteins depress the intensity of motion occurring over nanoseconds and increases the intensity on the tens of microseconds timescale.[190] This might be due to an effect on the "transverse waves" or ripples that may propagate through large segments of the fatty acyl chains.[190]

The effects of branched chain and *trans*-unsaturated fatty acids have been examined by another approach, which was to measure the order and dynamics of *Acholeplasma laidlawii* B membranes enriched with 8-^{19}F-palmitic acid plus the additional fatty acids of interest. [191] At a single temperature (37°), the order parameter was somewhat lower with linear saturated fatty acids than with branched and *trans*-unsaturated fatty acids. This correlated with the differences in phase transition temperatures for these membranes. When compared at comparable "fluidities" (i.e., at the Tm + 15° for each preparation), the effects of the differences among the fatty acids were greater and in the opposite order from these. Therefore, *trans*-double bonds induced a local decrease in the order of neighboring acyl chains, which was the opposite of the previously reported effect of *cis*-unsaturated fatty acyl chains.

Shift reagents have been used extensively to identify the phospholipid head groups in contact with the aqueous environment of one side of a bilayer.[192] These reagents include the lanthenide ions Pr^{3+}, Nd^{3+}, and Eu^{3+}, and other compounds such as free radicals and ferricyanide.

2. *Electron Paramagnetic Resonance (EPR) Spectroscopy*

EPR (also called ESR for electron spin resonance)[193-196] spectroscopy is based upon the alignment in a magnetic field of unpaired electrons due to their net magnetic moment. By varying the external magnetic field while holding the microwave frequency constant, transitions between energy states by different unpaired electrons can be detected and analyzed. It provides much the same information as NMR, such as peak position (the g value), line widths, splitting due to neighboring nuclei (termed "hyperfine splitting"), relaxation phenomena, and the ability of the probe to interact with other agents in solution or the membrane; e.g., addition of ascorbate reduces the radicals on the side of the bilayer accessible to this reagent. However, they differ in that EPR probes can be detected in low concentrations and EPR is typically insensitive to motions with a frequency of less than 10^7/sec, although saturation transfer EPR (ST-EPR) can detect motions in the range 10^3 to 10^7/sec. Since there

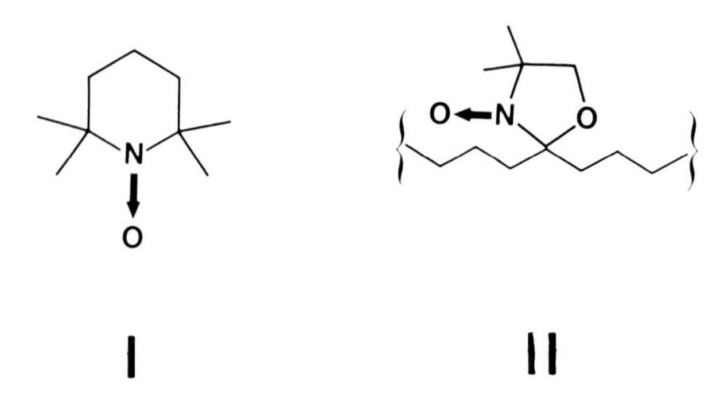

FIGURE 11. Structures of two types of spin probes frequently employed in EPR studies of phospholipid membranes. (I) 2,2,6,6-Tetramethylpiperidine nitroxide (TEMPO) and (II) dimethyloxazolidine nitroxide (DOXYL).

are relatively few natural free radicals in biomembranes (e.g., metalloproteins and flavin semiquinones), stable spin labels, such as the paramagnetic nitroxides TEMPO (Figure 11A) and *N*-oxyloxazolidines (Figure 11B), are covalently attached to the head groups or fatty acyl chains of phospholipids or to different positions of cholesterol.

Biradical probes have also been prepared since they are more sensitive to molecular motions than monoradicals.[197] The disadvantage of this, of course, is that each probe must be tested for its "unnatural" perturbation of the system of interest. The relative location of a number of nitroxide spin labels have been carefully determined.[198] Also, fortunately, the conclusions from EPR studies have often agreed with those derived by other methods, even though the introduction of a spin label presents some methodological problems.[199]

F. Differential Scanning Calorimetry

Differential scanning calorimetry[200-204] has been used extensively to measure thermal transitions of model and biological membranes. This technique is based upon the endothermic nature of transitions from more ordered (e.g., gel) to disordered (e.g., liquid-crystalline) states due to the standard enthalpy change of the process. Measurements of this change are made by gradually increasing the temperature of two cells (one of which contains the sample and the other is a reference) and by comparing the heat input necessary to increase the temperature of both cells equally. For a simple equilibrium between two states, differential scanning calorimetry will yield data like that shown in Figure 12. The area under the curve is related to the enthalpy of the process and the maximum is termed the "transition temperature" (Tm).

The advantage of differential scanning calorimetry is that it directly provides thermodynamic information without requiring the introduction of external probes that may perturb the system. Generally, the manipulations involved in differential scanning calorimetry do not alter the specimen, which can be reused in other experiments and the results of the two can be compared. In fact, additional information about the system can be obtained by determining whether or not the transition is reversible. The limitation of this method, of course, is that it provides thermodynamic, not mechanistic information about the changes.

Differential scanning calorimetric curves with synthetic phospholipids exhibit a main transition from gel to liquid crystalline states at Tm that depend on the polar head group and the acyl side chains (Table 1). Both Tm and the enthalpy changes increase upon increasing the acyl chain length in an homologous series of phospholipids with the same head group and different disaturated fatty acids. Introduction of *cis*-double bonds dramatically decrease

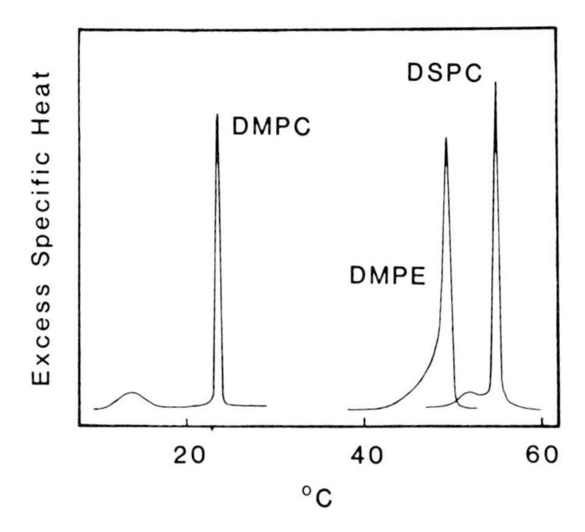

FIGURE 12. Examples of differential scanning calorimetric scans for different phospholipids (dimyristoylphosphatidylcholine, DMPC; dimyristoylphosphatidylethanolamine, DMPE; and distearoylphosphatidylcholine, DSPC), illustrating the variations in main transition temperature for phospholipids with different fatty acyl chains or headgroups. The smaller pretransition seen with the phosphatidylcholines is attributed to a tilt in the molecules resulting in a rippling of the bilayer. These data were modified from the review by Mabry-Gaud. [200]

the Tm due to introduction of greater disorder in the hydrophobic core of the membrane. Comparison of phosphatidylcholines and ethanolamines having identical fatty acyl groups reveal that the former have Tm values 20 to 30°C lower than the corresponding phosphatidylethanolamines. Phosphatidylserines of similar composition are usually intermediate between phosphatidylcholines and ethanolamines. If phospholipids with acyl chains of different lengths are mixed, the Tm may shift to a value between those for the individual phospholipids or exhibit a broader and more complex transition. In some complex systems (especially those containing glycolipids),[205] there can be multiple and distinct transitions due to the existence of several different lipid phases.

Some phospholipids (e.g., phosphatidylcholine) exhibit a smaller enthalpy change near the main transition. This is thought to reflect a reorientation of the gel state by a shift in the angle of the head groups and acyl chains.[206] The nomenclature for the different states[207] designates the type of lattice by capital letters: C for crystalline, L for one-dimensional lamellar, H for two-dimensional hexagonal, P for two-dimensional oblique, Q for cubic, and R for rhombohedral structures. The orientation of the acyl chains is designated by a Greek letter: α for liquid crystalline, β for relatively stiff acyl chains perpendicular to the plane of the lamellae, β' for tilted acyl chains, and δ for acyl chains coiled into helixes.

G. Chemical and Enzymatic Probes[208,210]

Most direct evidence for the asymmetric nature of biological membranes has been obtained by treating them with agents that interact with either the proteins or lipids.[208-210] Phospholipids have been examined by adding phospholipases (such as carefully purified phospholipase C), exchange proteins,[211,212] or reactive compounds that react with phospholipids. Examples of modifying reagents are trinitrobenzene sulfonic acid (TNBS), fluorodinitrobenzene (FDNB), 1,5-difluoro-2,4-dinitrobenzene (a cross-linking reagent[213] which reacts with the free amines of phosphatidylethanolamine and phosphatidylserine),[209] and periodate, which will only oxidize phosphatidylglycerol.[214] Proteins have been examined by protease digestion, enzyme assays under conditions where aqueous substrates will only have access to the active site if it is on the outer surface of the vesicle (or the vesicles have been disrupted — expression of activity after disruption of vesicles has been termed "latency"), reaction with antibodies, and examination of the effects of membrane impermeant inhibitors on the activity.[215] Glycolipids and glycoproteins have also been examined using various antibodies and lectins.[67] Antibodies have even been prepared against sphingomyelin and were shown to label the surface of red blood cells.[141]

In such studies, it is important to check that the agent reacts only with the intended side of the bilayer, does not itself perturb the asymmetric distribution of the component under study, and acts faster than the flip-flop rate of the lipid being studied. When carefully done, such agents can provide information about not only the asymmetric distribution of phospholipids, but also related processes, such as the rate of appearance of newly synthesized phosphatidylethanolamine at the plasma membrane[216] or the transbilayer movement of phospholipids during vesicle fusion.[217]

In a separate approach, fatty acids and phospholipids containing photoreactive groups (such as diazotrifluoropropionyl-and diazirinophenoxy compounds) have been synthesized.[218,219] When illuminated with high intensity light, they generate highly reactive nitrenes and carbenes that insert into single and double bonds, thereby cross-linking phospholipids with each other or proteins. The presumption is that these intermediates are so reactive that the structure of the cross-linked complex will provide information about their exact location in the bilayer (i.e., their nearest neighbors). The short half-lives of such intermediates (around 10^{-11} sec) allows them to react before undergoing much randomization in the membrane.

ACKNOWLEDGMENT

The authors are grateful to Drs. Jeffrey L. Browning, Dennis Chapman, David W. Deamer, and N. P. Franks for permission to use data in the figures and tables, and for helpful suggestions concerning the text.

REFERENCES

1. **Fulton, A. B.,** How crowded is the cytoplasm?, *Cell,* 30, 345, 1982.
2. **Cullis, P. R.,** Structural properties and functional roles of phospholipids in biological membranes, in *Phospholipids and Cellular Regulation,* Vol. I, Kuo, J. F., Ed., CRC Press, Boca Raton, Fla., 1985. chap. 1.
3. **Lambeth, J. D.,** On the role of phospholipids and membranes in the regulation of oxidative enzymes, in *Phospholipids and Cellular Regulation,* Vol. II, Kuo, J. F., Ed., CRC Press, Boca Raton, Fla., 1985. chap. 7.
4. **Raetz, C. R. H.,** Genetic control of phospholipid bilayer assembly, in *Phospholipids,* Hawthorne, J. N. and Ansell, G. B., Eds., Elsevier/North-Holland, Amsterdam, 1982, chap. 11.
5. **Nelson, D. R. and Robinson, N. C.,** Membrane proteins: a summary of known structural information, *Methods Enzymol.,* 97, 571, 1983.
6. IUPAC-IUB Commission on Biochemical Nomenclature. The nomenclature of lipids (1976), *J. Lipid Res.,* 12, 455, 1977.
7. **Hawthorne, J. N. and Ansell, G. B., Eds.,** *Phospholipids,* Elsevier/North-Holland, Amsterdam, 1982.
8. **Allan, D., Billah, M. M., Finean, J. B., and Michell, R. H.,** Release of diacylglycerol-enriched vesicles from erythrocytes with increased intracellular [Ca^{2+}], *Nature (London),* 261, 58, 1976.
9. **Marx, J. L.,** A new view of receptor action, *Science,* 224, 271, 1984.

9a. **Boggs, J. M.,** Intermolecular hydrogen bonding between lipids: influence on organization and function of lipids in membranes, *Can. J. Biochem.,* 58, 755, 1980.

10. **Kuksis, A., Ed.,** *Handbook of Lipid Research,* Vol. 1, *Fatty Acids and Glycerides,* Plenum Press, New York, 1978.

11. **Kummerow, F. A.,** Modification of cell membrane composition by dietary lipids and its implications for atherosclerosis, *Ann. N. Y. Acad. Sci.,* 414, 29, 1983.
12. **Snyder, F.,** *Ether Lipids: Chemistry and Biology,* Academic Press, New York, 1972.
13. **Paltauf, F.,** Ether lipids in biological and model membranes, in *Ether Lipids: Biochemical and Biomedical Aspects,* Mangold, H. K. and Paltauf, F., Eds., Academic Press, New York, 1983, chap. 17.
14. **Esko, J. D. and Matsuoka, K. Y.,** Biosynthesis of phosphatidylcholine from serum phospholipids in Chinese hamster ovary cells deprived of choline, *J. Biol. Chem.,* 258, 3051, 1983.
15. **Barenholtz, Y. and Thompson, T. E.,** Sphingomyelins in bilayers and biological membranes, *Biochim. Biophys. Acta,* 604, 129, 1980.
16. **Barenholtz, Y. and Gatt, S.,** Sphingomyelin: metabolism, chemical synthesis, chemical and physical properties, in *Phospholipids,* Hawthorne, J. N. and Ansell, G. B., Eds., Elsevier/North-Holland, Amsterdam, 1982, chap. 4.
17. **Hakomori, S.,** Glycosphingolipids in cellular interaction, differentiation, and oncogenesis, *Ann. Rev. Biochem.,* 50, 733, 1981.
18. **Rosenberg, H.,** Phosphonolipids, in *Form and Function of Phospholipids,* Ansell, G. B., Hawthorne, J. N., and Dawson, R. C. C., Eds., Elsevier, Amsterdam, 1973, chap. 12.
19. **Vanderkooi, J. M., Landsberg, R., Selick, H., II, and McDonald, G. G.,** Interaction of general anesthetics with phospholipid vesicles and biological membranes, *Biochim. Biophys. Acta,* 464, 1, 1977.
20. **Singer, S. J. and Nicholson, G. L.,** The fluid mosaic model of the structure of cell membranes, *Science,* 175, 720, 1972.
21. **De Kruijff, B., Cullis, P. R., and Verkleij, A. J.,** Non-bilayer lipid structures in model and biological membranes, *Trends Biochem. Sci.,* 5, 79, 1980.
22. **Rothman, J. E. and Lenard, J.,** Membrane asymmetry, *Science,* 195, 743, 1977.
23. **Chapman, D.,** Phase transitions and fluidity characteristics of lipids and cell membranes, *Q. Rev. Biophys.,* 8, 185, 1975.
24. **Quinn, P. J. and Chapman, D.,** The dynamics of membrane structure, *CRC Crit. Rev. Biochem.,* 8, 1, 1980.
25. **Edidin, M.,** Molecular motions and membrane organization and function, in *Membrane Structure,* Finean, J. B. and Michell, R. H., Eds., Elsevier/North-Holland, Amsterdam, 1981, chap. 2.
26. **Thompson, T. E. and Huang, C.,** Dynamics of lipids in biomembranes, in *Membrane Physiology,* Andreoli, T. E., Hoffman, J. F., and Fanestil, D. D., Eds., Plenum Press, New York, 1978, chap. 2.
27. **Tanford, C.,** *The Hydrophobic Effect: Formation of Micelles and Biological Membranes,* John Wiley & Sons, New York, 1973.
28. **Nagle, J. F.,** Theory of the main lipid bilayer phase transition, *Ann. Rev. Phys. Chem.,* 31, 157, 1980.
29. **Wilkinson, D. A. and Nagle, J. R.,** Thermodynamics of lipid bilayers, in *Liposomes: From Physical Structure to Therapeutic Applications,* Knight, C. G., Ed., Elsevier/North-Holland, Amsterdam, 1981, chap. 9.
30. **Bell, G. M., Combs, L. L., and Dunne, L. J.,** Theory of cooperative phenomena in lipid systems, *Chem. Rev.,* 81, 15, 1981.
31. **Klein, R. A.,** Thermodynamics and membrane processes, *Q. Rev. Biophys.,* 15, 667, 1982.
32. **Kreissler, M., Lemaire, B., and Bothorel, P.,** Theoretical conformational analysis of phospholipids. II. Role of hydration in the gel to liquid crystal transition of phospholipids, *Biochim. Biophys. Acta,* 735, 23, 1983.
33. **Finer, E. G. and Phillips, M. C.,** Factors affecting molecular packing in mixed phospholipid monolayers and bilayers, *Chem. Phys. Lipids,* 10, 237, 1973.
34. **Cullis, P. R. and De Kruijff, B.,** Lipid polymorphism and the functional roles of lipids in biological membranes, *Biochim. Biophys. Acta,* 559, 399, 1979.
35. **Mason, J. T. and Huang, C.,** Hydrodynamic analysis of egg phosphatidylcholine vesicles, Ann. N.Y. *Acad. Sci.,* 308, 29, 1978.
36. **Huang, C.-H.,** A structural model for the cholesterol-phosphatidylcholine complexes in biological membranes, *Lipids,* 12, 348, 1977.
37. **Seelig, J. and Seelig, A.,** Lipid conformation in model membranes and biological membranes, *Q. Rev. Biophys.,* 13, 19, 1980.
38. **Hauser, H., Pascher, I., Pearson, R. H., and Sundell, S.,** Preferred conformations and molecular packing of phosphatidylethanolamine and phosphatidylcholine, *Biochim. Biophys. Acta,* 650, 21, 1981.
39. **Demel, R. A. and DeKruiff, F. B.,** Functions of sterols in membranes, *Biochim. Biophys. Acta,* 457, 109, 1976.
40. **Cornell, B. A. and Separovic, F.,** Membrane thickness and acyl chain length, *Biochim. Biophys. Acta,* 733, 189, 1983.

41. **Macdonald, P. M., McDonough, B., Sykes, B. D., and McElhaney, R. N.,** Fluorine-19 nuclear magnetic resonance studies of lipid fatty acyl chain order and dynamics in *Acholeplasma laidlawii* B membranes. Effects of methyl-branch substitution and of *trans* unsaturation upon membrane acyl-chain orientational order, *Biochemistry*, 22, 5103, 1983.
42. **Lentz, B. R., Hoechli, M., and Barenholz, Y.,** Acyl chain order and lateral domain formation in mixed phosphatidylcholine-sphingomyelin multilamellar and unilamellar vesicles, *Biochemistry*, 20, 6803, 1981.
43. **Presti, F. T., Pace, R. J., and Chain, S. I.,** Cholesterol-phospholipid interactions in membranes. II. Stoichiometry and molecular packing of cholesterol-rich domains, *Biochemistry*, 21, 3831, 1982.
44. **Shapiro, E. and Ohki, S.,** The interaction energy between hydrocarbon chains, *J. Cell. Int. Sci.*, 47, 38, 1974.
45. **Seelig, J.,** ^{31}P Nuclear magnetic resonance and the head group structure of phospholipoids in membranes, *Biochim. Biophys. Acta*, 515, 105, 1978.
46. **Trauble, H.,** The movement of molecules across lipid membranes: a molecular theory, *J. Membrane Biol.*, 4, 193, 1971.
47. **Cherry, R. J.,** Rotational and lateral diffusion of membrane proteins, *Biochim. Biophys. Acta*, 559, 289, 1979.
48. **Backer, J. M. and Dawidowiz, E. A.,** Transmembrane movement of cholesterol in small unilamellar vesicles determined by cholesterol oxidase, *J. Biol. Chem.*, 256, 586, 1981.
49. **Op den Kamp, J. A. F.,** Membrane asymmetry, *Ann. Rev. Biochem.*, 48, 47, 1979.
50. **Nichols, J. W. and Pagano, R. E.,** Use of resonance energy transfer to study the kinetics of amphiphile transfer between vesicles, *Biochemistry*, 21, 1720, 1982.
51. **Duckwitz-Peterlin, G., Eilenberger, G., and Overath, P.,** Phospholipid exchange between bilayer membranes, *Biochim. Biophys. Acta*, 469, 311, 1977.
52. **Roseman, M. A. and Thompson, T. E.,** Mechanism of spontaneous transfer of phospholipids between bilayers, *Biochemistry*, 19, 444, 1980.
53. **Kader, J.-C., Douady, D., and Mazliak, P.,** Phospholipid transfer proteins, in *Phospholipids*, Hawthorne, J. N. and Ansell, G. B., Eds., Elsevier/North-Holland, Amsterdam, 1982, chap. 8.
54. **Zilversmit, D. B. and Hughes, M. E.,** Phospholipid exchange between membranes, *Meth. Membr. Biol.*, 7, 211, 1976.
55. **Wirtz, K. W. A.,** Phospholipid transfer proteins, in *Lipid-Protein Interactions*, Vol. 1, Jost, P. C. and Griffith, O. H., Eds., J. Wiley & Sons, New York, 1982, chap. 6.
56. **Nichols, J. W. and Pagano, R. E.,** Resonance energy transfer assay of protein-mediated lipid transfer between vesicles, *J. Biol. Chem.*, 258, 5368, 1983.
57. **Frank, A., Barenholz, Y., Lichtenberg, D., and Thompson, T. E.,** Spontaneous transfer of sphingomyelin between phospholipid bilayers, *Biochemistry*, 22, 5647, 1983.
58. **Pownall, H. J., Hickson D., Grotto, A. M., Jr., and Massey, J. B.,** Kinetics of spontaneous and plasma-mediated sphingomyelin transfer, *Biochim. Biophys. Acta*, 712, 169, 1982.
59. **Massey, J. B., Gotto, A. M., and Pownall, H. J.,** Kinetics and mechanism of the spontaneous transfer of fluorescent phospholipids between apolipoprotein-phospholipid recombinants. Effect of the polar headgroup, *J. Biol. Chem.*, 257, 5444, 1982.
60. **Steck, T. L.,** Membrane isolation, in *Membrane Molecular Biology*, Fox, C. F. and Keith, A. D., Eds., Sinauer Associates, Stamford, Conn., 1972, chap. 3.
61. **Fleischer, S. and Packer, L., Eds.,** *Methods in Enzymology*, Vol. 32, Academic Press, New York, 1974.
62. **Fleischer, S. and Packer, L., Eds.,** *Methods in Enzymology*, Vol. 55, Academic Press, New York, 1979.
63. **Finean, J. B. and Michell, R. H.,** Isolation, composition and general structure of membranes, in *Membrane Structure*, Finean, J. B. and Michell, R. H., Eds., Elsevier/North-Holland, Amsterdam, 1981, chap. 1.
64. **Beaufay, H. and Amar-Costesec, A.,** Cell fractionation techniques, *Methods Membrane Biol.*, 6, 1976.
65. **Evans, W. H.,** A biochemical dissection of the functional polarity of the plasma membrane of the hepatocyte, *Biochim. Biophys. Acta*, 604, 27, 1980.
66. **Gotlib, L. J. and Searls, D. B.,** Plasma membrane isolation on DEAE-Sephadex beads, *Biochim. Biophys. Acta*, 602, 207, 1980.
67. **Gahmberg, C. G,** Membrane glycoproteins and glycolipids: structure, localization, and function of the carbohydrate, in *Membrane Structure*, Finean, J. B.and Michell, R. H., Ed., Elsevier/North-Holland, Amsterdam, 1981. chap. 4.
68. **Havel, R. J., Eder, H. A., and Bragdon, J. H.,** The distribution and chemical composition of ultracentrifugally separated lipoproteins in human serum, *J. Clin. Invest.*, 34, 1345, 1955.
69. **Rudel, L. L., Lee, J. A., Morris, M. D., and Felts, J. M.,** Characterization of plasma lipoproteins separated and purified by agarose-column chromatography, *Biochem. J.*, 139, 89, 1974.
70. **Wieland, H. and Seidel, D.,** A simple specific method for precipitation of low density lipoproteins, *J. Lipid Res.*, 24, 904, 1983.

71. **Folch, J., Lees, M., and Stanley, G. H. S.,** A simple method for the isolation and purification of total phospholipids from animal tissues, *J. Biol. Chem.*, 226, 497, 1957.
72. **Bligh, E. A. and Dyer, W. J.,** A rapid method of total lipid extraction and purification, *Can. J. Biochem. Physiol.*, 37, 911, 1959.
73. **Spanner, S.,** Separation and analysis of phospholipids, in *Form and Function of Phospholipids*, Ansell, G. B., Hawthorne, J. N., and Dawson, R. M. C., Eds., Elsevier, Amsterdam, 1973, chap. 2.
74. **Rouser, G., Nelson, G. J., Fleischer, S., and Simon, G.,** Lipid composition of animal cell membranes, organelles, and organs, in *Biological Membranes*, Dennis, E. A. and Chapman, D., Eds., Academic Press, New York, 1968, chap. 2.
75. **White, D. A.,** The phospholipid composition of mammalian tissues, in *Form and Function of Phospholipids*, Ansell, G. B., Hawthorne, J. N., and Dawson, R. M. C., Ed., Elsevier, Amsterdam, 1973, chap. 16.
76. **Mangold, H. K., Ed.,** *Lipids and Technical Lipid Derivatives*, Vol. 1, CRC Press, Boca Raton, Fla., 1984.
77. **Radin, N. S.,** Extraction of tissue lipids with a solvent of low toxicity, *Methods Enzymol.*, 72, 5, 1981.
78. **Klein, R. A.,** The detection of oxidation in liposome preparations, *Biochim. Biophys. Acta*, 210, 486, 1970.
79. **Harrington, C. A., Fenimore, D. C., and Eichberg, J.,** Fluorimetric analysis of polyunsaturated phosphatidylinositol and other phospholipids in picomole range using high performance thin-layer chromatography, *Anal. Biochem.*, 106, 307, 1980.
80. **Saito, M., Tanaka, Y., and Andos, S.,** Thin-layer chromatography-densitometry of minor acidic phospholipids: application to lipids from erythrocytes, liver, and kidney, *Anal. Biochem.*, 132, 376, 1983.
81. **Porter, N. A. and Weenen, H.,** High-performance liquid chromatographic separations of phospholipids and phospholipid oxidation products, *Methods Enzymol.*, 72, 34, 1981.
82. **Compton, B. J. and Purdy, W. C.,** The high performance liquid chromatography and detection of phospholipids and triglycerides. I. Non-polar stationary phase chromatographic behavior in ultraviolet transparent mobile phases, *Anal. Chim. Acta*, 141, 405, 1982.
83. **Duck-Chong, C. G.** A rapid sensitive method for determining phospholipid phosphorus involving digestion with magnesium nitrate, *Lipids*, 14, 492, 1979.
84. **Stewart, J. C. M.,** Colorimetric determination of phospholipids with ammonium ferrothiocyanate, *Anal. Biochem.*, 104, 10, 1980.
85. **Shand, J. H. and Noble, R. C.,** Quantitation of lipid mass by a liquid scintillation counting procedure following charring on thin-layer plates, *Anal. Biochem.*, 101, 427, 1980.
86. **Helenius, A. and Simons, K.,** Solubilization of membranes by detergents, *Biochim. Biophys. Acta*, 415, 29, 1975.
87. **Lichtenberg, D., Robson, R. J., and Dennis, E. A.,** Solubilizatin of phospholipids by detergents. Structural and kinetic aspects, *Biochim. Biophys. Acta*, 737, 285, 1983.
88. **Helenius, A., McCaslin, D. P., Fries, E., and Tanford, C.,** Properties of detergents, *Methods Enzymol.*, 56, 734, 1979.
89. **Oku, N. and MacDonald, R. C.,** Solubilization of phospholipids by chaotrophic ion solutions, *J. Biol. Chem.*, 258, 8733, 1983.
90. **Gonenne, A. and Ernst, R.,** Solubilization of membrane proteins by sulfobetaines, novel zwitterionic surfactants, *Anal. Biochem.*, 87, 28, 1978.
91. **Womack, M. D., Kendall, D. A., and MacDonald, R. C.,** Detergent effects on enzyme activity and solubilization of lipid bilayer membranes, *Biochim. Biophys. Acta*, 733, 210, 1983.
92. **Tanford, C. and Reynolds, J. A.,** Characteristics of membrane proteins, in detergent solutions, *Biochim. Biophys. Acta*, 457, 133, 1976.
93. **Scheule, R.K. and Gaffney, B. J.,** Reconstitution of membranes with fractions of Triton® X-100 which are easily removed, *Anal. Biochem.*, 117, 61, 1981.
94. **Chang, H. W. and Bock, E.,** Pitfalls in the use of commercial nonionic detergents for the solubilization of integral membrane proteins: sulfhydryl oxidizing contaminants and their elimination, *Anal. Biochem.*, 104, 112, 1980.
95. **Szoka, F. and Papahadjopoulos, D.,** Comparative properties and methods of preparation of lipid vesicles (liposomes), *Ann. Rev. Biophys. Bioeng.*, 9, 467, 1980.
96. **Szoka, F. and Papahadjopoulos, D.,** Liposomes: preparation and characteristics, in *Liposomes: From Physical Structure to Therapeutic Applications*, Knight, G., Eds., Elsevier/North-Holland, Amsterdam, 1981, chap. 3.
97. **Deamer, D. W. and Uster, P. S.,** Liposome preparation: methods and mechanisms, in *Liposomes*, Ostro, M. J., Ed., Marcel Dekker, New York, 1983, chap. 2.
98. **Gregoriadis, G.,** The carrier potential of liposomes in biology and medicine. I and II., *N. Engl. J. Med.*, 295, 704, and 765, 1976.

99. **Gregoriadis, G.,** *Liposome Technology*, Vol. 1, *Preparation of Liposomes*, CRC Press, Boca Raton, Fla., 1984.
100. **Papahadjopoulos, D., Ed.,** *Liposomes and Their Uses in Biology and Medicine*, Annals. N. Y., Academy of Science, Sp. Publ. 308, 1978.
101. **Bangham, A. D., Hill, M. W., and Miller, N. G. A.,** Preparation and use of liposomes as models of biological membranes, *Methods Membrane Biol.*, 1, 1, 1974.
102. **Kitagawa, T., Inoue, K., and Nojima, S.,** Properties of liposomal membranes containing lysolecithin, *J. Biochem.*, 79, 1123, 1976.
103. **Hargreaves, W. R. and Deamer, D. W.,** Liposomes from ionic, single-chain amphiphiles, *Biochemistry*, 17, 3759, 1978.
104. **Gavino, V. C. and Deamer, D. W.,** Purification of acyl CoA: 1-acyl-sn glycero-3-phosphocholine acyltransferase, *J. Bioeng. Biomembrane*, 14, 513, 1982.
105. **Sakurai, I. and Kawamura, Y.,** Magnetic-field-induced orientation and bending of the myelin figures of phosphatidylcholine, *Biochim. Biophys. Acta*, 735, 189, 1983.
106. **Bangham, A. D., Standish, M. M., and Watkins, J. C.,** Diffusion of univalent ions across the lamellae of swollen phospholipids, *J. Mol. Biol.*, 13, 238, 1965.
107. **Kremer, J. M. H., v. d. Esker, M. W. I., Pathmamanohara, C., and Wiersema, P. H.,** Vesicles of variable diameter prepared by a modified injection method, *Biochemistry*, 16, 3932, 1977.
108. **Barenholz, Y., Gibbes, D., Litman, B. J., Goll, J., Thompson, T. E., and Carlson, F. D.,** A simple method for the preparation of homogeneous phospholipid vesicles, *Biochemistry*, 16, 2806, 1977.
109. **Barenholz, Y., Amselem, S., and Lichtenberg, D.,** A new method for preparation of phospholipid vesicles (liposomes) — French Press, *FEBS Lett.*, 99, 210, 1979.
110. **Deamer, D. W.,** Preparation and properties of ether-injection liposomes, *Ann. N. Y. Acad. Sci.*, 308, 250, 1978.
111. **Rhoden, V. and Goldin, S. M.,** Formation of unilamellar lipid vesicles of controllable dimensions by detergent dialysis, *Biochemistry*, 18, 4173, 1979.
112. **Racker, E.,** Reconstitution of membrane processes, *Methods Enzymol.*, 55, 699, 1979.
113. **Pick, U.,** Liposomes with a large trapping capacity prepared by freezing and thawing of sonicated phospholipid mixtures, *Arch. Biochem. Biophys.*, 212, 186, 1981.
114. **Olson, F., Hunt, C. A., Szoka, F., Vail. W. J., and Papahadjopoulos, D.,** Preparation of liposomes of defined size distributions by extrusion through polycarbonate membranes, *Biochim. Biophys. Acta*, 557, 9, 1979.
114a. **Papahadjopoulos, D., Vail, W. J., Jacobson, K., and Poste, G.,** Cochleate lipid cylinders: formation by fusion of unilamellar vesicles, *Biochim. Biophys. Acta*, 394, 483, 1975.
115. **Nozaki, Y., Lasic, D. D., Tranford, C., and Reynolds, J. A.,** Size analysis of phospholipid vesicle preparations, *Science*, 217, 366, 1982.
116. **Jain, M. K.,** *The Bimolecular Lipid Membrane: A System*, Van Nostrand Reinhold, New York, 1972, chap. 4 to 6.
117. **Finkelstein, A.,** Bilayers: formation, measurement, and incorporation of components, *Methods Enzymol.*, 32, 489, 1974.
118. **Andreoli, T. E.,** Planar lipid bilayer membranes, *Methods Enzymol.*, 32, 513, 1978.
119. **Fettiplace, R., Gordon, L. G. M., Hladky, S. B., Requena, J., Zingsheim, H. P., and Haydon, D. A.,** Techniques in the formation and examination of "Black" lipid bilayer membranes, *Meth. Membr. Biol.*, 4, 1975.
120. **Montal, M.,** Asymmetric lipid bilayers. Response to multivalent ions, *Biochem. Biophys. Acta*, 298, 750, 1973.
121. **Freeman, R. B.,** Membrane-bound enzymes, in *Membrane Structure*, Finean, J. B. and Michell, R. H., Eds., Elsevier/North-Holland, Amsterdam, 1981, chap. 5.
122. **Darszon, R., Vandenberg, C. A., Schonfeld, M., Ellisman, M. H., Spitzer, N. C., and Montal, M.,** Reassembly of protein-lipid complexes into large bilayer vesicles: perspectives for membrane reconstitution, *Proc. Natl. Acad. Sci. U.S.A.*, 77, 239, 1980.
123. **McIntyre, J. O., Holladay, L. A., Smigel, M., Puett, D., and Fleischer, S.,** Hydrodynamic properties of D-β-hydroxybutyrate dehydrogenase, a lipid-requiring enzyme, *Biochemistry*, 17, 4169, 1978.
124. **Hopkins, C. R.,** *Structure and Function of Cells*, W. B. Saunders, Philadelphia, 1978, chap. 1 to 3.
125. **Jain, M. K. and Wagner, R. C.,** *Introduction to Biological Membranes*, John Wiley & Sons, New York, chap. 2.
126. **Munn, E. A.,** The application of the negative staining technique to the study of membranes, *Methods Enzymol.*, 32, 20, 1974.
127. **Verkleij, A. J. and de Gier, J.,** Freeze fracture studies on aqueous dispersions of membrane lipids, in *Liposomes: From Physical Structure to Therapeutic Applications*, Knight, C. J., Ed., Elsevier/North-Holland, Amsterdam, 1981, chap. 4.

128. **Verkleij, A. J. and Ververgaert, P. H. J. Th.,** Freeze-fracture morphology of biological membranes, *Biochim. Biophys. Acta,* 515, 303, 1978.
129. **Zingsheim, H. P.,** Membrane structure and electron microscopy. The significance of physical problems and techniques (freeze etching), *Biochim. Biophys. Acta,* 265, 339, 1972.
130. **Deamer, D. W., Leonard, R., Tardieu, A., and Branton, D.,** Lamellar and hexagonal lipid phases visualized by freeze-etching, *Biochim. Biophys. Acta,* 219, 47, 1970.
131. **Bachmann, L. and Smidt, W. W.,** Improved cryofixation applicable to freeze etching, *Proc. Natl. Acad. Sci., U.S.A.,* 68, 2149, 1971.
132. **Ververgaert, P. H. J. Th., Verhoeven, J. J., and Elbers, P. F.,** Spray-freezing of liposomes, *Biochim. Biophys. Acta,* 311, 651, 1973.
133. **Heuser, J. E., Reese, T. S., Dennis, M. J., Jan, Y., Jan, L., and Evans, L.,** Synaptic vesicle exocytosis captured by quick freezing and correlated with quantal transmitter release, *J. Cell Biol.,* 81, 275, 1979.
134. **Burnell, E., Van Alphen, L., Verkleij, A., and De Druijff, B.,** ^{31}P nuclear magnetic resonance and freeze-fracture electon microscopy studies on *Escherichia coli.* I. Cytoplasmic membrane and total phospholipids, *Biochim. Biophys. Acta,* 597, 492, 1980.
135. **Severs, N. J. and Robenek, H.,** Detection of microdomains in biomembranes. An appraisal of recent developments in freeze-fracture cytochemistry, *Biochim. Biophys. Acta,* 737, 373, 1983.
136. **Kisher, K. A.,** Monolayer freeze-fracture autoradiography: quantitative analysis of the transmembrane distribution of radioiodinated concanavalin A, *J. Cell Biol.,* 93, 155, 1982.
137. **Poste, G., Koppel, D. E., Schlessinger, J., Elson, E., and Webb, W. W.,** Identification of a potential artifact in the use of electron microscope autoradiography to localize saturated phospholipids in cells, *Biochim. Biophys. Acta,* 510, 256, 1978.
138. **Pinto de Silva, P., Kachar, B., Torrisi, M. R., Brown, C., and Parkinson, C.,** Freeze-fracture cytochemistry: replicas of critical point-dryed cells and tissues after fracture-label, *Science,* 213, 230, 1981.
139. **Wischnitzer, S.,** *Introduction to Electron Microscopy,* Pergamon Press, New York, 1981, 405.
140. **Elias, P. M., Goerke, J., and Friend, D. S.,** Freeze-fracture identification of sterol-digitonin complexes in cell and liposome membranes, *J. Cell. Biol.,* 78, 577, 1978.
141. **Teitelbaum, D., Arnon, R., Sela, M., Rabinosohn, Y., and Shapiro, D.,** Sphingomyelin specific antibodies elicited by synthetic conjugates, *Immunochemistry,* 10, 735, 1973.
142. **Demel, R. A.,** Monolayers. description of use and interaction, *Methods Enzymol.,* 32, 539, 1974.
143. **Parsegian, V. A., Fuller, N., and Rand, R. P.,** Measured work of deformation and repulsion of lecithin bilayers, *Proc. Natl. Acad. Sci., U.S.A.,* 76, 2750, 1979.
144. **Franks, N. P. and Lieb, W. R.,** X-Ray and neutron diffraction studies of lipid bilayers, in *Liposomes: From Physical Structure to Therapeutic Applications,* Knight, C. G., Ed., Elsevier/North-Holland, Amsterdam, 1981. chap. 8.
145. **Shipley, G. G.,** Recent X-ray diffraction studies of biological membranes and membrane components, in *Biological Membranes,* Vol. 2, Chapman, D. E. and Wallach, D. F. H., Eds., Academic Press, New York, 1973, chap. 1.
146. **Franks, N. P. and Levine, Y. K.,** Low-angle X-ray diffraction, in *Membrane Spectroscopy,* Grell, E., Ed., Springer-Verlag, Heidelberg, 1981, chap. 9.
147. **Blaurock, A. E.,** Evidence of bilayer structure and of membrane interactions from X-ray diffraction analysis, *Biochim. Biophys. Acta,* 650, 167, 1982.
148. **Cantor, C.R. and Schimmel, P. R.,** Absorption spectroscopy and other optical techniques, in *Biophysical Chemistry, Part II: Techniques for the Study of Biological Structure and Function,* W. H. Freeman, San Francisco, 1980. chap. 7 and 8.
149. **Urry, D. W. and Long, M. M.,** Circular dichroism and absorption studies on biomembranes, *Meth. Membr. Biol.,* 1, 1976.
150. **Urry, D. W. and Long, M. M.,** Ultraviolet absorption, circular dichroism, and optical rotatory dispersion in biomembrane studies, in *Membrane Physiology,* Andreoli, T. E., Hoffman, J. F., and Fanestil, D. D., Eds., Plenum Medical, New York, 1978, chap. 6.
151. **Holzwarth, G.,** Ultraviolet spectroscopy of biological membranes, in *Membrane Molecular Biology,* Fox. C. F. and Keith, A. D., Eds., Sinauer Associates, Stamford, Conn., 1972, chap. 9.
152. **Long, M. M. and Urry, D. W.,** Absorption and circular dichroism spectroscopies, in *Membrane Spectroscopy,* Grell, E., Springer-Verlag, Berlin, 1981, chap. 3.
153. **Heesemann, J. and Zingsheim, H. P.,** Optical spectroscopy of monolayers, multilayers, multilayer assemblies, and single model membranes, in *Membrane Spectroscopy,* Grell, E., Springer-Verlag, Berlin, 1981, chap. 4.
154. **Johansson, L. B. and Lindblom, G.,** Orientation and mobility of molecules in membranes studied by polarized light spectroscopy, *Q. Rev. Biophys.,* 13, 63, 1980.
155. **Littman, B. J. and Barenholtz, Y.,** The optical activity of D-erythro-sphingomyelin and its contribution to the circular dichroism of sphingomyelin-containing systems, *Biochim. Biophys. Acta,* 394, 166, 1975.

156. **Chen, G. and Kane, J. P.,** Temperature dependence of the optical activity of human serum low density lipoprotein. The role of lipids, *Biochemistry,* 14, 3357, 1975.
157. **Azzi, A.,** The use of fluorescent probes for the study of membranes, *Methods Enzymol.,* 32, 234, 1974.
158. **Yguerabide, J. and Foster, M. C.,** Fluorescence spectroscopy of biological membranes, in *Membrane Spectroscopy,* Grell, E., Springer-Verlag, Berlin, 1981, chap. 5.
159. **Radda, G. K. and Vanderkooi, J.,** Can fluorescent probes tell us anything about membranes?, *Biochim. Biophys. Acta,* 265, 509, 1972.
160. **Skylar, L. A., Hudson, B. S., and Simoni, R. D.,** Conjugated polyene fatty acids as fluorescent probes: synthetic phospholipid membrane studies, *Biochemistry,* 16, 819, 1977.
161. **Thulborn, K. R. and Sawyer, W. H.,** Properties and location of a set of fluorescent probes sensitive to the fluidity gradient of the lipid bilayer, *Biochim. Biophys. Acta,* 511, 125, 1978.
162. **London, E.,** Investigation of membrane structure using fluorescence quenching by spin-labels, *Mol. Cell. Biochem.,* 45, 181, 1982.
163. **Fung, B. K.-K. and Stryer, L. L.,** Surface density determination in membranes by fluorescence energy transfer, *Biochemistry,* 17, 5341, 1978.
164. **Shinitzky, M. and Barenholtz, Y.,** Dynamics of the hydrocarbon region of micelles and membranes determined with fluorescence probes, *J. Biol. Chem.,* 249, 2651, 1974.
165. **Laskowicz, J. R.,** *Spectroscopy in Biochemistry,* Bell, J. E., Ed., Vol. 1, CRC Press, Boca Raton, Fla., 1981, 195.
166. **Pal, R., Wiener, J. R., Barenholz, Y., and Wagner, R. R.,** Influence of membrane glycoprotein and cholesterol of vesicular stomatitis virus on the dynamics of viral and model membranes: fluorescence studies, *Biochemistry,* 22, 3624, 1983.
167. **Elson, E. L. and Webb, W. W.,** Concentration correlation spectroscopy: a new biophysical probe based on occupation number fluctuations, *Annu. Rev. Biophys. Bioeng.,* 4, 311, 1975.
168. **Koppel, D. E., Axelrod, D., Schlessinger, J., Elson, E. L., and Webb, W. W.,** Dynamics of fluorescence marker concentrations as a probe of mobility, *Biophys, J.,* 16, 1315, 1976.
169. **Gall, W. E. and Edelman, G. M.,** Lateral diffusion of surface molecules in animal cells and tissues, *Science,* 213, 903, 1981.
170. **Fringeli, U. P. and Gunthard, Hs. H.,** Infrared membrane spectroscopy, in *Membrane Spectroscopy,* Grell, E., Springer-Verlag, Berlin, 1981, chap. 7.
171. **Wallach, D. F. H. and Oseroff, A. R.,** Infrared and laser Raman spectroscopy, *Methods Enzymol.,* 32, 247, 1974.
172. **Wallach, D. F. H., Verma, S. P., and Fookson, J.,** Application of laser Raman and infrared spectroscopy to the analysis of membrane structure, *Biochim. Biophys. Acta,* 559, 153, 1979.
173. **Lord, R. C.and Mendelsohn, R.,** Raman spectroscopy of membrane constituents and related molecules, in *Membrane Spectroscopy,* Grell, E., Springer-Verlag, Berlin, 1981, chap 8.
174. **Larsson, K.,** Conformation-dependent features in the Raman spectra of simple lipids, *Chem. Phys. Lipids,* 10, 165, 1973.
175. **Bush, S. F., Levin, H., and Levin, I. W.,** Cholesterol-lipid interactions: an infrared and Raman spectroscopic study of the carbonyl stretching mode region of 1,2-dipalmitoyl phosphatidylcholine bilayers, *Chem. Phys. Lipids,* 27, 101, 1980.
176. **Cantor, C. R. and Schimmel, P. R.,** Introduction to magnetic resonance, in *Biophysical Chemistry. Part II: Techniques for the Study of Biological Structure and Function,* W. H. Freeman, San Francisco, 1980, chap. 9.
177. **Browning, J. L.,** NMR studies of the structural and motional properties of phospholipids in membranes, in *Liposomes: From Physical Structure to Therapeutic Applications,* Knight, C. G., Ed., Elsevier/North-Holland, Amsterdam, 1981, chap. 7.
178. **Seelig, J.,** ^{31}P Nuclear magnetic resonance and the head group structure of phospholipids in membranes, *Biochim. Biophys. Acta,* 515, 105, 1978.
179. **Wennerstrom, H. and Lindblom, G.,** Biological and model membranes studied by nuclear magnetic resonance of spin one half nuclei, *Q. Rev. Biophys.,* 10, 67, 1977.
180. **Chan, S. I., Bocian, D. F., and Peterson, N. O.,** Nuclear magnetic resonance studies of the phospholipid bilayer membrane, in *Membrane Spectroscopy,* Grell, E., Ed., Springer-Verlag, Berlin, 1981, chap. 1.
181. **Seelig, J.,** Deuterium magnetic resonance: theory and application to lipid membranes, *Q. Rev. Biophys.,* 10, 353, 1977.
182. **Burnell, E., Van Alphen, L., Verkleij, A., and De Kruijff, B.,** ^{31}P nuclear magnetic resonance and freeze-fracture electron microscopy studies on *Escherichia coli.* I. Cytoplasmic membrane and total phospholipids, *Biochim. Biophys. Acta,* 597, 492, 1980.
183. **Burns, R. A., Jr., Stark, R. E., Vidusek, D. A., and Roberts, M. F.,** Dependence of phosphatidylcholine phosphorus-31 relaxations times and ^{31}P[^{1}H]nuclear overhauser effect distribution on aggregate structure, *Biochemistry,* 22, 5084, 1983.

184. **London, E. and Feigenson, G. W.,** Phosphorus NMR analysis of phospholipids in detergents, *J. Lipid Res.*, 20, 408, 1979.
185. **Seelig, A. and Seelig, J.,** The dynamic structure of fatty acyl chains in a phospholipid bilayer measured by deuterium magnetic resonance, *Biochemistry*, 13, 4839, 1974.
186. **Levine, Y. K., Birdsall, N. J. M., Lee, A. G., and Metcalf, J. C.,** ^{14}C Nuclear magnetic resonance relaxation measurements of synthetic lecithins and the effect of spin-labeled lipids, *Biochemistry*, 11, 1416, 1972.
187. **Gally, H. U., Pluschke, G., Overath, P., and Seelig, J.,** Structure of *Escherichia coli* membranes. Phospholipid conformation in model membranes and cells as studied by deuterium magnetic resonance, *Biochemistry*, 18, 5605, 1979.
188. **Brown, M. F., Ribeiro, A. A., and Williams, G. D.,** New view of lipid bilayer dynamics from ^{2}H and ^{13}C NMR relaxation time measurements, *Proc. Natl. Acad. Sci. U.S.A.*, 80, 4325, 1983.
189. **Gally, H. U., Pluschke, G., Overath, P., and Seelig, J.,** Structure of *Escherichia coli* membranes. Glycerol auxotrophs as a tool for the analysis of the phospholipid head-group region by deuterium magnetic resonance, *Biochemistry*, 20, 1826, 1981.
190. **Cornell, B. A., Hiller, R. G., Raison, J., Separovic, F., Smith, R., Vary, J. C., and Morris, C.,** Biological membranes are rich in low-frequency motion, *Biochim. Biophys. Acta*, 732, 473, 1983.
191. **Macdonald, P. M., McDonough, B., Sykes, B. D., and McElhaney, R. N.,** Fluorine-19 nuclear magnetic resonance studies of lipid fatty acyl chain order and dynamics in *Acholeplasma laidlawii* B membranes. Effects of methyl-branch substitution and of trans unsaturation upon membrane acyl-chain orientational order, *Biochemistry*, 22, 5103, 1983.
192. **Bergelson, L. D.,** Paramagnetic hydrophilic probes in NMR investigations of membrane systems, *Meth. Membr. Biol.*, 9, 1978.
193. **Marsh, D. and Watts, A.,** ESR spin label studies of liposomes, in *Liposomes: From Physical Structure to Therapeutic Applications*, Knight, C. G., Ed., Elsevier/North-Holland, Amsterdam, 1981, chap. 6.
194. **Gaffney, B. J.,** Spin-label measurements in membranes, *Methods Enzymol.* 32, 161, 1974.
195. **Keith, A. D., Sharnoff, M., and Cohn, G. E.,** A summary and evaluation of spin labels used as probes for biological membrane structure, *Biochim. Biophys. Acta*, 300, 379, 1973.
196. **Marsh, D.,** Electron spin resonance: spin labels, in *Membrane Spectroscopy*, Grell, E., Ed., Springer-Verlag, Berlin, 1981, chap. 2.
197. **Meier, P., Blume, A., Ohmes, E., Neugebauer, F. A., and Kothe, G.,** Structure and dynamics of phospholipid membranes: an electron spin resonance study employing biradical probes, *Biochemistry*, 21, 526, 1982.
198. **Schreier-Muccillo, S., Marsh, D., and Smith, I. C. P.,** Monitoring the permeability profile of lipid membranes with spin probes, *Arch. Biochem. Biophys.*, 172, 1, 1976.
199. **Tiller, et al.,** *Anal. Biochem.*, 141, 262, 1984.
200. **Mabrey-Guad, S.,** Differential scanning calorimetry of liposomes, in *Liposomes: From Physical Structure to Therapeutic Applications*, Knight, C. G., Ed., Elsevier/North-Holland, Amsterdam, 1981, chap. 5.
201. **Mabrey, S. and Sturtevant, J. M.,** High-sensitivity differential scanning calorimetry in the study of biomembranes and related model systems, *Meth. Membr. Biol.*, 9, 237, 1978.
202. **Biltonen, R. L. and Friere, E.,** Thermodynamic characterization of conformational states of biological macromolecules using differential scanning calorimetry, *CRC Crit. Rev. Biochem.*, 5, 85, 1978.
203. **Steim, J. M.,** Differential scanning calorimetry, *Methods Enzymol.*, 32, 262, 1974.
204. **Scheidler, P. J. and Steim, J. M.,** Differential scanning calorimetry of biological membranes: instrumentation, *Meth. Membr. Biol.*, 4, 1976.
205. **Correa-Freire, M. C., Freire, E., Barenholz, Y., Biltonen, R. L., and Thompson, T. E.,** Thermotropic behavior of monoglucocerebroside-dipalmitoylphosphatidylcholine multilamellar liposomes, *Biochemistry*, 18, 422, 1979.
206. **Serrallach, E. N., de Haas, G. H., and Shipley, G. G.,** Structure and thermodynamic properties of mixed-chain phosphatidylcholine bilayer membranes, *Biochemistry*, 23, 713, 1984.
207. **Tardieu, A., Luzzati, V. V., and Reman, F. C.,** Structure and polymorphism of the hydrocarbon chains of lipids: a study of lecithin-water phases, *J. Mol. Biol.*, 75, 711, 1973.
208. **Op den Kamp, J. A. F.,** The asymmetric architecture of membranes, in *Membrane Structure*, Finean, J. B. and Michell, R. H., Eds., Elsevier/North-Holland, Amsterdam, 1981, chap. 3.
209. **Op den Kamp, J. A. F.,** Lipid asymmetry in membranes, *Ann. Rev. Biochem.*, 48, 47, 1979.
210. **Etemandi, A.-H.,** Membrane asymmetry. A survey and critical appraisal of the methodology, *Biochim. Biophys. Acta*, 604, 423, 1980.
211. **Rothman, J. E. and Dawidowicz, E. A.,** Asymmetric exchange of vesicle phospholipids catalyzed by the phospholipid exchange proteins. Measurement of inside-outside transitions, *Biochemistry*, 14, 2809, 1975.

212. **Johnson, L. W., Hughes, M. E., and Zilversmit, D. B.,** Use of phospholipid exchange proteins to measure inside-outside transposition in phosphatidylcholine liposomes, *Biochim. Biophys. Acta,* 375, 176, 1975.
213. **Valtersson, C. and Dallner, G.,** Compartmentalization of phosphatidylethanolamine in microsomal membranes from rat liver, *J. Lipid Res.,* 23, 868, 1982.
214. **Lentz, B. R., Alford, D. R., and Dombrose, F. A.,** Determination of phosphatidylglycerol asymmetry in small, unilamellar vesicles by chemical modification, *Biochemistry,* 19, 2555, 1980.
215. **Bell, R. M., Ballas, L. M., and Coleman, R. A.,** Lipid topogenesis, *J. Lipid Res.,* 22, 391, 1980.
216. **Sleigh, R. G. and Pagano, R. E.,** Rapid appearance of newly synthesized phosphatidylethanolamine at the plasma membrane, *J. Biol. Chem.,* 258, 9050, 1983.
217. **Hoekstra, D. and Martin, O. C.,** Transbilayer redistribution of phosphatidylethanolamine during fusion of phospholipid vesicles. Dependence on fusion rate, lipid phase separation, and formation of nonbilayer structures, *Biochemistry,* 24, 6097, 1982.
218. **Khorana, H. G.,** Chemical studies of biological membranes, *Bioorg. Chem.,* 9, 363, 1980.
219. **Stoffel, W.,** Studies on lipid-lipid and lipid-protein interactions with physical and chemical methods, in *Structure and Function of Biomembranes,* Yagi, K., Ed., Japan Scientific Societies Press, Tokyo, 1979, 1.
220. **Mabrey, S. and Sturtevant, J. M.,** Investigation of phase transitions of lipids and lipid mixtures by high sensitivity differential scanning calorimetry, *Proc. Natl. Acad. Sci. U.S.A.,* 73, 3862, 1976.
221. **Keough, K. M. W. and Davis, P. J.,** Gel to liquid-crystalline phase transitions in water dispersions of saturated mixed-acid phosphatidylcholines, *Biochemistry,* 18, 1453, 1979.
222. **Browning, J. L. and Seelig, J.,** Bilayers of phosphatidylserine: a deuterium and phosphorus nuclear magnetic resonance study, *Biochemistry,* 19, 1262, 1980.
223. **Barenholz, Y., Suurkuusk, J., Mountcastle, D., Thompson, T. E., and Biltonen, R. L.,** A calorimetric study of the thermotropic natural and synthetic sphingomyelins, *Biochemistry,* 15, 2441, 1976.
224. **Calhoun, W. I. and Shipley, G. G.,** Fatty acid composition and thermal behavior of natural sphingomyelins, *Biochim. Biophys. Acta,* 555, 436, 1979.
225. **O'Brien, J. S.,** Cell membranes: composition, structure, function, *J. Theor. Biol.,* 15, 307, 1967.
226. **Phillips, M. C. and Chapman, D.,** Monolayer characteristics of saturated 1,2-diacylphosphatidylcholines (lecithins) and phosphatidylethanolamines at the air-water interface, *Biochim. Biophys. Acta,* 163, 301, 1968.

Chapter 3

PHARMACOLOGICAL, DEVELOPMENTAL, AND PHYSIOLOGICAL REGULATION OF SYNAPTIC MEMBRANE PHOSPHOLIPIDS*

Robert J. Hitzemann, R. Adron Harris, and Horace H. Loh

TABLE OF CONTENTS

* Abbreviations used: Phosphatidylcholine (PC), lysophosphatidylcholine (LPC), phosphatidylethanolamine (PE), phosphatidylserine (PS), phosphatidylinositol (PI), diphosphatidylinositol (DPI), triphosphatidylinositol (TPI), ethanolamine plasmalogen (EP), sphingomyelin (SM), differential scanning calorimetry (DSC), differential thermal analysis (DTA), synaptic plasma membrane (SPM), electron spin resonance spectroscopy (ESR), nuclear magnetic resonance spectroscopy (NMR).

I. SYNAPTIC MEMBRANES

This chapter focuses on the modulation of phospholipid structure and function relationships in nerve membranes and in particular, synaptic plasma membranes (SPMs). For those unfamiliar with the SPM preparation, there are several excellent articles which can be consulted.[1-3] Briefly, purified nerve ending particles (synaptosomes) are isolated from a crude brain homogenate by differential centrifugation-density gradient techniques. The nerve endings are then lysed and the plasma membranes are separated from intrasynaptic mitochondria, synaptic vesicles, and other contaminating membranes (e.g., myelin membranes) by an additional density gradient step. In some procedures, the density of the mitochondria is increased to minimize contamination from this organelle.[4] The purity of the SPM preparation is especially critical since environmental and pharmacological manipulations[5,6] have markedly different effects on SPMs as compared to other membranes. The most common contaminant of SPM preparations are so-called "microsomal membranes", a fraction in brain composed largely of smooth endoplasmic reticulum and plasma membrane from nonsynaptic sites. Contamination from this source is particularly difficult to determine but a conservative estimate would be 5 to 10% of the total membrane present.[2,3] Contamination from these membranes may be particularly troublesome due to their similarity to SPMs in terms of lipid and protein composition and functional activities (e.g., receptor-mediated processes).[7,8] SPMs, when derived from asymmetric synapses, can be differentiated from other plasma membranes on the basis of the asymmetric synaptic junction region which includes the postsynaptic membrane apparatus;[9] however, some SPM preparations are scarce in asymmetric complexes.[10] These preparations may represent a selective isolation from symmetrical synapses, which in some brain regions are primarily associated with axo-somatic inhibitory synapses.[11] Overall, it should be recognized that SPM preparations are heterogeneous. Despite this problem, the information gained from SPM preparations has contributed greatly both to our understanding of the role(s) phospholipids play in regulating synaptic activities and to our understanding of how pharmacological, developmental, and physiological manipulations change synaptic activities by modifying phospholipid composition and turnover.

II. SYNAPTIC MEMBRANE PHOSPHOLIPID COMPOSITION

Despite differences in isolation techniques, there is general agreement among laboratories as to the general lipid composition of SPMs. Data from our laboratories on mouse and rat brain SPMs are representative of the results obtained (Table 1).[12,13] Compared to other plasma membranes, SPMs have a lower cholesterol to phospholipid ratio, a higher lipid to protein ratio, and a higher ganglioside level.[14-16] In SPMs, gangliosides have largely replaced sphingomyelin (SM), a phospholipid, as the major sphingolipid present (see Table 2). While SM may contribute as much as 20 to 30% of the total phospholipid in other plasma membrane preparations, in SPMs the percentage is quite low (e.g., 1.8% in the rat). Decreasing SM and increasing ganglioside content will have important biophysical consequences. Data obtained both with fluorescence depolarization[17] and electron spin resonance spectroscopy (ESR) techniques[18] illustrate that gangliosides are markedly more effective than SM to increase membrane order(s). Gangliosides are also effective in inhibiting membrane fusion.[19] Given their asymmetric distribution[20,21], and relatively high concentration, gangliosides could account for as much as 30% of the lipid in the external half of the membrane.

There is greater diversity among SPM preparations in terms of phospholipid composition. The strains of rat and mouse used in our laboratories show marked compositional differences which may reflect species differences and/or differences in SPM heterogeneity (Table 2). Ethanolamine plasmalogen (EP) constitutes 24.4% of the total membrane phospholipid in the mouse but only 5.7% in the rat. However, the total amount of ethanolamine phospho-

Table 1
LIPID COMPOSITION OF RAT AND MOUSE SPMs

Parameter	Rat[a]	Mouse[b]
Lipid phosphate (μmol/mg protein)	0.96 ± 0.08	0.86 ± 0.11
Cholesterol/phospholipid (molar ratio)	0.51 ± 0.08	0.72 ± 0.03
Gangliosides/phospholipid (molar ratio)	0.17 ± 0.02	0.12 ± 0.01

[a] N = 8. See Reference 12.
[b] N = 6. See Reference 114.

Table 2
PHOSPHOLIPID COMPOSITION IN RAT AND MOUSE SPMs

(% total phospholipid)

	PC	PS	EP	PE	SM	Other
Rat[a]	40.7 ± 2.3	16.1 ± 0.8	5.7 ± 0.5	34.2 ± 1.9	1.8 ± 0.3	1.5 ± 1.0
Mouse[b]	39.0 ± 2.0	8.1 ± 0.7	24.4 ± 0.7	20.3 ± 0.8	5.2 ± 0.3	3.7 ± 1.5

[a] N = 8. See Reference 12 and unpublished observations.
[b] N = 6. See Reference 114.

glyceride is relatively equal in the mouse (44.7%) and in the rat (39.9%). Furthermore, in comparison to the mouse the rat SPM preparation has nearly twice the phosphatidylserine (PS) (16.1 vs. 8.1%). The significance of such compositional differences is not readily evident. The large difference in EP is particularly intriguing given our current lack of understanding regarding a unique functional role for this phospholipid. Plasmalogens are most prominent in skeletal muscle, cardiac muscle, and nerve tissue, which suggests some special function of plasmalogens in excitable cells.[22]

III. MEMBRANE ASYMMETRY

The amino phospholipids of murine SPMs have been shown to be arranged asymmetrically.[23,24] Using TNBS to label intact nerve endings, Fontaine et al.[23] demonstrated that a maximum of 10 to 15% of the SPM phosphotidylethanolamine (PE) and approximately 20% of the SPM PS resides on the outer half of the membrane. These figure may be somewhat high since some of the TNBS will react with the aminophospholipids of the cytoplasmic face of the post-synaptic complex which remains attached to the nerve ending. In addition, small amounts of TNBS probably leak into the nerve ending. The fatty acid composition of PE is also asymmetric. Phosphatidylethanolamine in the outer half of the membrane is richer in 16:10 and 18:0 and poorer in 18:1 and long-chain polyunsaturates as compared to PE in the cytoplasmic surface. The cause of the asymmetry is unknown, but two possible causes can be eliminated. First, serum lipoprotein-phospholipid exchange processes which probably contribute to the enrichment of phosphatidylcholine (PC) and SM in the external face of some plasma membranes.[25] do not operate to any significant degree in the brain. Second, phospholipid methylation, although present in synaptic membranes,[26] is of such low specific activity as to make its probable contribution to membrane asymmetry negligible. The role of base-exchange, a fairly active process in SPMs[27] in creating phospholipid asymmetry, is unknown. It is not known if the base-exchange enzymes for choline, ethanolamine, or serine are arranged asymmetrically. However, it is known that plasmalogens are poor substitutes for exchange,[28] which may contribute to the asymmetry of the ethanolamine phospholipids (EPs). Interestingly, both short- and long-term morphine administration has been shown to alter serine exchange.[29] One may speculate that the response of these exchange reactions to drug treatment may be to ensure specific phospholipid arrangements, including asymmetry, within specific membrane domains.

The asymmetry of cholinergic synaptic vesicles has been determined.[30,31] Michaelson et al.[30] noted that TNBS permeated the vesicle membrane and thus used a phospholipase *c* digestion technique to measure asymmetry. For Torpedo cholinergic vesicles they found that 100% of the phosphatidylinositol (PI), 77% of the PE, 48% of the EP, and 58% of the PC was located in the outer leaflet. The distribution of PS could not be accurately determined. Some speculations as to the function of the vesicle asymmetry in membrane fusion have been presented.[31]

IV. FATTY ACID COMPOSITION

The fatty acid composition of PC, PE, EP, and PS for the mouse and rat SPM preparations are shown in Table 3. For both preparations, PC has a low unsaturation index. PC is markedly enriched in palmitic acid (16:0) and some work suggests that a significant amount of the PC 16:0 is found in dipalmitoyl-PC (DPPC).[33] Crawford and Wells[34] have noted that appreciable amounts of DPPC are found only in erythrocytes, brain, and lung tissue, suggesting that perhaps this phospholipid plays an important role in O_2 uptake, transport, and or utilization. In contrast to PC, the unsaturation index of PE is quite high. As noted by Israelachvili et al.,[35] the packing shape of such highly unsaturated PE molecules will be that of an inverted truncated cone which facilitates the formation of inverted micelles (hexagonal-II phase). Interestingly, the more saturated PE molecules in the external side of the membrane will have a more cylindrical packing shape and will be more similar to the packing shape of the highly saturated PC and SM molecules found in this part of the membrane. The highly unsaturated PE molecules and those of EP and PS, which are relatively unique to brain tissue, will also have an interesting effect on the gradient of membrane order and fluidity along the hydrocarbon chain. As noted by Robertson[36] for a phospholipid such as PC where the first unsaturation generally will not appear, if at all until carbon 8, the movement of the initial segment of the hydrocarbon chain would be restricted (depending in part on the mutual attraction or repulsion of the head groups of the surrounding phospholipid molecules). The more interior portion of the molecule will remain more fluid due to the "bend" caused by the double bond. However, for a molecule like PE where the first unsaturation may appear at carbon 3, the dimension of the region of restricted movement will be markedly reduced. This arrangement may have great functional importance since it will reduce the ability of the membrane to trap laterally diffusive molecules in the center of the membrane.

V. SPM MEMBRANE ASSEMBLY

The assembly of biological membranes and the role(s) of the individual phospholipids in this assembly has been described in detail in Chapter 1. In general, SPMs appear to fit the general pattern of the fluid mosaic model described by Singer and Nicolson.[37] Although considerable domains of the SPM may exist as fluid bilayers, obviously other structures are possible and even probable. The diagram in Figure 1 illustrates some of the possible vari-

Table 3
FATTY ACID COMPOSITION OF RAT AND MOUSE SPM PHOSPHOLIPIDS[a]

	PC		PE		EP		PS	
Acyl group	**Rat**	**Mouse**	**Rat**	**Mouse**	**Rat**	**Mouse**	**Rat**	**Mouse**
16:0	50.4 ± 2.1	43.1 ± 2.2	11.6 ± 0.9	6.5 ± 0.8	1.5 ± 0.3	0.6 ± 0.1	3.1 ± 0.4	3.2 ± 0.9
18:0	9.8 ± 0.1	14.6 ± 0.6	16.3 ± 1.7	33.3 ± 0.8	1.5 ± 0.1	1.1 ± 0.2	48.8 ± 1.7	38.7 ± 0.9
18:1	24.4 ± 2.3	31.8 ± 1.4	9.9 ± 0.4	14.7 ± 0.5	5.2 ± 0.9	11.2 ± 0.5	9.1 ± 0.6	17.7 ± 1.4
18:2	0.6 ± 0.1	0.5 ± 0.1	0.8 ± 0.3	0.4 ± 0.1	ND	0.2 ± 0.1	ND	ND
18:3(W-3 + W-6)	TR	1.6 ± 0.1	TR	1.3 ± 0.2	ND	4.7 ± 0.5	ND	1.5 ± 0.2
20:3(W-6)	TR	0.3 ± 0.1	TR	0.4 ± 0.1	1.2 ± 0.1	0.5 ± 0.1	TR	1.2 ± 0.2
20:4(W-6)	6.8 ± 1.1	3.8 ± 0.1	20.6 ± 2.1	12.1 ± 0.2	10.8 ± 1.0	7.5 ± 0.1	1.8 ± 0.3	5.0 ± 0.6
20:5(W-3 + W-6)	ND	0.4 ± 0.1	1.1 ± 0.2	TR	ND	0.5 ± 0.1	ND	ND
22:4(W-6)	0.8 ± 0.2	0.4 ± 0.1	4.7 ± 0.6	4.1 ± 0.4	6.4 ± 0.2	7.1 ± 0.2	2.5 ± 0.3	4.7 ± 0.6
22:5(W-6 + W-3)	0.9 ± 0.3	ND	2.8 ± 0.6	ND	1.6 ± 0.4	0.5 ± 0.1	1.7 ± 0.7	ND
22:6(W-3)	3.8	3.3 ± 0.2	30.3 ± 2.2	26.9 ± 0.6	21.6 ± 1.9	17.2 ± 1.1	32.2 ± 1.9	27.9 ± 1.8
UI	83	77	314	246	215	194	228	232

[a] Data is for acyl linkages only. For EP it is assumed that the akyl-1-enyl linkages contribute 50% to the total and contain only saturated fatty acids. Data expressed as percent of total and are adapted from References 12 and 114.

ND = Not detected.
TR = Trace.
UI = Unsaturation index.

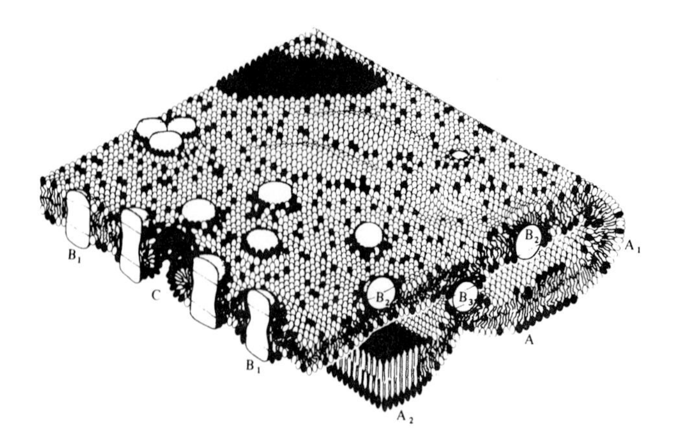

FIGURE 1. The variations in membrane structure.[35] (A) Typical bilayer structure (A_1) folded bilayer, (A_2) gel region, (B_1) proteins spanning the membrane with hydrophalic portions on both ends, (B_2) proteins embedded to different depths, (B_3) protein trapped between two folds, (C) lipid pore. (From Israelachvili, J. N., *Light Transducing Membranes: Structure, Function and Evolution*, Deamer, D. W., Ed., Academic Press, New York, 1978. With permission.)

ations. For SPMs this diagram should also include the synaptic junction region in which the mobility of the membrane components is restricted by the cytoskeletal elements present. Assuming a specialized arrangement of lipids and proteins is necessary at this region for maximally effective transmitter release, one might predict an unique phospholipid composition and structure in the synaptic junction. However, determination of the precise lipid composition in the junctional region is hindered in two ways. First, isolated synaptic junctions contain both pre- and post-synaptic elements. Second, the synaptic junctions are prepared by a detergent treatment (Triton® X-100) technique. MacDonald[38] has examined the effects of Triton® X-100 treatment on the extractability of lipid from intact erythrocyte ghosts and from vesicles prepared from ghost lipids. The data obtained may be useful in understanding the resistance of the junctional membrane region to extraction: (1) the presence of cholesterol reduces Triton® X-100-induced solubilization; (2) the types of proteins present will influence solubilization (in general, intrinsic membrane proteins promote whereas extrinsic membrane proteins inhibit solubilization); (3) at the relatively low Triton® X-100 concentrations used to prepare synaptic junctions (e.g., Triton® lipid ratios of 1 or 2:1), SM is selectively retained in the membrane residue as long as cholesterol is present. Thus, one may predict that the lipids of the junctional region will be enriched in SM and cholesterol, which, due to their complimentary packing structure and the low level of unsaturation in SM, would ensure a relatively rigid domain. However, SM has been largely replaced in SPMs by gangliosides as the major sphingolipid.[32] In the presence of 3 mm $CaCl_2$, which is used in combination with Triton® X-100 to increase the recovery of synaptic junctions,[9] gangliosides are selectively retained over phospholipid in the junctional fraction.[9] Furthermore, GM_1, the ganglioside which has the greatest ordering effect on PC membranes,[17,18] has been reported to be enriched in the junctional region as compared to the surrounding membrane.[39]

The highly active transport systems of neuronal membranes in general and SPMs in particular are undoubtedly associated with unique membrane lipids and lipid structures. Recent studies from our laboratories[40] have shown that membrane perturbants, e.g., free fatty acids or ethanol, have marked effects on the thermodynamics of glutamate and GABA transport — effects which are most easily explained in terms of altering membrane structure. For example, the incorporation of oleic acid, which Klausner et al.[41] suggest partitions selectively into fluid membrane domains, decreased the enthalpy of activation ($\Delta H\ddagger$) for glutamate transport into nerve endings and increased the activation entropy ($\Delta S\ddagger$). These data suggest that oleic acid caused a change in the reacting components such that glutamate, the carrier(s), Na^+, and the membrane had a greater requirement to orient themselves properly in three-dimensional space and to achieve an appropriate spatial proximity before translocation could occur. Interestingly, saturated fatty acids such as palmitic and stearic acids, which can stabilize membrane structures, had the opposite effect on the activation thermodynamics. As will be discussed in a subsequent section, during development there is a marked increase in the cholesterol to phospholipid ratio in SPMs and a concominant increase in membrane order. Developmental differences in the specificity and rate of GABA and glutamate transport, especially during the latter stages of development, are most easily explained by this change in membrane properties.[42]

Nerve endings are also secretory organelles for which specific lipid structures are required. The mechanism(s) by which neurotransmitters are released in the CNS are unclear and may not involve exocytosis as in peripheral secretory cells.[43] Cooper et al.[43] suggest that if we discount exocytosis, the simplest explanation for explaining the release of a transmitter such as acetylcholine (ACh) would be that ACh is released through a gated channel: a voltage-dependent entrance of Ca^{++} into the terminal would open a gate for a finite period of time, allowing ACh to be liberated into the synaptic cleft. The physical nature of such gates is unknown. A transitory local increase in lysophospholipids, especially lysolecithin could form a membrane pore which would result in a large increase in permeability for small molecules such as neurotransmitters. Evidence supporting the view that lysophosphatidylcholine (LPC) has a role in transmitter release (either by facilitating exocytosis or pore formation) can be summarized:

1. LPC can disrupt membrane preparations.[44]
2. Small amounts of LPC have been found in brain.[45]
3. LPC can promote the fusion of erythrocytes or fibroblasts in vitro.[46]
4. High levels of LPC are found in chromaffin granules isolated from the adrenal medulla.[47]
5. Phospholipase A activity is found in brain synaptic vesicle preparations.[48]

Evidence which suggests LPC may not be involved in transmitter release may also be summarized:

1. Outside of the chromaffin granules only low levels of LPC are found in other vesicle preparations, including synaptic vesicles from the rat brain[48] and the splenic nerve[49] and serotonin granules from blood platelets.[50]
2. Other lipids besides LPC can induce cell fusion.[51]
3. The identification of a truly vesicular phospholipiase A is difficult because of contaminating membranes.

In an attempt to more fully examine the role of LPC in transmitter release, Baker et al.[52] studied the effects of chemical and electrical stimulation on the lipid content of synaptic vesicles prepared from the electric organs of *Torpedo marmorata* and from guinea pig cerebral cortex. LPC was the only lysophospholipid present in either vesicle preparation and its content did not increase as the result of stimulation. In fact, a small but not significant decrease in the LPC levels was found. Similarly, we have found that an increase in the firing of the nigrostriatal neurons is associated with a decrease in synaptic vesicle LPC levels.[53] As shown in Table 4, haloperidol, which is known to increase the firing of nigrostriatal neurons, induced a significant decrease in the ratio of LPC:PC.

Ashley and Brammer[54] have examined the effect of depolarization on the physical prop-

Table 4
EFFECT OF HALOPERIDOL ON [^{3}H]LPC to [^{3}H]PC RATIO IN STRIATAL SYNAPTIC VESICLES[a]

	dpm/sample		
Group	**[^{3}H]LPC**	**[^{3}H]PC**	**[^{3}H]LPC/[^{3}H]PC**
Saline	412 ± 94	5821 ± 644	7.3 ± 1.3
Haloperidol	264 ± 48	5830 ± 270	4.4 ± 0.6[b]

[a] Rats were injected unilaterally with 50 μCi [^{3}H] choline into the substantia nigra. 1 week later animals were administered saline or haloperidol, 1 mg/kg, i.p. 1 hr later the animals were sacrificed, synaptic vesicles were isolated from the caudate nucleus, and the levels of [^{3}H]LPC and [^{3}H]PC were determined. N = Mean + S.E. of 8 pooled samples.

[b] Significantly different from control (Mann Whitney U-test), $p < 0.05$.

From Hitzemann, R. and Loh, H., *Res. Commun. Chem. Pathol. Pharmacol.*, 35(2), 209, 1982. With permission.

erties of nerve endings. Using the fluorescence polarization technique and 1,6-diphenyl-1,3,5 hexatriene (DPH) as the probe, these authors found that high concentrations of veratridine (150 μm), even in the presence of 2 m*M* $CaCl_2$, had no effect on the steady-state anisotropy. However, these authors did observe that Ca^{++} decreased the mobility of a second probe, dansyl-dipalmitoyl-PE (dns-DPPE). On the basis of these results, the authors suggested that Ca^{++} may cause a phase separation in SPM acidic lipids involving either lipid aggregation or transition to a gel-like phase.

VI. BIOPHYSICAL PROPERTIES OF SPM METHODOLOGY

A variety of techniques have been used to measure the physical properties of biological membranes including DSC, DTA, X-ray crystallography, Fourier transform infrared spectroscopy, NMR, ESR, and fluorescence polarization. Only ESR and fluorescence polarization have been widely used in the study of SPMs, although it is likely that in the near future NMR will be more widely employed. All three techniques can be used to measure two general types of parameters regarding lipid, primarily acyl chain rotational mobility: the *rate of motion* and the *range of motion*. The rate of motion may be expressed in terms of the rotational diffusion constant (R) or the wobbling diffusion constant (Dw) or the dynamic component of anisotropy (r_f), while the range of motion is generally expressed as the order parameter (S) and less frequently in terms of the cone angle (Θ) or the limiting anisotropy (r_∞).

In comparison to ESR and fluorescence polarization, NMR is the least perturbing technique and the theoretical analyses of NMR spectra are generally more rigorous and based on a sounder theoretical framework.[55] Furthermore, the time scale for the molecular motions measured by NMR (10^4 to 10^5 sec) is considerably slower than that for fluorescence (10^8 to 10^9 sec) or ESR (10^7 to 10^9 sec). Motions in the NMR time domain are most likely to be of biological significance.[55,56]

By the use of ^{2}H-NMR, it has been possible to determine the range of motion along the entire length of the hydrocarbon chain in the membrane phospholipids. For example, in DPPC membranes, ^{2}H-NMR reveals a plateau of membrane order from C_2 to C_{10}, with progressively increasing disorder after C_{10}.[57] The spectra obtained from ^{2}H-NMR in natural membranes often reveal a broad gel to liquid crystalline phase transition, which is also seen by more direct techniques, such as DSC.[58-60] In contrast, ESR probes such as 5-doxyl stearic acid may, under identical conditions, reveal a sharp phase transition and not report the peripheral boundaries of the gel to liquid crystalline transition.[55,58] Furthermore, on the NMR time scale, protein does not perturb lipid order, indicating a rapid exchange of the bulk and protein boundary lipids.[61,62] In different ways ESR and fluorescence polarization report that proteins immobilize membrane lipids and that the exchange between boundary and bulk lipids is slow in reference to the time scale being used.[63] The major disadvantage of ^{2}H-NMR is that the technique has a low sensitivity which requires the presence of relatively high probe concentrations (e.g., >50 mol%).[64] Furthermore, the synthesis of the ^{2}H-fatty acids is difficult and tedious. Finally, the time for obtaining NMR spectra is long, which hinders the use of NMR with notoriously unstable membranes, such as SPMs.

Due to the high natural abundance of the ^{31}P isotope and the few chemically equivalent sites which exist for phosphorus in membranes, ^{31}P-NMR spectra are easy to acquire and interpret.[65] The major disadvantage of ^{31}P-NMR is the large amount of membrane required (often several μmol of lipid-P) and the long acquisition time (several minutes to more than 1 hr). Nonetheless, under some circumstances, ^{31}P-NMR may prove useful for studies of SPMs and their isolated lipid components. In this latter regard, Cullis and associates have extensively published on ^{31}P-NMR of model membranes, describing the characteristics which promote bilayer and hexagonal-II packing arrangements.[66,67] A theoretical analysis of such spectra has been published by Thayer and Kohler.[65] ^{13}C-NMR spectra can also be used to detect molecular ordering of the membrane lipids. Recently, Brown et al.[68] have suggested that ^{13}C-NMR may be particularly useful in obtaining spin-lattice relaxation times, and thus a measure of the rate of molecular motion. Overall, we conclude that NMR spectroscopy will eventually surface as a dominant measure of lipid motion in SPMs; however, its widespread application awaits significant increases in sensitivity. An example of where the necessary sensitivity can be obtained by using reconstituted vesicles is provided by Seelig et al.[69]

ESR has been widely used to study lipid motions in natural membranes, including SPMs, and has provided important information on the order of these membranes and their response to various pharmacological or environmental manipulations, e.g., anesthetics or ethanol.[70-73] Basically, two types of probes have been employed: (1) the fatty acid derivatives, e.g., 5 or 12-doxyl-stearic acid, and (2) the small, nonpolar nitroxide molecules such as 2,2,6,6-tetramethylpeperidone-*N*-oxyl (TEMPO). The latter group may be particularly useful in monitoring gel to liquid crystalline phase transitions.[55] The scope and limitations of the ESR technique have been reviewed.[74] Some of the obvious limitations include the time scale of the measurements, the measurements are generally confined to the range of motion — order parameter, and the effect of the probe itself on the immediate membrane environment, an effect which cannot be completely eliminated by using low concentrations. Historically, the quantitative differences between results obtained from ESR and NMR spectra have been assumed to result from differences in the time scales of the two measurements. However, in a recent study, Taylor and Smith[75] suggest that one of the fundamental tenets of evaluating some ESR spectra may be in error. In analyzing the spectra of a probe such as 5-doxyl-stearic acid, it is assumed that the five-membered oxazolidine ring system is oriented perpendicularly to the long molecular axes of the probe. Taylor and Smith[75] tested this assumption by examining the NMR spectra of ^{2}H-labeled doxyl spin probes. These authors found that the nitroxide probe did not display this assumed geometry in the membrane, with a more pronounced departure for position 5 relative to position 12.

Another problem in the interpretation of ESR data arises from the breaks in the Arrhenius plots for nitroxide-labeled fatty acid probes. These discontinuities have been widely assumed

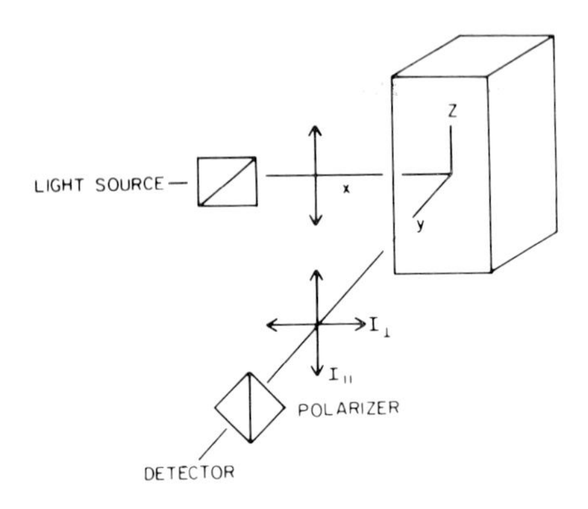

FIGURE 2. Schematic diagram of the principle for measurement of fluorescence anisotropies. (From Lakowicz, J. R., Ed., *Principles of Fluorescence Spectroscopy*, Plenum Press, New York, 1983. With permission.)

to represent gel to liquid crystalline phase transitions. Independent verification of these breaks may suggest otherwise. For example, in *A. laidalwii* membranes of differing lipid composition, ESR often reports quite sharp breaks while the DTA-determined phase transition temperature is quite broad.[55] Possibly, the ESR probes are partitioning into quite selective membrane domains and in fact some data[63] suggest that nitroxide fatty acids are excluded from gel-like domains. Despite the limitations with ESR probes, this methodology has proven useful in a variety of SPM studies. By using ESR probes it has been possible to examine the effects of drugs (e.g., ethanol and anesthetics[70-73]) and aging[76] on SPM order.

In the last 15 years fluorescence polarization has been widely used to analyze membrane structure; the significance of the approach has strengthened with the advent of time resolved techniques.[77,78] The basic principles of measuring fluorescence polarization are illustrated in Figure 2. The sample is excited with vertically polarized light and emission is measured through polarizers both parallel and perpendicular to the excitation polarizer. For measurements of the rate and or range of membrane motion an apolar fluorophore such as DPH is partitioned into the lipid matrix of the membrane. DPH is a particularly ideal probe in that it has a high extinction coefficient, but does not fluoresce in aqueous solutions.[79] The system is excited by polarized monochromatic light. For "steady-state" measurements, emission is measured under continuous illumination; emission is measured parallel (||) and perpendicular (⊥) to the polarized exciting light. Historically, the data have been more frequently reported as polarization (P) where

$$P = \frac{I_{\parallel} - I_{\perp}}{I_{\parallel} + I_{\perp}}$$

However, from a mathematical and theoretical viewpoint,[80] the preferable expression is in terms of anisotropy (r) where

$$r = \frac{I_{\parallel} - I_{\perp}}{I_{\parallel} + 2I_{\perp}}$$

The limiting anisotropy (r_o) refers to the anisotropy in the absence of depolarizing processes such as rotational diffusion or energy transfer.

$$r_o = 2/5\ (3\ \cos\alpha^2 - 1/2)$$

where (2/5) is the loss of anisotropy due to photoselection and α equals the angle of displacement between the emission and absorption dipoles. In an isotropic solution, i.e., an homogeneous oil, the range of r_o values must be $-0.2 \leq r_o \leq 0.4$ (P_o ranges from $-0.33 \leq P_o \leq 0.50$). When $\alpha = 54.7°$ (the "magic angle"), $r_o = 0$.

Until recently, fluorescence polarization (anisotropy) data were interpreted in terms of microviscosity using the Perrin equation:[81]

$$(1/r - 1/3) = (1/r_o - 1/3)\ (1 + 3\ \tau/\bar{\rho})$$

where τ is fluorescence lifetime and $\bar{\rho}$ is the rotational relaxation time. Alternatively, the Perrin equation may be expressed as

$$r = r_o/1 + (\tau/T_c)$$

where T_c is rotational correlation time. Viscosity (η) can be determined from

$$\eta = T_c\ RT/V$$

where T is absolute temperature, V is the volume of the rotating unit and R is the gas constant. When τ is constant and homogenous, η can be calculated directly from the steady state anisotropy. Changes in τ can be measured by pulse lifetime or phase-modulation measurements. The relative merits of both approaches have been discussed by Lackowicz.[80] In our experience with SPMs,[6,82] neither drugs nor development appreciably affect the time or homogeneity of τ. Thus, one could calculate T_c or η directly from the steady state anisotropy as long as the fluoropore, e.g., DPH had the same freedom (isotropy) of depolarizing rotations as in a homogeneous reference oil solution. A variety of data indicate this is not the case. Figure 3, taken from an article by van Blitterswijk et al.,[83] illustrates the problem. DPH is incorporated into a membrane (e.g., SPM) and anisotropy is measured as a function of time after a brief pulse of polarized light. Initially, the emitting dipoles of DPH are parallel to the direction of polarization and r approaches r_o. With time, DPH rotates and the anisotropy decreases. In a isotropic oil, the anisotropy falls to zero, as indicated by the dotted line. However, in real membranes, r reaches a limiting value r_∞ because the motion of the probe is hindered. The steady state anisotropy r_s is then composed of two components: r_f, the fast decaying component related to rotational diffusion, and r_∞, the slowly decaying component related to membrane order (the S parameter)

$$r_s = r_\infty + r_f$$

van Blitterswijk et al.[83] found on the basis of their data and that of others that there exists an empirical relationship between r_s and r_∞ for DPH such that:

$$r_\infty = 4/3\ r_s - 0.10, \qquad 0.13 < r_s < 0.28$$

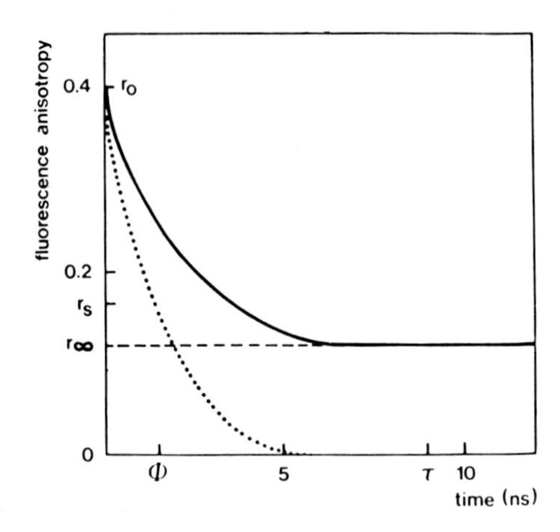

FIGURE 3. Time-dependent fluorescence anisotropy decay in lipid membranes (solid line) and in an isotropic reference oil (dotted line). The indicated values are typical for a rather fluid biomembrane, labeled with DPH. (From Van Blitterswijk, W. J., Van Hoeven, R. P., and van der Meer, B. W., *Biochim. Biophys. Acta*, 644, 1981. With permission.)

The range of permissible r_s values is similar to that encountered with most natural membranes. Pottel et al.[84] have developed a similar semiempirical relationship. They found that

$$r_\infty/r_o = 4/3\ r_s/r_o - 0.28, \qquad 0.33 < r_s/r_o < 0.70$$

When $r_o = 0.4$, the difference in r_∞ from van Blitterwijk et al.[83] is 0.01. Kinosita et al.[85] suggested that the rotational diffusion of DPH is limited to a cone formed by the surrounding membrane structures. The cone angle, θ, reflects the membrane order and is related to r_∞ by

$$r_\infty/r_o = [1/2 \cos\theta\ (1 + \cos\theta)]^2$$

The ratio r_∞/r_o has been shown on theoretical grounds[86-88] to be related to the square of the order parameter(s). Thus,

$$r_\infty/r_o = S^2, \qquad 0 \leq S \leq 1$$

S is the same orientational order parameter measured in NMR and ESR. The mathematical similarities between S_{DPH} and S_{13CNMR} have been elegantly described by Lipari and Szabo.[88] The question of whether or not it will be possible to develop similar relationships for other fluorescence probes remains to be answered. To some degree this issue is dealt with in a recent article by Hare.[89]

The significance of this "new" interpretation of r_s is readily evident when examining some data recently acquired in our laboratory on developmental changes in SPM anisotropy (Table 5).[82] There is an increase in r_s from day 3 to day 120 (adult) which roughly corresponds to the effect of a 7°C decrease in temperature.[12] Fluorescence lifetime was found to slightly increase during development as determined by both phase and modulation techniques at 6, 18, and 30 mHz. Furthermore, only one lifetime was detected, i.e., the phase and modulation measures gave identical results.[80] A direct application of the Perrin equation gives an increase in the rotational correlation time from 11.3 to 15.3 nsec. Factoring r_s into r_∞ and r_f reveals

Table 5
DEVELOPMENTAL CHANGES IN POLARIZATION, ANISOTROPY, FLUORESCENCE LIFETIME, AND RELATED PARAMETERS IN DEVELOPING SPMs[a,82]

Age	Polarization (P)	Anisotropy (r_s)	Lifetime (τ)	Rotational correlation time (Tc)	Anisotropy static (r_∞)	Anisotropy free (r_f)	Order parameter (S_{DPH})	Rotational correlation (Tc′)
3	0.265 ± 0.002	0.194 ± 0.002	9.82 ± 0.02	11.33 ± 0.24	0.159 ± 0.003	0.035 ± 0.001	0.662 ± 0.005	2.06 ± 0.01
7	0.267	0.195	10.30	12.06	0.160	0.035	0.666	2.16
10	0.270	0.198	9.85	11.86	0.164	0.034	0.672	2.04
14	0.274	0.201	10.00	12.46	0.168	0.033	0.681	2.05
21	0.286	0.211	10.30	14.33	0.181	0.030	0.707	2.02
28	0.290	0.214	10.20	14.73	0.185	0.029	0.716	1.98
Adult	0.294 ± 0.001[b]	0.217 ± 0.001[b]	10.27 ± 0.16[b]	15.33 ± 0.20[b]	0.190 ± 0.001[b]	0.028 ± 0.001[b]	0.723 ± 0.001[b]	1.96 ± 0.03[b]

[a] The determination of P, r_s, and τ were performed using DPH as the probe and a SLM-4800 spectrofluorometer. N = 3 separate preparations for day 3 and adult animals and one preparation for the other days. Lifetime data were determined by modulation at 18 mHz. Similar results were obtained by both phase and modulation techniques at 6, 18, and 30 mHz. The rotational correlation time, Tc, was determined from $r_s = r_o/1 + \tau/Tc$ where r_o = .362. The order parameter was calculated from $S = \sqrt{r_\infty/r_o}$ $r_f = r_s - r_\infty$. Tc′ was calculated from $r_f = r_o - r_\infty/1 + \tau/Tc'$.

[b] Significantly different from day 3 value.

that during development there is a significant increase in membrane order. In contrast, the kinetic component, r_f, actually decrased during development, as reflected by the small but significant decrease in T_c'. This decrease in T_c' is in marked contrast to the increase in T_c obtained by direct application of the Perrin equation. The major change in SPM lipids during development is the doubling of the molar sterol/phospholipid molar ratio. Using the approach described here, Pottel et al.[84] found that cholesterol does not make membranes more viscous, but contrarily seems to lower the true microviscosity, which agrees with the earlier conclusions of Heyn.[90] Cholesterol, however, increases the order parameter, and since S is a measure of lipid packing, cholesterol is in fact a condensor of membranes; this is in agreement with its effects on monolayers. Our data on SPMs and development seem to perfectly reflect these changes in membrane cholesterol.

Due to its unique geometry, it has been possible to analyze in great detail DPH anisotropy. There are, however, a number of other fluorescence probes which have specific uses, although a precise interpretation of the data is more difficult. Those include probes of the membrane surface, e.g., TMA-DPH and 1,8-ANS, and *cis*- and *trans*-parinaric acids which may selectively partition repeatedly into the more liquid crystalline and gel-like domains of the membrane.[91-93] *cis*-Parinaric acid can be directly incorporated into membrane phospholipids, including SPMs.[94] Finally, the greatest advantage of the fluorescence depolarization technique is its exquisite sensitivity. In our laboratories it has been possible to obtain data on as little as 5 to 10 μg membrane (as protein).

The authors hope that the previous sections have provided the necessary framework for evaluating the role(s) of phospholipids in regulating SPM function. Our attempt in the next sections will not be a comprehensive review of the effects of pharmacological, developmental, and physiological interventions on SPM phospholipids. Rather, we will quite selectively focus on those areas which have been of particular interest in our laboratories over the past decade. Three main types of research strategies have been employed. In one strategy we have related pharmacological or physiological intervention to change in SPM lipid content which will affect the physical properties of the membranes. For the second strategy, we have probed the effects of various pharmacological interventions on the synthesis, metabolism, and turnover of SPM phospholipids. Our final strategy has been to elucidate the role(s) of specific membrane lipids as coparticipants in membrane activities, e.g., hormone receptor binding.

VII. DRUGS AND SPM LIPIDS

A. Ethanol

The mechanism(s) of ethanol action and the processes underlying the development of ethanol tolerance and dependence have been areas of active interest in our laboratories. Our approach, which has largely focused on the biophysical properties of ethanol, has its historical antecedent in the studies by Overton[95] and Meyer and Hemmi,[96] showing that the anesthesia produced by a series of primary alcohols in tadpoles is related to the oil/water partition coefficient. Butanol will be more potent than ethanol and so on. Thus, it is the degree of partitioning of the alcohol into the nerve membrane which determines potency. This theory has been questioned on both theoretical and empirical grounds,[97,98] but it has served to focus our attention on ethanol and SPM lipids.

The potential acute effects of ethanol (and related anesthetics) on SPMs may be divided into two broad categories: (1) change in phase transitions and (2) changes in the rate and/or range of motion of the membrane lipids. A variety of studies using differential scanning calorimetry (DSC) absorbance, and fluorescence[99-104] have shown that alcohols decrease the transition temperature (T_m) of gel phases in both synthetic and biological membranes. However, the significance of these findings may be questioned on several grounds. First, the concentrations of ethanol required to produce significant effects are more than 300 m*M* and are thus not pharmacologically relevant. Second, short-chain alcohols decrease T_m values while long-chain alcohols increase T_m values; however, both groups inhibit nerve conduction.[98] Third *cis*- and *trans*-tetradecanol, respectively, decrease and increase the T_m of PC vesicles, but both isomers are equipotent as general and local anesthetics.[103] There is an interesting variation of the T_m hypothesis, which suggests that ethanol may alter the lipid configuration. For example, ethanol at a high concentration (400 m*M*) stabilizes synthetic PE membranes in the bilayer conformation and reduces the conversion to hexagonal phase II.[104] The hexagonal phase facilitates the transport of calcium across membranes. These data are of interest given the finding by Harris and Hood[105] that ethanol inhibits calcium transport across synaptic membranes and also given the hypothesis that synaptic calcium transport may be regulated by hexagonal-phase phosphatidic acid.[106]

Alcohol may act by altering the rate and/or range of motion of the SPM lipids. It is the range of motion (membrane order) which has been shown by ESR techniques to be sensitive to ethanol. Chin and Goldstein[71,72] found that ethanol decreased synaptic membrane order and that animals administered ethanol chronically were resistant to the disordering. More recently, these same authors have found that ethanol is less effective in disordering the membranes of short sleep (SS) when compared to long sleep (LS) mice.[107] The genetic characteristics of the LS-SS mice have been described elsewhere.[108] Interestingly, the difference in membrane order between the LS and SS mice has not been detected by the DPH-fluorescence depolarization technique.[109] The difference in the ESR and fluorescence data may reflect a difference in the parameter measured. The ESR order parameter is likely to be associated with the fast acyl chain isomerization motion.[110] Differently, DPH, a relatively rigid molecule, may be insensitive to conformational disorders associated with chain isomerizations and may instead sense a "rigid-body" order of the lipid chains.[110-112] However, fluorescence depolarization does detect an ethanol-induced decrease in SPM lipid order at pharmacologically relevant concentrations.[6,113,114] The ethanol effect is greatest in the hydrophobic core and is less pronounced at the membrane surface. In addition, myelin membranes in comparison to synaptic membranes are insensitive to the effect of ethanol.[6] Analyzing the DPH steady-state anisotropy[113] as described in Section V reveals not only that ethanol decreases r_∞ (and hence membrane order), but also increases r_f (and hence increases membrane viscosity).[109] Thus, the DPH data suggest that ethanol has effects on the rate and range of membrane order opposite to that of cholesterol.

This point is important since it has been suggested that during ethanol tolerance development there is an increase in SPM cholesterol content.[115] Johnson et al.[116] have suggested that the small increase in cholesterol is not sufficient to account for the decrease in ethanol sensitivity. However, other cells may be able to increase membrane cholesterol to attenuate the effects of ethanol and anesthetics. For example, Koblin and Wang[117] studied the effects of halothane or cyclopropane on *A. laidlawii* B cells grown in a cholesterol-containing medium. Continuous anesthetic exposure did not alter fatty acid composition, but did increase membrane cholesterol content, which led to an increase in membrane order as measured by 5-doxyl stearic acid.

The ESR probe 12-doxylstearate detects a difference in SPM membrane order between control and ethanol-adapted membranes,[71] a difference that is consistent with DPH fluorescence polarization.[114] In addition to increased cholesterol content, SPMs could adapt to ethanol exposure by decreasing the acyl group unsaturation in the membrane phospholipids. The effects of ethanol on acyl group composition have been conflicting. Smith and Gerhart[118] and Wing et al.[119] found no change in SPM acyl group composition. In contrast, Littleton and John,[120] Sun and Sun,[122] and Alling et al.[121] have found significant changes. For example, Sun and Sun[122] observed a small decrease in PC 18:1 from 28.1 to 25.9%. Recently, we have examined this parameter.[114] The experimental design was such that SPMs from ethanol-exposed and calorically paired controls were prepared by one of us (Harris) and then shipped blind to another laboratory (Hitzemann) for analysis. As shown in Table 6, the only significant changes in acyl group composition were found in PS. In the ethanol-treated SPMs, palmitic acid (16:0) markedly increased from 3.2 to 11.6%, while docashexanoic acid (22:6, W-3) decreased from 27.9 to 18.8%. Since PS comprises only 8% of the SPM lipid, these changes in PS acyl group composition would have virtually undetectable effects in the total SPM acyl group composition. Other lipid parameters were also examined. As shown in Table 7, there is no effect of chronic ethanol treatment on SPM phospholipid composition. Similarly, we found no significant effect on the protein/lipid-P or cholesterol/lipid-P ratios and membrane ganglioside content or composition (data not shown). Overall, we conclude that chronic ethanol exposure does not induce a significant effect on SPM lipid composition. These data are not surprising given that vesicles prepared from SPM lipid extracts of control and ethanol-treated animals show no significant difference in physical properties.[114] These SPM lipid extracts are also relatively insensitive to the disordering effects of ethanol, suggesting the SPM protein components are required for the full expression of ethanol effects.

B. Nicotine

Studies on role(s) of the phosphoinositides in the regulation of membrane activities, especially those of neuronal membranes, have enjoyed a renaissance. Understanding the significance of the "PI" effect has been an area of particularly intensive activity. Some studies have focused on the formation of phosphatidic acid (PA) during the PI cycle. For example, in our laboratories we have been able to show that PA is able to support a rapid Ca^{++} uptake in synaptosomes, which induces neurotransmitter release.[106] The metabolites of PI, DPI, and TPI, namely diacylglycerol and the inositol phosphates, are suggested to be involved in the regulation of protein kinase C activity and internal Ca^{++} mobilization.[123-126]

Work in our laboratories focused on the effects of nicotine on DPI and TPI synthesis and metabolism.[127,128] DPI and TPI may regulate membrane-bound Ca^{++} levels.[129] The simple dephosphorylation of TPI to DPI could significantly lower bound Ca^{++} levels due to the difference in Ca^{++} affinity.[130,131] It has been suggested that a TPI to DPI conversion may be involved in the generation of action potentials at nicotinic cholinergic synapses on the CNS.[132] To test this hypothesis we examined the effects of various cholinergic compounds on the incorporation of [^{3}H]glycerol and ^{32}Pi into DPI and TPI. As shown in Table 8, both eserine and nicotine significantly decreased while mecamylamine increased the labeled TPI/DPI ratio. The nicotine effect was dose-dependent and was reversed by mecamylamine pretreatment. Interestingly, nicotine in doses as low as 0.3 mg/kg significantly increased the incorporation of ^{32}Pi and [^{3}H]glycerol into PI. Overall, we interpreted these data to illustrate that the TPI to DPI interconversion is indeed sensitive to the level of nicotinic receptor activation.

TPI is also an excellent "binding site" for nicotinic ligands.[128,133] We observed that cholinergic ligands changed the partitioning of TPI between an aqueous and organic phase. The ability of such ligands to induce an aqueous-to-organic phase transfer was excellently correlated with the in vivo potency of these ligands to induce neuromuscular blockade and with the in vitro binding affinity of the ligand for nicotinic receptors.[128] It also is of interest to note that the proteolipid nicotinic cholinergic receptor isolated by DeRobertis[134] using Sephadex® LH-20 chromoatography contains large amounts of TPI.[133]

C. Opioid Narcotics

Paralleling our studies on TPI we observed that LH-20 chromatography of brain would isolate a fraction rich in opiate receptor-binding activity.[135] The active principle of this

Table 6
EFFECTS OF CHRONIC ETHANOL TREATMENT ON THE ACYL COMPOSITION OF SYNAPTIC MEMBRANE PHOSPHOLIPIDS

	Phosphatidylserine		Phosphatidylcholine		Phosphatidylethanolamine		Ethanolamine Plasmalogen	
Acyl group	Control	Ethanol	Control	Ethanol	Control	Ethanol	Control	Ethanol
16:0	3.2 ± 0.9[a]	11.6 ± 3.6[b]	43.1 ± 2.2	44.6 ± 0.4	6.5 ± 0.8	7.7 ± 0.4	1.2 ± 0.3	2.7 ± 0.4
18:0	38.7 ± 0.9	35.8 ± 3.6	14.6 ± 0.6	14.0 ± 0.2	33.3 ± 0.8	32.6 ± 0.5	2.2 ± 0.4	2.7 ± 0.6
18:1	17.7 ± 1.4	22.2 ± 2.1	31.8 ± 1.4	30.2 ± 0.6	14.7 ± 0.5	14.1 ± 0.5	22.3 ± 0.9	21.9 ± 2.1
18.2	ND[c]	ND	0.5 ± 0.1	0.5 ± 0.1	0.4 ± 0.1	0.4 ± 0.1	0.3 ± 0.1	0.3 ± 0.1
18:3(W-3 ± W-6)	1.5 ± 0.2	1.1 ± 0.1	1.6 ± 0.1	1.6 ± 0.1	1.3 ± 0.2	1.2 ± 0.1	9.5 ± 1.0	8.5 ± 0.6
20:3(W-6)	1.2 ± 0.2	1.1 ± 0.1	0.3 ± 0.1	0.3 ± 0.1	0.4 ± 0.1	0.4 ± 0.1	0.9 ± 0.1	0.9 ± 0.1
20:4(W-6)	5.0 ± 0.6	4.9 ± 0.4	3.8 ± 0.1	3.8 ± 0.1	12.1 ± 0.2	11.9 ± 0.2	15.0 ± 0.2	14.7 ± 0.2
20:5(W-3 + W-6)	ND	ND	0.4 ± 0.1	0.4 ± 0.1	>0.5	>0.5	0.9 ± 0.1	0.6 ± 0.1
22:4(W-6)	4.7 ± 0.6	3.8 ± 0.4	0.4 ± 0.1	0.4 ± 0.1	4.1 ± 0.4	4.5 ± 0.6	14.2 ± 0.5	13.3 ± 0.3
22:5(W-3)	ND	ND	ND	ND	ND	ND	1.0 ± 0.1	0.9 ± 0.1
22:6(W-3)	27.9 ± 1.8	18.8 ± 2.3[d]	3.3 ± 0.2	3.2 ± 0.1	26.9 ± 0.6	26.9 ± 1.0	34.3 ± 2.2	34.0 ± 2.9

[a] Values represent percent composition and are the mean ± S.E.M., n = 6. Brain SPM were obtained from DBA/2 mice fed an ethanol-containing diet or an isocaloric control diet for 7 days.
[b] Significantly different from control, $p < 0.05$.
[c] Not detectable.
[d] Significantly different from control, $p < 0.01$.

From Harris, R. A., Mitchell, M. A., and Hitzemann, R. J., *Mol. Pharmacol.*, 25, 401, 1984. With permission.

Table 7
EFFECT OF CHRONIC ETHANOL TREATMENT ON SYNAPTIC MEMBRANE PHOSPHOLIPID COMPOSITION[114]

Treatment	PC[a]	PS	EP	PE	SM	Other[b]
Ethanol	41.1 ± 0.8[c]	6.9 ± 0.5	24.6 ± 2.2	20.7 ± 0.7	5.6 ± 0.2	4.3 ± 1.4
Control	39.0 ± 2.0	8.1 ± 0.7	24.4 ± 0.7	20.3 ± 0.8	5.2 ± 0.3	3.7 ± 1.5

[a] Abbreviations: PC, phosphatidylcholine; PS, phosphatidylserine; EP, ethanolamine plasmalogen; PE, phosphatidylethanolamine; SM, sphingomyelin.
[b] Includes phosphatidic acid, phosphatidylinositol, and polyphosphatidylinositols.
[c] Values represent percent composition and are the mean ± S.E.M., r = 6.

From Harris, R. A., Mitchell, M. A., and Hitzemann, R. J., *Mol. Pharmacol.*, 25, 401, 1984. With permission.

Table 8
EFFECTS OF CHOLINERGIC DRUGS ON THE ACCUMULATION OF ^{32}P and ^{3}H-LABELED TPI AND DPI IN THE BRAINSTEM MICROSOMAL FRACTION[a,b 127]

	% Controls ± S.E.					
Group	[^{3}H]TPI	[^{3}H]DPI	[^{3}]TPI, [^{3}H]DPI	[^{32}P]TPI	[^{32}P]DPI	[^{32}P]TPI, [^{32}P]DPI
Control	100 ± 13	100 ± 9	100 ± 11	100 ± 10	100 ± 8	100 ± 9
Eserine	60 ± 13[c]	105 ± 15	57 ± 22[b]	52 ± 7	85 ± 7	61 ± 10[c]
Mecamylamine	231 ± 19[c]	111 ± 10	208 ± 6[c]	202 ± 9	110 ± 6	184 ± 13
Atropine	103 ± 10	126 ± 14	82 ± 10	123 ± 9	112 ± 11	111 ± 8
Nicotine	55 ± 6[c]	94 ± 8	59 ± 7[c]	51 ± 4	94 ± 7	54 ± 6[c]

[a] Rats were given 0.5 mg/kg eserine, 2 mg/kg mecamylamine, 30 mg/kg atropine, 1 mg/kg nicotine or saline, i.p., 15 min prior to the intraventricular injection of 200 μCi $^{32}P_i$ and 50 μCi [^{3}H]glycerol. The animals were sacrificed 1 hr later, brains were removed and dissected, and subcellular fractions were prepared under alkaline conditions. TPI and DPI were extracted from the microsomes and the specific activities of the phospholipids were determined. Data are the mean ± S.E. of at least 10 experiments.
[b] The brainstem refers to the midbrain-pons-medulla.
[c] Significantly different from control, $p < 0.05$.

From Hitzemann, R. J., Natsuki, R., and Loh, H. H., *Biochem. Pharmacol.*, 27, 2519, 1978. With permission.

fraction proved to be the acidic glycolipid, cerebroside sulfate (CS). In the ensuing years, we have accumulated a variety of evidence which indicate that opiate-CS interactions may have an important role in opiate pharmacology.[136-138] CS binds opiates stereospecifically and with a high affinity, and CS-opiate interactions show a Na^{+} effect to discriminate agonist-antagonist interactions. The level of CS in SPMs is low and probably undetectable by conventional methods. However, SPM opiate receptor binding can be inhibited by Azure A, a cationic dye which has a high affinity for CS,[139] and by the sulfatide degradative enzyme, cerebroside sulfatase (E.C.3.1.6.8).[40] Furthermore, the injection of a specific CS antibody into the periaqueductal gray area, the center for opiate analgetic activity, antagonizes the actions of morphine.[141]

Our work in cell culture also supports a role for CS in opiate mechanisms. The presence of opiate receptors in the neuroblastoma X glioma NG108-15 cell line and the inhibition by opiates of basal and PGE_1-stimulated adenylate cyclase are well documented.[142-145] In the parent cell line, neuroblastoma N18TG2, which contains opiate binding sites,[146] opiate agonists, and opiate peptides, have little or no effect on the adenylate cyclase activity.[144,146] After the incorporation of the acidic lipid CS by N18TG2, the opiate ligands inhibited the adenylate cyclase activity in the parent cells with a potency similar to that found in the daughter cells, NG108-15.[147,148] This lipid potentiation effect could not be observed in cell lines which do not possess opiate receptor.[148] These results suggest that the addition of the lipid to the parent plasma membrane promoted coupling between receptors and adenylate cyclase by changing either the lipid composition of the membrane and thus the packing order of the membrane lipid or the membrane fluidity. The effect of galactocerebrosides on the physical properties of phospholipid bilayers has been examined not long ago.[149]

We have succeeded in partially purifying an opiate receptor preparation from rat brain using affinity chromatography. In our initial work,[195] two fractions were isolated from this column, one containing protein and the other acidic lipids. While neither alone bound opiates, when combined, the two fractions contained most of the binding activity originally loaded onto the column. The opiate-binding properties of this reconstituted lipid-protein mixture exhibited many properties similar to those of membrane bound receptors of the u-type, including high affinity, stereospecificity, Na-effect, and rank order in affinity for opiates. More recently, we have succeeded in isolating opiate binding activity as a single peak from the affinity column[196] in a fraction containing both the acidic lipids and the protein. The opiate-binding properties of this "receptor" fraction are the same as those of membrane-bound receptor. Furthermore, this receptor fraction is highly sensitive to both trypsin and *N*-ethylmaleimide. Based on the protein content of this isolated receptor fraction, a 3200-fold purification over the original brain P_2 fraction was achieved.

The issue of how chronic drug exposure affects SPM phospholipid synthesis, metabolism, and turnover has been examined for d-amphetamine, pentobarbital, and morphine.[150-154] The data obtained with morphine are illustrative of the types of information that may be obtained. These studies derived from our earlier work on SPM protein synthesis and were part of a coordinated attempt to find a specific membrane component that increased in parallel with tolerance-dependence development. The pioneering studies of Mulé[155-157] clearly demonstrated that morphine and its cogeners have marked effect on brain phospholipid biosynthesis. During a period of rapid tolerance development, we examined the accumulation of [^{32}P]phospholipids in subcortical SPMs.[150] For reasons described in the original publication, measurements were made at 3 and 24 hr after ^{32}Pi administration. In one fraction of SPMs, chronic morphinization increased the synthesis of both PS and PI at 3 but not at 24 hr after administration (Table 9). The increased accumulation of [^{32}P]PS and PI was also found in animals chronically administered pentobarbital, suggesting the increase is rather nonspecific. It should be noted that only a small percentage of the total SPMs are directly associated with opiate mechanism. Thus, the chance of finding an opiate specific effect unless it is greatly amplified is relatively small. This same argument (or interpretation) would seem to supply to the recent report by Herron et al.[158] These authors noted that chronic morphinization-like denervation increased the cholesterol/phospholipid ratio and increased apparent microviscosity (most probably membrane order) in a crude brain membrane preparation. The experimental design of this study deserves some comment. First, the "chronic" morphine group was implanted with a single morphine pellet and sacrificed 11 days later. At this time the pellet is completely encapsulated and in fact, such animals are probably undergoing withdrawal. Secondly, the membrane preparation used is so heterogeneous that multiple fluorescence lifetimes are almost a certainty. Apparently, this point was not considered in the analysis. Furthermore, this preparation cannot be correlated with functional activity. Thirdly, the changes in microviscosity were measured at an unphysiological temperature (25°C). Fourth, the changes in viscosity are very small and would represent a change in anisotropy of 0.0025 units, a change generally at the limit of instrumental capability. In our

Table 9
EFFECT OF CHRONIC MORPHINE OR PENTOBARBITAL TREATMENT ON INCORPORATION OF ^{32}P, [^{3}H] CHOLINE INTO SYNAPTIC PLASMA MEMBRANE-H-PHOSPHOLIPIDS[a150]

		Activity (cpm/μmol lipid P) × 10^{-3}			
		$^{32}P_i$			[^{3}H] choline
Group	Time (hr)	PS + PI	PE	PC	PC
Control	3	67 ± 7	65 ± 8	82 ± 11	162 ± 14
Pentobarbital	3	95 ± 9(141)	35 ± 4(54)	86 ± 7	168 ± 16
Morphine	3	86 ± 8(128)	65 ± 7	68 ± 9	139 ± 17
Control	24	384 ± 31	117 ± 14	154 ± 13	433 ± 36
Pentobarbital	24	484 ± 26(126)	111 ± 16	126 ± 9	336 ± 24(78)
Morphine	24	369 ± 24	103 ± 10	133 ± 14	318 ± 17(73)

[a] Animals were implanted with 2 placebo, 75 mg morphine or 200 mg pentobarbital pellets, s.c.; 24 hr later, the animals were given 100 μCi $^{32}P_i$ (phosphoric acid) and 20 μCi [^{3}H]choline intraventricularly. The animals were sacrificed 3 and 24 hr later and synaptic plasma membranes were prepared from the subcortex. Phospholipids were extracted from the membranes, separated, and their specific activities determined by conventional techniques. Data are the mean ± S.E. of 4 to 6 experiments; a pooled sample of 3 rat brains per experiment was used. The percent change from control when significant differences ($P < 0.05$) were found is given in parentheses. Abbreviations used are PS + PI, phosphatidylserine plus phosphatidylinositol; PC, phosphatidylcholine; PE, phosphatidylethanolamine; and H, heavy, referring to synaptic plasma membranes derived from a heavy population of nerve ending particles.[150]

From Hitzemann, R. J. and Loh, H. H., *Biochem. Pharmacol.*, 26, 1087, 1977. With permission.

laboratories we have attempted to find an effect of morphine and/or naloxone on brain membranes or vesicles prepared from lipid extracts. Only at a very high concentration of naloxone (10^{-5}) was a somewhat small, but probably insignificant decrease in DPH polarization observed for intact SPMs.[6]

VII. PHYSIOLOGY, PATHOLOGY, AND SPMs

A. Temperature Acclimation

There have been numerous studies showing that temperature acclimation is associated with changes in membrane composition and function. Generally, reducing temperature results in an increase in phospholipid acyl group unsaturation.[159] Such changes have been described in terms of homeoviscous adaptation.[160] However, as noted in previous sections, the adaptive response is more likely to involve a change in membrane order rather than viscosity. Such changes have been observed in *E. coli*,[160] *Bacillus stearothermophilus*,[161] *Tetrahymena pyriformis*,[162] and teleost fish.[163,164] In this section we will briefly review the work of Cossins on teleost fish SPMs; this work has been discussed in detail elsewhere.[159] Goldfish were acclimated at 5, 15, and 25°C and SPMs were prepared. At all temperatures in the range examined (2 to 37°C) DPH polarization (P) was highest in the 25°C SPMs and lowest in the 5°C SPMs. The difference between the 5 and 25°C preparations was not the result of a higher fluorescence lifetime in the 5°C SPMs. Rather, at all temperatures, τ was somewhat higher in the 25°C SPMs. Furthermore, τ was homogeneous. Based on our earlier arguments, we can conclude that changes in P result from changes in the range of probe motion. The changes in P were also present in vesicles prepared from purified SPM phospholipids, indicating that the differences between the warm and cold acclimated fish can to a large degree be explained by differences in phospholipid composition. In this regard, the 5°C SPM phospholipids have significantly more unsaturated and less saturated fatty acids. Recently, Cosins[159] has used differential polarized phase flourometry to directly determine both the rotational diffusion coefficient (R) and the limiting anisotropy, r_∞. This technique measures the differential lifetime ($\Delta\tau$), or the difference in τ of the parallel and perpendicular components of emission. A difference in lifetimes will exist since the parallel component will select those fluorophors which have emitted their photon before significant rotation has occured, while the perpendicular component will select those fluorophors which have rotated significantly during their fluorescence lifetime. By combining both measurements of $\Delta\tau$ and r_s, it is possible to calculate a rotational diffusion constant from which r_∞ can be simply calculated from

$$r = r_s + (r_s - r_o)/6R$$

Overall, this approach allows the determination of $r\infty$ for any fluorescence probe and one is not limited to using DPH and the empirical equations derived for this probe.[83,84] The results obtained by Cossins unequivocally demonstrate that the difference in r_s between the membranes of cold- and warm-acclimated fish more specifically relates to a change in membrane order rather than membrane fluidity. Overall, the results of Cossins[159] indicate that thermal adaptation is a very active process in fish SPM involving marked changes in phospholipid composition. High pressure can also induce a significant adaptive response.[165] It is interesting to note the robustness of the thermal response in comparison to the small response caused by chronic ethanol intoxication. This difference further illustrates the unwiseness of drawing a similarity between the actions of ethanol and the effect of increasing temperature.

B. Growth, Differentiation, and Development

There is accumulating evidence that cell division is associated with changes in plasma membrane order and fluidity.[166-170] Furthermore, when untransformed cells in culture reach confluence, DPH polarization increases.[171] Interestingly, oncogenically transformed cells, e.g., with Rous sarcoma virus, do not show an increase in polarization at confluence.[172] The issue of whether or not transformed and untransformed subconfluent cells have a difference in membrane order and/or fluidity remains controversial.[172] For nerve cells the most extensive studies of changes on membrane order and fluidity relating to growth and development are found for neuroblastoma cells in culture.[170] Some of the salient points are as follows. (1) During the cell cycle there are marked changes in DPH polarization. Polarization is highest during mitosis and drops dramatically as cells enter G_1, and decrease which is maintained throughout the rest of interphase.[173] (2) Upon the addition of differentiating-inducing agents, e.g., dibutyryl cAMP, there is a rapid decrease in DPH polarization to a level never seen during the cell cycle. These changes are not simply the result of neurite extension, but are rather related with the initiation of morphological differentiation. The addition of DPPC vesicles to the culture medium (0.6 μmol/mℓ) prevents the decrease in polarization and neurite extension. Dioleoyl-PC vesicles were without effect. (3) There is a topographical asymmetry during differentiation. By using the fluorescence photobleaching recovery method, De Laat et al.[174] were able to show that the most fluid domains were located in the existing neurites. De Laat and van der Saag[170] have concluded that these local differences in membrane order and fluidity may be related to the process of membrane growth since Pfenninger[175] has demonstrated by freeze-fracture electron microscopy of developing neurites that membrane growth occurs through a predominant insertion of membrane lipid in specialized areas of the growth cone. The change(s) during differentiation in membrane lipid composition related to change in DPH polarization appears to relate to a preferential (over phospholipid) decrease in cholesterol biosynthesis.[170] Finally, we have recently observed in a crude plasma membrane preparation that there is a significant increase in DPH polarization when comparing undifferentiated fetal rat brain neurons in culture to differentiated neurons (data not sown). The cells were grown in defined medium.

Studies in our laboratories have focused on the changes in SPM order and fluidity that occur during the final stages of synaptic differentiation.[12,82] As described in Table 3, there is a marked increase in membrane order during maturation. For these studies we used SPMs prepared from nerve endings which could be identified on the basis of numerous intrasynaptic mitochondria and vesicles. Furthermore, the nerve endings were isolated only from the cortex, a brain region in the rat which develops postnatally.[176] As previously noted the increase in membrane order which is maintained in vesicles formed from lipid extracts is substantial and is roughly equivalent to a 7°C decrease in temperature. The increase in membrane order is most likely to result from the marked increase in the membrane sterol to phospholipid ratio.[12] The sterol to phospholipid ratio increases nearly 100% from day 7 to 120 (adult) with the bulk of the change occurring before day 28 (unpublished observation). Desmosterol, which is present in the young SPMs, disappears with maturation. The functional significance of this change is unclear since desmosterol and cholesterol have equivalent effects on membrane order.[12] Overall, our data and those of de Laat and van der Saag[170] indicate that both the regulation of cholesterol biosynthesis and the incorporation of cholesterol into the terminal membrane are important in controlling nerve differentiation and synaptic maturation. However, there are other changes in the membrane lipids which may act as ''fine-tuning'' on these processes.

For example, we have observed that during development there is a significant change on SPM phospholipid composition. Most notably, the proportion of PC significantly decreases[12] (Table 10). The decrease in PC was not offset by the increase in SM, so during development there is a net loss of lipid-bound choline. This is largely compensated for by the significant increases in PE and EP. The reasons for this decrease in PC may involve a developmental decrease in phospholipid methylation activity which we have suggested is related to the increase in membrane order.[177]

In addition to the changes in phospholipid composition, the fatty acid composition of the SPM phospholipids changes during development. As shown in Table 11 there is a marked (−36%) decrease in the short-chain saturates (14:0 + 16:0). In contrast to this decrease, oleic acid (18:1) and the long-chain polyunsaturates 22:4 (W-6) and 22:6 (W-3) increase during development. Overall, there was a significant increase in the unsaturation index from day 7 to 14. Data on changes in fatty acid composition of the individual phospholipids are presented elsewhere.[178] Overall, we conclude that development is a period of intensive changes in SPM lipid composition, changes which are necessary elements of the process of synaptic maturation. Interestingly, many of the lipid changes which occur during development continue at a slower pace after maturity is reached and probably eventually contribute to synaptic aging and deterioration.

C. Aging

The data described in the previous sections clearly illustrate that there is an optimal membrane composition, order, and fluidity for the regulation of synaptic activities, e.g., receptor binding or neurotransmitter release. Even small deviations from this optimum, as those induced by ethanol, can lead to severe functional deficits. In recent years, it has been suggested that at least some of the manifestations of aging result from changes in the lipid

Table 10
PHOSPHOLIPID COMPOSITION IN DEVELOPING SPMs[a]

Phospholipid	Age (days) 7	14	120 (Adult)
PC	50.4 ± 3.2	47.8 ± 2.7	40.7 ± 2.3[b]
PE	29.6 ± 0.9	31.1 ± 1.3	34.2 ± 1.9[b]
PS	16.2 ± 0.5	16.2 ± 0.9	17.6 ± 0.7
EP	3.6 ± 0.4	4.2 ± 0.5	5.7 ± 0.5[b]
SM	0.2 ± 0.1	0.6 ± 0.2	1.8 ± 0.3[b,c]

[a] Phospholipids were isolated from 7, 14, and 120 day SPMs by conventional techniques. Data are the mean ± S.E. of 4 separate experiments, each performed in duplicate (n = 8). PS = PS + PI. Data expressed as percent of total lipid P recovered.

[b] Significantly different from day 7 value, $p < 0.05$.

[c] Significantly different from day 14 value, $p < 0.05$.

Table 11
FATTY ACID COMPOSITION (ACYL LINKAGES) IN DEVELOPING MEMBRANES

Fatty acid	Age (days) 7	14	120 (Adult)
14:0 + 16:0	35.2 ± 3.1	27.8 ± 2.5	22.6 ± 2.4[a]
16:1	1.6 ± 0.2	0.6 ± 0.1[a]	0.5 ± 0.1[a]
18:0	16.8 ± 1.7	17.4 ± 1.9	23.2 ± 1.6[a]
18:1	12.7 ± 0.8	12.5 ± 1.3	18.9 ± 1.1[a,b]
18:2	1.5 ± 0.1	1.3 ± 0.2	0.8 ± 0.1[a]
18:3(W-3 + W-6)	0.3 ± 0.1	0.3 ± 0.1	0.3 ± 0.1
20:0 (20:1)	0.6 ± 0.1	0.4 ± 0.1	0.6 ± 0.1
20:4(W-6)	13.6 ± 1.3	14.2 ± 1.6	11.7 ± 0.9
20:5(W-3 + W-6)	0.2 ± 0.1	0.9 ± 0.2	0.1 ± 0.1
24:0; 24:1	<0.1	<0.1	<0.1
22:4(W-6)	2.5 ± 0.2	6.1 ± 0.5[a]	4.1 ± 0.4[a,b]
22:5(W-6)	3.3 ± 0.3	2.6 ± 0.3	1.4 ± 0.1[a,b]
22:6(W-3)	9.9 ± 0.8	12.8 ± 0.9[a]	13.3 ± 1.2[a]
Unsaturation index[c]	163 ± 11	206 ± 15[a]	185 ± 9

Note: The total partitioned lipid extracts from the 7, 14, and 120 day membrane preparations were analyzed for fatty acids (acyl linkages) using a gas chromatographic technique. The data are the mean of the 3 experiments, each of which was performed in duplicate (n = 6). The data are expressed as the percent of the total comprised by each fatty acid.

[a] Significantly different from day 7 value, $p < 0.05$.

[b] Significantly different from day 14 value, $p < 0.05$.

[c] The unsaturation index is computed as the sum of the percent weight multiplied by the number of olefinic bonds for each unsaturated fatty acid in the mixture.

matrix. While data in the area are not without controversy, there are some emerging trends which merit comment. There is an increase in SPM cholesterol content with aging.[179-181] In one report[179] the magnitude of this change was quite large (0.22 to 0.34 μmol/mg protein from 2 to 24 months). Unfortunately, it was not reported whether the rats used in this study, a local breed, were of determinate or indeterminate size (some rat strains continue to grow throughout life). Other groups have reported more moderate changes[180,181] in SPM cholesterol. In lieu of other compensatory factors, an increase in membrane cholesterol should increase membrane order, and indeed, these workers[179-181] have reported an increase in DPH polarization or anisotropy (r). Nagy et al.[179] found that the changes in r were small, but due to the large Ns used (>50 per group), were highly significant. Contrasting these results, Wood and Armbrecht[182] could find no age-related difference in SPM lipid order of C57BL/NNIA male mice using the ESR probe 5-doxyl stearic acid.

In one study[180] the increase in membrane lipid order was associated with an increase in serotonin receptor (5-HT) binding. This change was reversed by feeding the rats a special diet containing an "activated lipid mixture". The diet has been described elsewhere.[183] Sun and Samorajaski[184] examined the effects of ethanol on SPM Na^+, K^+ ATPase prepared from C57BL/10 mice, and human brain at autopsy. Ethanol was more effective to inhibit Na^+, K^+ ATPase in old mice (or old humans) as compared to younger mice. These authors concluded that the difference in sensitivity was the result of differences on membrane cholesterol and/or fatty acid composition.

Lipid peroxidation has been suggested to be involved in the aging process.[185] It is well known that peroxidation of membrane lipids leads to marked changes in the structural and functional properties of membranes.[186] Decreased fluidity (membrane order) has been reported in both natural and artificial membranes using the fluorescent probes pyrene, perylene, 4-dimethylchalcone, and DPH.[186-191] Using a series of ESR probes, Bruck and Thayer[192] found the effect of peroxidation on the range of probe motions to be greatest for 12-doxyl stearic acid. Seven (7-) and 10-doxyl stearic acid were less affected and 5- and 16-doxyl stearic were essentially unaffected. Peroxidation significantly, but somewhat differently, decreased the rate of motion for all probes with the 10- and 12-doxyl probes showing the greatest effect. Overall, these data suggest that peroxidation may lead to an alteration of the transbilayer fluidity gradient. One interesting aspect of the peroxidation hypothesis in aging is that one can artificially introduce peroxidation and, thus, for some studies at least, it is not necessary to use aged animals. A caveat to the lipid peroxidation hypothesis, as pointed out by Rothstein,[193] is that membrane components undergo continual turnover. For this reason, it seems unlikely that peroxidation could affect membrane order or fluidity until it reaches a very advanced stage, far outstripping the biosynthetic capacity. On the other hand, the peroxidized lipids might react with proteins to form lipofuscin pigments which have been associated with various disease states.[194]

VIII. CONCLUSION

In this chapter we have attempted to review various aspects of SPM lipid composition and structure. The data presented clearly portray the dynamic qualities of the SPM phospholipids in regulating synaptic activities, especially during development and also illustrate the importance of biophysical studies in understanding synaptic phenomena. Phospholipids not only participate passively in synaptic activities, but also directly participate in some functions as indicated by the putative role(s) for the phosphoinositides.

ACKNOWLEDGMENTS

The authors wish to acknowledge the editorial and typing assistance of Ruth Rupley. This work was supported by PHS grants MH-37377 (RJH), AA-06399 (RAH), DA-00564 and 01583 (HHL), Career Research Scientist Award K2-DA-70554 (HHL), and funds from the Veterans Administration (RAH).

REFERENCES

1. **Cotman, C. W.,** Isolation of synaptosomal and synaptic plasma membrane fractions, in *Methods in Enzymology*, Vol. 31, Fleischer, S. and Packer, L., Eds., Academic Press, New York, 1974, 445.
2. **Gurd, J. W., Jones, L. R., Mahler, H. R., and Moore, W. J.,** Isolation and partial characterization of rat brain synaptic plasma membranes, *J. Neurochem.*, 22, 281, 1974.
3. **Morgan, I. G., Wolfe, L. S., Mandell, P., and Gombos, G.,** Isolation of plasma membranes from rat brain, *Biochim. Biophys. Acta*, 241, 737, 1971.
4. **Davis, G. and Bloom, F. E.,** Isolation of synaptic junctional complexes from rat brain, *Brain Res.*, 62, 135, 1973.
5. **Cossins, A. R.,** Adaptation of biological membranes to temperature: the effect of temperature acclimation of goldfish upon the viscosity of synaptosomal membranes, *Biochim. Biophys. Acta*, 470, 395, 1977.
6. **Harris, R. A. and Schroeder, F.,** Effects of barbiturates and ethanol on the physical properties of brain membranes, *J. Pharmacol. Exp. Ther.*, 223, 424, 1982.
7. **Smith, A. P. and Loh, H. H.,** The subcellular localization of stereospecific opiate binding in mouse brain, *Res. Commun. Chem. Pathol. Pharmacol.*, 15, 205, 1976.
8. **Skrivanek, J. A., Ledeen, R. W., Margolis, R. U., and Margolis, R. K.,** Gangliosides associated with microsomal subfractions of brain: comparison with synaptic plasma membranes, *J. Neurobiol.*, 13, 95, 1982.
9. **Cotman, C. W. and Matthews, D. A.,** Synaptic plasma membranes from rat brain synaptosomes: isolation and partial characterization, *Biochim. Biophys. Acta*, 249, 380, 1971.
10. **Morgan, I. G., Wolfe, L. S., Marinari, U., Breckenridge, W. C., and Gombos, G.,** The isolation and characterization of synaptosomal plasma membranes, *Adv. Exp. Med. Biol.*, 25, 209, 1972.
11. **Gray, E. G.,** Electron microscopy of excitatory and inhibitory synapses: a brief review, *Prog. Brain Res.*, 31, 141, 1969.
12. **Hitzemann, R. J. and Johnson, D. A.,** Developmental changes in synaptic membrane lipid composition and fluidity, *Neurochem. Res.*, 8, 121, 1983.
13. **Hitzemann, R. J., Harris, R. A., and Mitchell, M. A.,** Lipid composition of brain membranes from ethanol tolerant-dependent mice, *Soc. Neurosci.*, 9, 1234, 1983.
14. **Dodge, J. T. and Phillips, G. B.,** Composition of phospholipids and of phospholipid fatty acids and aldehydes in human red cells, *J. Lipid Res.*, 8, 667, 1967.
15. **Marcus, A. J., Ullman, H. L., and Safrer, L. B.,** Lipid composition of subcellular particles of human blood platelet, *J. Lipid Res.*, 10, 108, 1969.
16. **Pfleger, R. C., Anderson, N. G., and Snyder, F.,** Lipid class and fatty acid composition of rat liver plasma membranes isolated by zonal centrifugation, *Biochemistry*, 7, 2826, 1968.
17. **Harris, R. A., Groh, G. I., Baxter, D. M., and Hitzemann, R. J.,** Gangliosides enhance the membrane actions of ethanol and pentobarbital, *Mol. Pharmacol.*, 25, 410, 1984.
18. **Bertoli, E., Masserini, M., Sonniso, S., Ghidoni, R., Cestaro, B., and Tettamanty, G.,** Electron paramagnetic resonance studies on the fluidity and surface dynamics of egg phosphatidylcholine vesicles containing gangliosides, *Biochim. Biophys. Acta*, 467, 196, 1981.
19. **Felgner, P. L., Freire, E., Barenholz, Y., and Thompson, T. E.,** Asymmetric incorporation of trisialoganglioside into dipalmitoylphosphatidylcholine vesicles, *Biochemistry*, 20, 2168, 1981.
20. **Steck, T. L. and Dawson, G.,** Topographical distribution of complex carbohydrates in the erythrocyte membrane, *J. Biol. Chem.*, 249, 2135, 1974.
21. **Stoffel, W., Anderson, R., and Stahl, J.,** Studies on the asymmetric arrangement of membrane lipid enveloped virions as a model system, *Hoppe-Seyler's Z. Physiol. Chem.*, 356, 1123, 1975.

22. **Ansell, G. B. and Spanner, S.,** Functional metabolism of brain phospholipids, *Adv. Neurobiol.*, 9, 1, 1977.
23. **Fontaine, R. N., Harris, R. A., and Schroeder, F.,** Aminophospholipid asymmetry in murine synaptosomal plasma membranes, *J. Neurochem.*, 34, 269, 1980.
24. **Smith, A. R. and Loh, H. H.,** The topographical distribution of phosphatidylethanolamine and phosphatidylserine in synaptosomal plasma membrane, *Proc. West. Pharmacol.*, 19, 147, 1976.
25. **De Medio, G. E., Trovarelli, G., Hamberger, A., and Porcellate, G.,** Synaptosomal phospholipid pool in rabbit brain and its effect on GABA uptake, *Neurochem. Res.*, 5, 171, 1980.
26. **Crews, F. F., Hirata, F., and Axelrod, J.,** Indentification and properties of two methyltransferases that synthesize phosphatidylcholine in rat brain synaptosomes, *Fed. Proc. Fed. Am. Soc. Exp. Biol.*, 8, 1517, 1979.
27. **Fontaine, R. N. and Schroeder F.,** Plasma membrane aminophospholipid distribution in transformed murine fibroblasts, *Biochim. Biophys. Acta*, 558, 1, 1979.
28. **Giati, A., De Medico, G. E., Brunetti, M., Amaducci, L., and Porcellati, G.,** Base-exchange enzymatic system for the synthesis of phospholipids in neuronal and glial cells and their subfractions: a possible marker for neuronal membranes, *J. Neurochem.*, 20, 1167, 1973.
29. **Natsuki, R., Hitzemann, R., and Loh, H.,** Effects of morphine on the incorporation of [^{14}C] serine into phospholipid via the base-exchange reaction, *Mol. Pharmacol.*, 14, 448, 1978.
30. **Michaelson, D. M., Barkai, G., and Barenholz, Y.,** Asymmetry of lipid organization in cholinergic synaptic vesicle membranes, *Biochim. J.*, 211, 155, 1983.
31. **Deutsch, J. W. and Kelley, R. B.,** Lipids of synaptic vesicles: relevance to the mechanism of membrane fusion, *Biochemistry*, 20, 378, 1981.
32. **Hitzemann, R. J.,** Developmental changes in the fatty acids of synaptic membrane phospholipids: effect of protein malnutrition, *Neurochem. Res.*, 6, 935, 1981.
33. **O'Brien, J. F. and Geison, R. L.,** The mass distribution of the phosphatidylcholine in subcellular fractions of rat brain, *J. Neurochem.*, 18, 1615, 1971.
34. **Crawford, C. G. and Wells, M. A.,** Fatty acid and molecular species composition of rat brain phosphatidylcholine and ethanolamine from birth to weaning, *Lipids*, 14, 757, 1979.
35. **Israelachvili, J. N.,** The packing of lipids and proteins in membranes, in *Light Transducing Membranes: Structure, Function and Evolution*, Deamer, D. W., Ed., Academic Press, New York, 1978, 91.
36. **Robertson, R. N.,** *The Lively Membranes*, Cambridge University Press, Cambridge, 1984.
37. **Singer, S. J. and Nicolson, G. C.,** The fluid mosaic model of the structure of cell membranes, *Science*, 175, 720, 1972.
38. **MacDonald, R.,** Action of detergents on membranes: differences between lipid extracted from red cell ghosts and from red cell lipid vesicles by Triton X-100, *Biochimie*, 19, 1916, 1980.
39. **Ledeen, R. W., Skrivanek, J. A., Tirri, L. J., Margolis, R. R., and Margolis, R. U.,** Gangliosides of the neuron: localization and origin, in *Ganglioside Function: Biochemical and Pharmacological Implications*, Porrellati, G., Ceccarelli, B., and Tettamante, G., Eds., Plenum Press, New York, 1973, 83.
40. **Hitzemann, R. J., Mark, C., and Panini, A.,** Effect of fatty acids, ethanol and development on a-aminobutyric acid and glutamate fluxes in rate nerve endings, *Biochem. Pharmacol.*, 31, 4039, 1982.
41. **Klausner, R. D., Kleinfeld, A. M., Hoover, R. L., and Karzowsky, M. J.,** Lipid domains in membranes. Evidence derived from structural perturbations induced by free fatty acids and lifetime heterogeneity analysis, *J. Biol. Chem.*, 255, 1286, 1980.
42. **Hitzemann, R., Mark, C., and Panini, A.,** Gaba and glutamate fluxes in developing nerve endings, *Neurochem. Int.*, 6(1), 133, 1984.
43. **Cooper, J. R., Bloom, F. E., and Roth, R. H.,** *The Biochemical Basis of Neuropharmacology* 3rd ed., Oxford University Press, New York, 1978, 87.
44. **Thompson, R. H. S.,** Lipolytic enzymes and demyelination, in *Metabolism and Physiological Significance of Lipids*, Dawson, R. M. C. and Rhodes, D. N., Eds., Wiley & Son, London, 1964, 541.
45. **Cotman, C., Blank, M. L., Moehl, A. et al.,** Lipid composition of synaptic plasma membranes isolated from rat brain by zonal centrifugation, *Biochemistry*, 8, 4606, 1969.
46. **Poole, A. R., Howell, J. I., and Lucy, J. A.,** Lysolecithin and cell fusion, *Nature (London)*, 227, 810, 1970.
47. **Blaschko, H., Firemark, H., Smith, A. D. et al.,** Lipids of the adrenal medulla: lysolecithin, a characteristic constituent of chromaffin granules, *Biochem. J.*, 104, 545, 1967.
48. **Breckenridge, W. C., Morgan, I. G., Zanetta, J. P. et al.,** Adult rat brain synaptic vesicles. II. Lipid composition, *Biochim. Biophys. Acta*, 3, 357, 1973.
49. **Lagercrantz, H.,** Lipids of the sympathetic nerve trunk vesicles, comparison with adrenomedullary vesicles, *Acta Physiol. Scand.*, 82, 567, 1971.
50. **De Prada, M., Pletscher, A., and Tranzer, J. P.,** Lipid composition of membranes of amine-storage granules, *Biochem. J.*, 127, 681, 1972.

51. **Ahkong, Q. F., Fisher, D., Tampin, W. et al.,** The fusion of erythrocytes by fatty acids, esters, retinol and α-tocopherol, *Biochem. J.*, 136, 147, 1973.
52. **Baker, R. R., Dowdall, M. J., and Whittaker, V. P.,** The involvement of lysophosphoglycerides in neurotransmitter release; the compositional and turnover of phospholipids of synaptic vesicles of guinea pig cerebral cortex and torpedo electric organ and the effect of stimulation, *Brain Res.*, 100, 629, 1975.
53. **Hitzemann, R. and Loh, H.,** The transport and turnover of phospholipids in the rat nigrostriatal system: effects of d-amphetamine and haloperidol, *Res. Commun. Chem. Pathol. Pharmacol.*, 35(2), 209, 1982.
54. **Ashley, R. H. and Brammer, M. J.,** A fluorescence polarization study of calcium and phase behavior in synaptosomal lipids, *Biochim. Biophys. Acta*, 769, 363, 1984.
55. **McElhaney, R. N.,** The structure and function of the *Acholeplasm laidlawii* plasma membrane, *Biochim. Biophys. Acta*, 779, 1, 1984.
56. **Cornell, B. A., Heller, R. G., Raison, J., Separovic, F., Smith, R., Vary, J. C., and Morris, C.,** Biological membranes are rich in low frequency motion, *Biochim. Biophys. Acta*, 732, 473, 1983.
57. **Stockton, G. W., Johnson, K. G., Butler, K. W., Tullock, A. P., Boulanger, Y., Smith, I. C. P., Davis, J. H., and Bloom, M.,** Deuterium NMR study of lipid organization on *Acholeplasma laidlawii* membranes, *Nature (London)*, 269, 267, 1977.
58. **Stockton, G. W., Johnson, K. G., Butler, K. W., Panaszek, C. F., Carr, R., and Smith, I. C. P.,** Molecular order in *Acholeplasma laidlawii* membranes as determined by deuterium magnetic resonance of brosynthetically incorporated specifically labelled lipids, *Biochim. Biophys Acta*, 401, 535, 1975.
59. **McElhaney, R. N.,** The effect of membrane-lipid phase transitions on membrane structure and on the growth of *Acholeplasma laidlawii* B, *J. Supramol. Struct.*, 2, 617, 1974.
60. **McElhaney, R. N.,** The effect of alterations in the physical state of the membrane lipids on the ability of *Acholeplasma laidlawii* B to grow at various temperatures, *J. Mol. Biol.*, 84, 145, 1974.
61. **Smith, I. C. P., Butler, K. W., Tullock, A. P., Davis, J. H., and Bloom, M.,** The properties of gel state lipid in membranes of *Acholeplasma laidlawii* as observed by ^{3}H NMR, *FEBS Lett.*, 100, 57, 1979.
62. **Jarrell, H. C., Butler, K. W., Byrd, R. A., Deslauriers, R., Ekiel, I., and Smith, I. C. P.,** A ^{2}H-NMR study of *Acholeplasma laidlawii* membranes highly enriched in myristic acid, *Biochim. Biophys. Acta*, 688, 622, 1982.
63. **Jost, P. C., Griffith, O. H., Capaldi, R. A., and Vanderkooi, G.,** Evidence for boundary lipid in membranes, *Proc. Natl. Acad. Sci. U.S.A.*, 70, 480, 1973.
64. **Oldfield, E., James, N., Kensey, R., Kintanar, A., Lee, R. W., Rothgeb, T. M., Schramm, S., Skayune, R., Smith, R., and Tsai, M. D.,** Protein crystals, membrane proteins and membrane lipids. Recent advances in the study of their static and dynamic structures using nuclear magnetic resonance spectroscopic techniques, *Biochem. Soc. Symp.*, 46, 155, 1981.
65. **Thayer, A. M. and Kohler, S. J.,** Phosphorus-31 nuclear magnetic resonance spectra characteristic of hexagonol and isotropic phases generated from phosphatidylethanolamine in the bilayer phase, *Biochemistry*, 20, 6831, 1981.
66. **Nakar, R., Schmid, S. L., Hope, M. J., and Cullis, R. R.,** Structural preferences of phosphatidylinositol and phosphatidylinositol-phosphatidylethanolamine model membranes, *Biochim. Biophys. Acta*, 688, 169, 1982.
67. **Cullis, P. R. and DeKruijff, B.,** Lipid polymorphism and the functional roles of lipids in biological membranes, *Biochim. Biophys. Acta*, 559, 399, 1979.
68. **Brown, M. F., Ribeiro, A. A., and Williams, G. D.,** New view of lipid bilayer dynamics from ^{2}H and ^{13}C NMR relation time measurements, *Proc. Natl. Acad. Sci., U.S.A.*, 80, 4325, 1983.
69. **Seelig, J., Tamm, L., Hymol, L., and Fleischer, S.,** Deuterium and phosphorus nuclear magnetic resonance and fluorescence depolarization studies of functional reconstituted sarcoplasmic reticulum membrane vesicles, *Biochemistry*, 20, 3922, 1981.
70. **Michaelis, E. K., Chang, H. H., Roy, S., McFaul, J. A. and Zembreck, J. D.,** Ethanol effects on synaptic glutamate receptor function and on membrane lipid organization, *Pharmacol. Biochim. Behav.*, 18, 1, 1983.
71. **Lyon, R. C. and Goldstein, D. B.,** Changes in synaptic membrane order associated with chronic ethanol treatment in mice, *Mol. Pharmacol.*, 28, 86, 1983.
72. **Chin, J. H. and Goldstein, D. B.,** Effects of low concentrations of ethanol on the fluidity of spin-labelled erythrocyte and brain membranes, *Mol. Pharmacol.*, 13, 435, 1977.
73. **Lenaz, G., Curatoloa, G., Mazzanti, L., Bertoli, L., and Pastusko, A.,** Spin label studies on the effect of anesthetic in synaptic membranes, *J. Neurochem.*, 32, 1689, 1979.
74. **Schrier, S., Polnaszek, C. F., and Smith, I. C. P.,** Spin labels in membranes. Problems in practice, *Biochim. Biophys. Acta*, 515, 375, 1978.
75. **Taylor, M. G. and Smith, I. C. P.,** The conformations of nitroxide-labelled fatty acid probes of membrane structure as studied by ^{2}H-NMR, *Biochim. Biophys. Acta*, 733, 256, 1978.

76. **Wood, W. G. and Armbrecht, H. J.,** Effect of ethanol or fluidity of brain microsomal membranes isolated from young and old mice, *Alcohol Clin. Exp. Res.*, 5, 172, 1981.
77. **Weber, G.,** Resolution of the fluorescence lifetimes in a heterogeneous system by phase and modulation measurements, *J. Phys. Chem.*, 85, 749, 1981.
78. **Barkley, M. D., Kowalezyk, A. A., and Brand, L.,** Fluorescence decay studies of anisotropic motions of small molecules, *J. Chem. Phys.*, 75, 3581, 1978.
79. **Shinitsky, M. and Barenholz, Y.,** Dynamics of the hydrocarbon layer in liposomes of lecithin and sphingomyelin containing dicetylphosphate, *J. Biol. Chem.*, 249, 2652, 1974.
80. **Lackowicz, J. R.,** *Principles of Fluorescence Spectroscopy*, Plenum Press, New York, 1983.
81. **Perrin, F.,** Polarization of light of fluorescence, average life of molecules in the excited state, *J. Phys. Radium*, 7, 390, 1926.
82. **Hitzemann, R. J. and Harris, R. A.,** Developmental changes in synaptic membrane fluidity: a comparison of DPH and TMA-DPH, *Dev. Brain Res.*, 14, 113, 1984.
83. **van Blitterswijk, W. J., Van Hoeven, R. P., and van der Meer, B. W.,** Lipid structural order parameters (reciprocal of fluidity) in biomembranes derived from steady-state fluorescence polarization measurements, *Biochim. Biphys. Acta*, 644, 323, 1981.
84. **Pottel, H., van der Meer, W., and Herreman, W.,** Correlation between the order parameter and the steady-state fluorescence anisotropy of 1,6-diphenyl-1,3,5-hexatriene and an evaluation of membrane fluidity, *Biochim. Biophys. Acta*, 730, 181, 1983.
85. **Kinosita, K., Kawato, S., and Ikegami, A.,** A theory of fluorescence decay in membranes, *Biophys. J.*, 20, 829, 1977.
86. **Dale, R. E., Chen, L. A., and Brand, L.,** Rotational relaxation of the "microvisity" probe diphenylhexatriene in parathin oil and egg lechithin vesicles, *J. Biol. Chem.*, 252, 7500, 1977.
87. **Jahnig, F.,** Structural order of lipids and proteins in membranes: evaluation of fluorescence anisotropy data, *Proc. Natl. Acad. Sci. U.S.A.*, 76, 6361, 1979.
88. **Lipari, G. and Szato, A.,** Effect of liberational motion on fluorescence depolarization and nuclear magnetic resonance relaxation in macromolecules and membranes, *Biophys. J.*, 30, 489, 1980.
89. **Hare, F.,** Simplified derivation of angular order and dynamics of rodlike fluorophores in models and membranes, *Biophys. J.*, 42, 205, 1983.
90. **Heyn, M. P.,** Determination of lipid order parameters and rotational correlation times from fluorescence depolarization experiments, *FEBS Lett.*, 108, 359, 1979.
91. **Prendergast, F. G., Gaugland, R. P., and Callahan, P. J.,** 1-[4-(trimethyl-amino)phenyl]-6-phenylhexa-1,3,5-triene: synthesis, fluorescence properties, and use as fluorescence probe of lipid bilayers, *Biochemistry*, 20, 7333, 1981.
92. **Radda, G. K.,** Fluorescent probes in membrane studies in *Methods in Membrane Biology, Vol. 4*, Korns, E. D., Plenum Press, New York, 1979, 97.
93. **Schroeder, F.,** Fluorescence probes as monitors of surface membrane fluidity gradients in murine fibroblasts, *Eur. J. Biochem.*, 112, 293, 1980.
94. **Harris, W. E. and Stahl, W. L.,** Incorporation of cis-parinaric acid, a fluorescent fatty acid, into synaptosomal phospholipids by an acyl-CoA acyltransferase, *Biochim. Biophys. Acta*, 736, 79, 1983.
95. **Overton, E.,** Studien uber die Narkose zugleich un beitag zur allgemeinen pharmakolologic, *Jena, Gustav Fischer*, p. 101, 1901.
96. **Meyer, K. H. and Hemmi H.,** Beitrage zur theorie der Narkose. III, *Biochem. Z.*, 277, 39, 1935.
97. **Goldstein, A., Aronow, L., and Kolman, S.,** *Principles of Drug Action*, Harper & Row, New York, 1968, 154.
98. **Richards, C. D., Martin, K., Gregory, S., Keightley, C. A., Hesketh, T. R., Smith, G. A., Warren, G. B., and Metcalfe, J. C.,** Degenerate perturbations of protein structure as the mechanism of anesthetic action, *Nature (London)*, 276, 775, 1978.
99. **Jain, M. K. and Wu, N. M.,** Effect of small molecules on the dipalmitoyl liposomal bilayer. III. Phase transition in lipid bilayer, *J. Membrane Biol.*, 34, 157, 1977.
100. **Kirshnan, K. S. and Brandts, J. F.,** Interaction of phenothiazines and lower alipatic alcohols with erythrocyte membranes. A scanning calorimetric study, *Mol. Pharmacol.*, 16, 181, 1979.
101. **Hill, M. W.,** The effect of anesthetic-like molecules on the phase transition in smectic mesophases of dipalmitoyllecithin I. The normal alcohol up to C-9 and three inhalation anesthetics, *Biochim. Biophys. Acta*, 356, 117, 1974.
102. **Rowe, E. S.,** Effect of ethanol on the phase transition temperature of dimyristoylphosphatidylcholine. *Alcoholism, Clin. Exp. Res.*, 4, 227, 1980.
103. **Pringle, M. J. and Miller, K. W.,** Structural isomers of tetradecenol disemminate between the lipid fluidity and phase transition theories of anesthesia, *Biochem. Biophys. Res. Commun.*, 85, 1192, 1978.
104. **Cullis, P. R., Hornby, A. P., and Hope, M. J.,** Effects of anesthetics on lipid polymorphism, in *Molecular Mechanisms of Anesthesia*, Fink, B. R., Ed., Raven Press, New York, 1980, 397.

105. **Harris, R. A. and Hood, W. F.,** Inhibition of synaptosomal clacium uptake by ethanol, *J. Pharmacol. Exp. Ther.*, 213, 562, 1980.
106. **Harris, R. A., Schmidt, J., Hitzemann, B. A., Hitzemann, R. J.,** Phosphatidate as a molecular link between depolarization and neurotransmitter release in brain, *Science*, 1981.
107. **Goldstein, D. B., Chin, J. H., and Lyon, R. C.,** Ethanol disordering of spin-labelled mouse brain membranes: correlation with genetically determined ethanol sensitivity of mice, *Proc. Natl. Acad. Sci. U.S.A.*, 79, 4231, 1982.
108. **McClearn, G. E. and Kakihana, R,** Selective breeding for ethanol sensitivity: short-sleep and long-sleep mice, in *Development of Animal Models as Pharmacogenetic Tools*, Res. Monogr. No. 6, McClearn, G. E., Deitrich, R. A., and Erwin, V. G., Eds., National Institute on Alcohol Abuse and Alcoholism, Rockville, MD: 1981, 147.
109. **Harris, R. A.,** Unpublished observations.
110. **Stubbs, C. D. and Smith, A. D.,** The modification of mammalian membrane polyunsaturated fatty acid composition in relation to membrane fluidity and function, *Biochim. Biophys. Acta*, 779, 89, 1984.
111. **Hoffmanh, W., Pink, D. A., Restall, C., and Chapman, D.,** Intrinsic molecules in fluid phospholipid bilayers, *Eur. J. Biochem.*, 114, 585, 1981.
112. **Jahnig, F., Vogel, H., and Best, L.,** Unifying description of the effect of membrane proteins on lipid order. Verification for the nelettin/myristoylphsphatidylcholine system, *Biochemistry*, 21, 6790, 1982.
113. **Harris, R. A. and Schroeder, F.,** Ethanol and the physical properties of brain membranes: fluorescence studies, *Mol. Pharmacol.*, 20, 128, 1981.
114. **Harris, R. A., Mitchell, M. A., and Hitzemann, R. J.,** Physical properties and lipid composition of brain membranes from ethanol tolerant-dependent mice, *Mol. Pharmacol.*, 25, 401, 1984.
115. **Chin, J. H., Parsons, L. M., and Goldstein, D. B.,** Increased cholesterol content of erythrocyte and brain membranes in ethanol tolerant mice, *Biochim. Biophys. Acta*, 513, 358, 1978.
116. **Johnson, D. A. Lee, N. M., Cooke, R., and Loh, H. H.,** Ethanol-induced fluidization of the brain lipid bilayers: required presence of cholesterol in membranes for the expression of tolerance, *Mol. Pharmacol.*, 15, 739, 1979.
117. **Koblin, D. D. and Wang, H. H.,** Chronic exposure to inhaled anesthetics increases cholesterol content in *Acholeplasma laidlawii*, *Biochim. Biophys. Acta*, 649, 717, 1981.
118. **Smith, T. L. and Gerhart, M. J.,** Alterations in brain lipid composition of mice made physically dependent to ethanol, *Life Sci.*, 31, 1419, 1982.
119. **Wing, D. R., Harvey, D. J., Hughes, J., Dunbar, P. G., McPherson, K. A., and Paton, W. D. M.,** Effects of chronic ethanol administration on the composition of membrane lipids in the mouse, *Biochem. Pharmacol.*, 31, 3431, 1982.
120. **Littleton, J. M. and John, G.,** Synaptosomal membrane lipid of mice during continuous exposure to ethanol, *J. Pharm. Pharmacol.*, 29, 579, 1977.
121. **Alling, C., Liljequist, S., and Engel, J.,** The effect of chronic ethanol administration on lipids and fatty acids in subcellular fractions of rat brain, *Med. Biol.*, 60, 149, 1982.
122. **Sun, G. Y. and Sun, A. Y.,** Effects of chronic ethanol administration on phospholipid acyl groups of synaptic plasma membrane fraction isolated from guinea pig brain, *Res. Commun. Chem. Pathol. Pharmacol.*, 24, 405, 1979.
123. **Streb, H., Irvine, R. F., Berridge, M. J., and Schulz, I.,** Inositol-1,4-5-trisphosphate releases Ca^{2+} from a non-mitochondrial intracellular Ca^{2+} store in pancreatic acinar cells, *Nature (London)*, 306, 67, 1983.
124. **Nishizuka, Y.,** Phospholipid degradation and signal translation for protein phosphorylation, *Trends Biochem. Sci.*, 8, 13, 1983.
125. **Castagna, M., Takai, Y., Kaibuchi, K., Sano, K., Kikkawa, U., and Nishizuka, Y.,** Direct activation of Ca^{++} activated, phospholipid dependent protein kinase by tumor-promoting phorbol esters, *J. Biol. Chem.*, 257, 7847, 1982.
126. **Thomas, A. P. and Marks, J. S., Coll, K. E., and Williamson, J. R.,** Quantitation and early kinetics of inositol lipid changes induced by vasopressin in isolated and cultured hepatocytes, *J. Biol. Chem.*, 258, 5716, 1983.
127. **Hitzemann, R. J., Natsuki, R., and Loh, H. H.,** Effects of nicotine on brain 1-phosphatidylinositol-4-phosphate and 1-phosphatidylinositol-3,4-biphosphate synthesis and metabolism — possible relationship to nicotine induced behaviors, *Biochem. Pharmacol.*, 27, 2519, 1978.
128. **Wu, Y. C., Cho, T. M., and Loh, H. H.,** Binding of dimethyltubocurarine to triphosphoinositide, *J. Neurochem.*, 29, 598, 1977.
129. **Yamamoto, H. A., Harris, R. A., Loh, H. H., and Way, E. L.,** Calcium binding to brain membranes: effects of morphine tolerance, *Proc. West. Pharmacol. Soc.*, 19, 71, 1976.
130. **Hendrickson, H. S. and Reinersten, J. L.,** Comparison of metal binding properties of trans-1, 2-cyclohexanediol diphosphate and deacylated phosphoinositides, *Biochemistry*, 8, 4855, 1969.

131. **Hendrickson, H. S. and Reinersten, J. L.,** Phosphoinositide interconversion: a model for control of Na^+ and K^+ permeability in the nerve axon membrane, *Biochem. Biophys. Res. Commun.*, 44, 1258, 1971.
132. **Torda, C.,** *A Depolarization-Hyperpolarization Cycle, A Molecular Model,* Torda, New York, 1972.
133. **Cho, T. M., Cho, J. S., and Loh, H. H.,** Alteration of the physiochemical properties of triphosphoinositide by nocotinic ligands, *Proc. Natl. Acad. Sci. U.S.A.*, 75, 784, 1978.
134. **De Robertis, E., Lunt, G. S., and La Torre, J. L.,** Multiple binding sites for acetylcholine in a proteolipid from electric tissue, *Mol. Pharmacol.*, 7, 97, 1971.
135. **Loh, H. H., Cho, T. M., Wu, Y. C., Way, E. L.,** Stereospecific binding of narcotics to brain cerebrosides, *Life Sci.*, 14, 2231, 1974.
136. **Loh, H. H., Cho, T. M., Wu, Y. C., Harris, R. A., and Way, E. L.,** Opiate binding to cerebroside sulfate: a model system for opiate-receptor interaction, *Life Sci.*, 16, 1811, 1975.
137. **Cho, T. M., Cho, J. S., and Loh, H. H.,** ^3H-cerebroside sulfate redistribution induced by cations, opiate or phosphatidylserine, *Life Sci.*, 19, 117, 1976.
138. **Cho, T. M., Cho, J. S., and Loh, H. H.,** A model system for opiate-receptor interaction: mechanism of opiate-cerebroside sulfate interaction, *Life Sci.*, 18, 231, 1978.
139. **Law, P. Y., Harris, R. A., Loh, H. H., and Way, E. L.,** Evidence for the involvement of cerebroside sulfate in opiate receptor binding. Studies with Azure A and Jimpy mutant mice, *J. Pharmacol. Exp. Ther.*, 207, 458, 1978.
140. **Law, P. Y., Fischer, G., Loh, H. H., and Herz, A.,** Inhibition of specific opiate binding to synaptic membrane by cerebroside sulfatase, *Biochem. Pharmacol.*, 28, 2557, 1979.
141. **Craves, F., Leybin, L., Loh, H. H., Zalc, B., and Baumann, N.,** Cerebroside sulfate antibodies inhibit the effects of morphine and B-endorphin, *Science*, 207, 75, 1980.
142. **Traber, J., Fischer, K., Latzin, S., and Hamprecht, B.,** Morphine antagonises action of prostaglandin in neuroblastoma and glioma hybrid cells, *Nature, (London)* 253, 120, 1975.
143. **Sharma, S. K., Klee, W. A., and Nirenberg, M.,** Opiate-dependent modulation of adenylate cyclase, *Proc. Natl. Acad. Sci. U.S.A.*, 74, 3365, 1977.
144. **Sharma, S. K., Nirenberg, M., and Klee, W. A.,** Morphine receptor as regulator of adenylate cyclase activity, *Proc. Natl. Acad. Sci. U.S.A.*, 72, 590, 1975.
145. **Hamprecht, B.,** Structural electrophysiological, biochemical and pharmacological, biochemical and pharmacological properties of neuroblastoma-glioma cell hybrids in cell culture, *Int. Rev. Cytol.*, 49, 99, 1977.
146. **Law, P. Y., Herz, A., and Loh, H. H.,** Demonstration and characterization of a stereospecific opiate receptor in the neuroblastoma N18TG2 cells, *J. Neurochem.*, 33, 1177, 1979.
147. **Loh, H. H., Law, P. Y., Fischer, G., and Herz, A.,** Possible role of cerebroside sulfate in opiate receptor mechanism, *7th Int. Congr. Pharmacol.*, Paris, 1978, 563.
148. **Nicksic, T. D., Law, P. Y., Herz, A., and Loh, H. H.,** Potentiation of opiate effects in the neuroblastoma cell line N18TG2 by sulfatide incorporation, in *Endogeneous and Exogenous Opiate Agonists and Antagonists,* Way, E. L., Ed., Pergamon Press, New York, 1980, 275.
149. **Ruocco, M. J. and Shipley, G. G.,** Galactocerebroside-phospholipid interactions in bilayer membranes, *Biophys. J.*, 43, 91, 1983.
150. **Hitzemann, R. J. and Loh, H. H.,** Influence of chronic pentobarbital or morphine treatment on the incorporation of ^{32}Pi and [^3H]choline into rat synaptic plasma membranes, *Biochem. Pharmacol.*, 26, 1087, 1977.
151. **Natsuki, R., Hitzemann, R. J., Hitzemann, B. A.,** et al., Effect of pentobarbital on regional brain phospholipid synthesis, *Biochem. Pharmacol.*, 26, 2095, 1977.
152. **Natsuki, R., Hitzemann, R. J., and Loh, H. H.,** Influence of morphine, B-endorphin and naloxone on the synthesis of phosphoinositides in the rat midbrain, *Res. Commun. Chem. Pathol. Pharmacol.*, 24(2), 233, 1979.
153. **Hitzemann, R. J. and Loh, H. H.,** Effect of d-amphetamine on the turnover, synthesis and metabolism of brain phosphatidylcholine, *Biochem. Pharmacol.*, 22, 2731, 1973.
154. **Hitzemann, R. J., Natsuki, R.- Ohizum, Y., and Johnson, D.,** Influence of morphine on membrane turnover and function in *Neurochemical Mechanisms of Opiates and Endorphines,* Loh, H. and Ross, D., Eds., Raven Press, New York, 1978, 495.
155. **Mulé, S. J.,** Effect of morphine and nalorphine on the metabolism of phospholipids in guinea pig cerebral cortex slices, *J. Pharmcol. Exp. Ther.*, 154, 370, 1966.
156. **Mulé, S. J.,** Morphine and the incorporation of ^{32}Pi into brain phospholipids of non-tolerant, tolerant and abstinent guinea pigs, *J. Pharmacol. Exp. Ther.*, 156, 92, 1967.
157. **Mulé, S. J.,** Morphine and the incorporation of ^{32}P-orthophosphate in vivo into phospholipids of the guinea pig cerebral cortex, liver and subcellular fractions, *Biochem. Pharmacol.*, 19, 581, 1970.
158. **Herron, D. S., Shinitzky, M., Zamir, N., and Samuel, D.,** Adaptive modulation of brain membrane lipid fluidity in drug addiction and denervation supersensitivity, *Biochem. Pharmacol.*, 31, 2435, 1982.

159. **Cossins, A. R.,** Steady-state and dynamic fluorescence studies of the adaption of cellular membranes to temperature, in *Fluorescent Probes*, Beddard, G. S. and West, M. A., Eds., Academic Press, London, 1981, 39.
160. **Sinensky, M.,** Homeoviscous adaptation — a homeostatic process that regulates the viscosity of membranes in *Escherechia coli*, *Proc. Natl. Acad. Sci. U.S.A.*, 71, 522, 1974.
161. **Esser, A. F. and Souza, K. A.** Correlation between thermal death and membrane fluidity in *Bacillus stearothermophilus*, *Proc. Natl. Acad. Sci. U.S.A.*, 71, 4111, 1974.
162. **Nowaza, Y., Iida, H., Fukushima, H., Okki, K., and Ohnishi, S.,** Studies on tetrahymena membranes: temperature induced alterations in fatty acid composition of various membrane fractions in tetrahymena pyreformis and its effect on membrane fluidity as inferred by spin label study, *Biochim. Biophys. Acta*, 367, 134, 1974.
163. **Cossins, A. R.,** Changes in muscle lipid composition and resistance adaptation to temperature in the freshwater crayfish, *Austropotamobuis pallipes*, *Lipids*, 11, 306, 1976.
164. **Cossins, A. R. and Prosser, C. L.,** Evolutionary adaptation of membranes to temperature, *Proc. Natl. Acad. Sci. U.S.A.*, 75, 2040, 1978.
165. **Chong, P. L. G., Cossins, A. R., and Weber, G.,** A differential polarized phase fluorometer study of the effects of high hydrostatic pressure upon the fluidity of cellular membranes, *Biochemistry*, 22, 409, 1982.
166. **Inbar, M. and Shinitzky, M.,** Decrease in microviscosity of lymphocyte surface membrane associated with stimulation induced by concanavalin A, *Eur. J. Immunol.*, 5, 166, 1975.
167. **Collard, J. G., de Wildt, A., Oomen-Meulemens, E. P. M., Smeekins, J., Emmelot, P., and Inbar, M.,** Increase in fluidity of membrane lipids in lymphocytes, fibroblasts and liver cells for growth, *FEBS Lett.*, 77, 173, 1977.
168. **Chen, S. and Levy, D.,** The effects of cell proliferations on the lipid composition and fluidity of hepatocyte plasma membranes, *Arch. Biochem. Biophys.*, 196, 424, 1979.
169. **Lai, C.-S., Hopwood, L. E., and Swartz, H. M.,** Electron spin resonance studies of changes in membrane fluidity of chinese hamster ovary cells during the cell cycle, *Biochim. Biophys. Acta*, 602, 117, 1980.
170. **De Latt, S. and van der Saag, P. T.,** The plasma membrane as a regulatory site in growth and differentiation of neuroblastoma cells, *Int. Rev. Cytol.*, 74, 1, 1982.
171. **Nicolau, C., Hildenbrand, K., Reimann, A., Johnson, S. M., Vaheri, A., and Freus, R.,** *Exp. Cell Res.*, 113, 63, 1978.
172. **Johnson, S. M.,** Steady state diphenyl hexatriene fluorescence polarization in the study of cells and cell membranes, in *Fluorescent Probes*, Beddard, G. S. and West, M. A., Eds., Academic Press, London, 1981, 143.
173. **De Laat, S. W., van deer Saag, P. T., Ad Nelemans, S., and Shinitzky, M.,** Microviscosity changes during differentiation of neuroblastoma cells, *Biochim. Biophys. Acta*, 509, 188, 1978.
174. **De Laat, S. W., van der Saag, P. T., Elson, E. L., and Schlessinger, J.,** Lateral diffusion of membrane lipids and proteins is increased specifically in neurites of differentiating neuroblastoma cells, *Biochim. Biophys. Acta*, 558, 247, 1979.
175. **Pfenninger, K. H.,** in *Neuroscience Fourth Study Program*, Schmidt, F. O. and Worden, F. G., Eds., MIT Press, Cambridge, 19, 779, 1968.
176. **Caley, D. W. and Maxwell, D. S.,** Ultrastructure of the developing cerebral cortex in the rat, in *Brain Development and Behavior*, Sterman, M. B., McGinty, D. S., and Adenolfi, A. M., Eds., Academic Press, New York, 1971, 91.
177. **Hitzemann, R. J.,** Developmental regulation of phospholipid methylation in rat brain synaptosomes, *Life Sci.*, 30, 1297, 1982.
178. **Hitzemann, R. J.,** Developmental changes in the fatty acids of synaptic membrane phospholipids: effect of protein malnutrition, *Neurochem. Res.*, 6, 935, 1981.
179. **Nagy, K., Nagy, V., Bertoni-Freddari, C., and Zs.-Nazy, I.,** Alterations of the synaptosomal membrane microviscosity in the brain cortex of rats during aging and cerebrophenoxine treatment, *Arch. Gerentol. Geriatr.*, 2, 23, 1983.
180. **Shinitsky, M., Heron, D. S., and Samuel, D.,** Restoration of membrane fluidity and serotonin receptors in the aged mouse brain, in *Aging of the Brain*, Samuel, D., Algeri, S., Gershon, S., Grimm, V., and Toffano, G., Eds., Raven Press, New York, 1983, 329.
181. **Bonetti, A. C., Battestella, A., Calderini, G., Teolato, S., Crews, F. T., Gatti, A., Algeri, S., and Toffano, G.,** Biochemical alterations in the mechanisms of synaptic transmission in aging brain, in *Aging of the Brain*, Samuel, D., Algeri, S., Gershon, S., Grimm, V., and Toffano, G., Eds., Raven Press, New York, 1983, 171.
182. **Wood, W. G. and Armbrecht, H. J.,** Effect of ethanol on fluidity of brain microsomal membranes isolated from young and old mice, *Alcohol Clin. Exp. Res.*, 5, (Abstr.), 172, 19, 1981.

183. **Herron, D. S., Hershkowitz, M., Shinitzky, M., and Samuel, D.,** In *Neurotransmitters and Their Receptors,* Teichberg, V.I. and Vogel, Z., Eds., Wiley, New York, 1980, 125.
184. **Sun, A. Y., and Samorajski, T.,** The effects of age and alcohol on Na^+ + K^+)-ATPase activity of whole homogenate and synaptosomes prepared from mouse and human brain, *J. Neurochem.,* 24, 161, 1975.
185. **Harmon, D.,** *Proc. Natl. Acad. Sci. U.S.A.,* 78, 7124, 1981.
186. **Vladimitov, Y. A., Olenev, V. I., Suslowa, T. B., and Cheremisina, Z. P.,** Lipid peroxidation in mitochondrial membrane, *Adv. Lipid Res.,* 17, 173, 1980.
187. **Dobretsov, G. E., Borschevskaya, T. A., Petrov, V. A., and Vladimirov, Y. A.,** The increase of phospholipid bilayer rigidity after lipid peroxidation, *FEBS Lett.,* 84, 125, 1977.
188. **Barrow, D. A. and Lentz, B. R.,** A model for the effect of lipid oxidation on diphenylhexatriene fluorescence in phospholipid vesicles, *Biochim. Biophys. Acta,* 645, 17, 1981.
189. **Rice-Evans, C. and Hochstein, P.,** Alterations in erythrocyte membrane fluidity by phenylhydrozine-induced peroxidation of lipids, *Biochem. Biophys. Res. Commun.,* 100, 1537, 1981.
190. **Eichenberger, K., Bohni, P., Winterhalter, K. H., Kawato, S., and Richter, C.,** Microsomal lipid peroxidation causes an increase in the order of the membrane lipid domain, *FEBS Lett.,* 142, 59, 1982.
191. **Shinitzky, M. and Barenholz, Y.,** Fluidity parameters of lipid regions determined by fluorescence polarization, *Biochim. Biophys. Acta,* 515, 367, 1978.
192. **Bruck, R. C. and Thayer, W. S.,** Differential effect of lipid peroxidation on membrane fluidity as determined by electron spin resonance probes, *Biochim. Biophys. Acta,* 733, 216, 1983.
193. **Rothstein, M.,** *Biochemical Approaches to Aging,* Academic Press, New York, 1982.
194. **Zeman, W.,** The neuronal ceroid-lipofuscinoses-Batten-Vogt Syndrome: a model for human aging? *Adv. Gerontol. Res.,* 3, 147, 1971.
195. **Cho, T. M., Go, B. L., Yamato, L., Smith, A. P., and Loh, H. H.,** Isolation of opiate binding components by affinity chromatography and reconstitution of binding activities, *Proc. Natl. Acad. Sci. U.S.A.,* 80, 5176, 1983.
196. **Cho, T. M., Go, B. L., and Loh, H. H.,** Isolation and purification of morphine receptor by affinity chromatography, *Proc. Natl. Acad. Sci., U.S.A.,* in press.
197. **Hitzemann, R. J.,** unpublished observations.

APPENDIX: REFERENCES

In the course of research for this chapter we reviewed a great many excellent references. Unfortunately, it was not possible to include all of these references in the text. For the reader's convenience we have arranged these "extra" references according to topic.

Drug-Membrane Interactions

Akagi, M., Mio, M., and Taska, K., Histamine release inhibition and prevention of the decrease in membrane fluidity induced by certain anti-allergic drugs: analysis of the inhibitory mechanism of NCO-650, *Agents Actions,* 13, 2, 1983.

Browning, J. L. and Akutsu, H., Local anesthetics and divalent cations have the same effect on the headgroups of phosphatidylcholine and phosphatidylethanolamine, *Biochim. Biophys. Acta,* 684, 172, 1982.

Burgess, G. M., Giraud, F., Poggioli, J., and Claret, M., Adrenergically mediated changes in membranes lipid fluidity and Ca^{2+} binding in isolated rat liver plasma membranes, *Biochim. Biophys. Acta,* 731, 387, 1983.

Crews, F. T., Majchrowicz, E., and Meeks, R., Changes in cortical synaptosomal plasma membrane fluidity and composition in ethanol-dependent rats, *Psychopharmacology,* 81, 208, 1983.

Domber, B. M. and Ingram, L. A., Effects of ethanol on the *Escherichia coli* plasma membrane, *J. Bacteriol.,* 157(1), 233, 1984.

Fenn, G. C., Lynch, M. A., Nhamburo, P. T., Caberos, L., and Littleton, J. M., Comparison of effects of ethanol on platelet function and synaptic transmission, *Pharmacol. Biochem. Behav.,* 18, 37, 1983.

Fleuret-Balter, C., Beauge, F., Barin, F., Nordmann, J., and Nordmann, R., Brain membrane disordering by administration of a single ethanol dose, *Pharmacol. Biochem. Behav.,* 18, 25, 1983.

Garda, H. A. and Brenner, R. R., Short-chain aliphatic alcohols increase rat liver microsomal membrane fluidity and affect the activities of some microsomal membrane-bound enzymes, *Biochim. Biophys. Acta,* 769, 160, 1984.

Herbette, L., Katz, A. M., and Sturtevant, J. M., Comparisons of the interaction of propranolol and timolol with model and biological membrane systems, *Mol. Pharmacol.,* 24, 259, 1983.

Hornby, A. P. and Cullis, P. R., Influence of local and neutral anaesthetics on the polymorphic phase preferences of egg yolk phosphatidylethanolamine, *Biochim. Biophys. Acta,* 647, 285, 1981.

Keegan, R., Wilce, P. A., Ruczkal-Pietrzak, E., and Shanley, B. C., Effect of ethanol on cholesterol and phospholipid composition of Hela cells, *Biochem. Biophys. Res. Commun.,* 114(3), 985, 1983.

Littleton, J. M., Fenn, C. G., Umney, N. D., and Yazdanbakhsh, M., Effects of ethanol administration on platelet function in the rat, *Alcoholism Clin. Exp. Res.,* 6(4), 512—519 (1982).

Logan, B. J., Laverty, R., and Peake, B. M., ESR measurements on the effects of ethanol on the lipid and protein conformation in biological membranes, *Pharmacol. Biochem. Behav.,* 18(1), 31, 1983.

Maher, P. and Singer, S. J., Structural changes in membranes produced by the binding of small amphipathic molecules, *Biochemistry,* 23, 232, 1984.

Ondrias, K., Balgavy, P., Stolc, S., and Horvath, L. I., A spin label study of the perturbation effect of tertiary amine anesthetics on brain lipid liposomes and synaptosomes, *Biochim. Biophys. Acta,* 732, 627, 1983.

Rubin, E. and Rottenberg, H., Ethanol and biological membranes: injury and adaptation, *Pharmacol. Biochem. Behav.,* 18(1), 7, 1983.

Sauerheber, R., Esgate, R. A., and Kuhn, C. R. E., Alcohols inhibit adipocyte basal and insulin-stimulated glucose uptake and increase the membrane lipid fluidity, *Biochim. Biophys. Acta,* 691, 115, 1982.

Schwalb, H., Dickstein, Y., and Heller, M., Interactions of cardiac glycosides with cardiac cells. III. Alterations in the sensitivity of (Na^+ + K^+)-ATPase to inhibition by ouabain in rat hearts, *Biochim. Biophys. Acta,* 689, 241, 1982.

Verkleij, A. J., DeMaagd, R., Leunissen-Bijvelt, J., and De Kruijff, B., Divalent cations and chlorpromazine can induce non-bilayer structures in phosphatidic acid containing model membranes, *Biochim. Biophys. Acta,* 684, 255, i982.

Environment and Membranes

Dickens, B. F. and Thompson, G. A., Phospholipid molecular species alterations in microsomal membranes as an initial key step during cellular acclimation to low temperature, *Biochemistry,* 21, 3604, 1982.

Durairaj, G. and Vijayakumar, I., Temperature acclimation and phospholipid phase transition in hypothalamic membrane phospholipids of garden lizard, calotes versicolor, *Biochim. Biophys. Acta,* 770, 7, 1984.

Otto, M. K., Krabichler, G., Thiele, J., Olkrug, D., and Schulzt, J. E., Effect of temperature on membrane fluidity and calcium conductance of the excitable ciliary membrane from paramecium, *Biochim. Biophys. Acta,* 769, 253, 1984.

Ramesha, C. S. and Thompson, G. A., Changes in the lipid composition and physical properties of tetrahymena ciliary membranes following low-temperature acclimation, *Biochemistry,* 21, 3612, 1982.

Ramesha, C. A. and Thompson, G. A., Cold stress induces *in situ* phospholipid molecular species changes in cell surface membranes, *Biochim. Biophys. Acta,* 731, 251, 1983.

Yager, P. and Chang, E. L., Destabilization of a lipid non-bilayer phase by high pressure, *Biochim. Biophys. Acta,* 731, 491, 1983.

Fluorescence Probes

Beechem, J. M., Knutson, J. R., Ross, J. B. A., Turner, B. W., and Brand, L., Global resolution of heterogeneous decay by phase/modulation fluorometry: mixtures and proteins, *Biochemistry,* 22, 6054, 1983.

Kleinfeld, A. M., Dragstein, P., Klausner, R. D., Pjura, W. J., and Matayoshi, E. D., The lack of relationship between fluorescence polarization and lateral diffusion in biological membranes, *Biochim. Biophys. Acta,* 649, 471, 1981.

Welti, R. and Silbert, D. F., Partition of parinaroyl phospholipid probes between solid and fluid phosphatidylcholine phases, *Biochemistry,* 21, 5685, 1982.

Welti, R., Partition of parinaroyl phospholipids in mixed head group systems, *Biochemistry,* 21, 5690, 1982.

General

Kinosita, K. and Ikegami, A., Reevaluation of the wobbling dynamics of diphenylhexatriene in phosphatidylcholine and cholesterol/phosphatidylcholine membranes, *Biochim. Biophys. Acta,* 769, 523, 1984.

Jahnig, F. and Bramhall, J., The origin of a break in arrhenius plots of membrane processes, *Biochim. Biophys. Acta,* 690, 310, 1982.

Madden, T. D. and Cullis, P. R., Stabilization of bilayer structure for unsaturated phosphatidylethanolamines by detergents, *Biochim. Biophys. Acta,* 684, 149, 1982.

Sessions, A. and Horwitz, A. F., Differentiation-related differences in the plasma membrane phospholipid asymmetry of myogenic and fibrogenic cells, *Biochim. Biophys. Acta,* 728, 103, 1983.

Wahle, K. W. J., Fatty acid modification and membrane lipids, *Proc. Nutr. Soc.,* 42, 273, 1983.

Phospholipid Methylation

Koch, T.K., Gordon, A. S., and Diamond, I., Phospholipid methylation in myogenic cells, *Biochem. Biophys. Res. Commun.*, 114(1), 339, 1983.

Smith, J. D., Effect of modification of membrane phospholipid composition on the activity of phosphatidylethanolamine N-methyltransferase of tetrahymena, *J. Biochem. Biophys.*, 223(1), 193, 1983.

Plasma Membranes

Crowe, L. M. and Crowe, J. H., Hydration-dependent hexagonal phase lipid in a biological membrane, *Arch. Biochem. Biophys.*, 217(2), 582, 1982.

Golan, D. E., Alecio, R., Veatch, W. R., and Rando, R. R., Lateral mobility of phospholipid and cholesterol in the human erythrocyte membrane: effects of protein-lipid interactions, *Biochemistry*, 23, 332, 1984.

Grunberger, D., Haimovitz, R., and Shinitzky, M., Resolution of plasma membrane lipid fluidity in intact cells labelled with diphenylhexatriene, *Biochim. Biophys. Acta*, 688, 764, 1982.

Kuhry, J. G., Poindron, P., and Laustriat, G., Evidence for early fluidity changes in the plasma membranes of enterferon treated L cells, from fluorexcence anisotropy data, *Biochem. Biophys. Res. Commun.*, 110(1), 88, 1983.

McKenzie, R. C. and Brophy, P. J., Lipid composition and physical properties of membranes from C-6 glial cells with altered phospholipid polar headgroups, *Biochim. Biophys. Acta*, 769, 357, 1984.

Poon, R. and Clark, W. R., The relationship between plasma membrane lipid composition and physical chemical properties. III. Detailed physical and biochemical analysis of fatty acid-substituted ELA plasma membranes, *Biochim. Biophys. Acta*, 689, 230, 1982.

Steiner, M. and Luscher, E. F., Fluorescence anisotropy changes in platelet membranes during activation, *Biochemistry*, 23, 247, 1984.

Storch, J., Schachter, D., Inoue, M., and Wolkoff, A. W., Lipid fluidity of hepatocyte plasma membrane subfraction and their differential regulation by calcium, *Biochim. Biophys. Acta*, 727, 209, 1983.

Reviews and Theoretical Articles

Axelrod, D., Lateral motion of membrane proteins and biological function, *J. Membrane Biol.*, 75, 1, 1983.

Blanquet, P. R., Regulation of surface membrane enzymes by lipid ordering, *Biochem. J.*, 213, 479, 1983.

Meyer, H. W., Lipid domain structures in biological membranes, *Exp. Pathol.*, 23, 3, 1983.

Reith, E. J., A model for transcellular transport of calcium based on membrane fluidity and movement of calcium carriers within the more fluid microdomains of the plasma membrane, *Calcif. Tissue Int.*, 35, 129, 1983.

Synaptic Plasma Membranes

Chang, H. H. and Michaelis, E. K., Depolarization induced release of L-glutamic acid from isolated resealed synaptic membrane vesicles, *Biochim. Biophys. Acta*, 769, 499, 1984.

Deliconstantinos, G., Phenobarbital modulates the (Na^+, K^+)-stimulated ATPase and Ca^{2+}-stimulated APTase activities by increasing the bilayer fluidity of dog brain synaptosomal plasma membranes, *Neurochem. Res.*, 8(9), 1143, 19.

Foot, M., Cruz, T. F., and Clandinin, M. T., Influence of dietary fat on the lipid composition of rat brain synaptosomal and microsomal membranes, *Biochem. J.*, 208, 631, 1982.

Hershkowitz, M., Zwiers, H., and Gispen, W. H., The effect of ACTH on rat brain synaptic plasma membrane lipid fluidity, *Biochim. Biophys. Acta*, 692, 495, 1982.

Fleischer, P. N. and Fleischer, S., Alteration of synaptic membrane cholesterol/phospholipid ratio using a lipid transfer protein, *J. Biol. Chem.*, 258(2), 1242, 1983.

Sterols and Membranes

Block, K. E., Sterol structure and membrane function, *CRC Crit. Rev. Biochem.*, 14(1), 47, 1982.

Duval, D., Durant, S., and Homo-delarche, F., Interactions of steroid molecules with membrane structures and functions, *Biochim. Biophys. Acta*, 737, 409, 1983.

Fisher, G. J., Freter, C. E., Ladenson, R. C., and Silbert, D. F., Effect of membrane sterol content on the susceptibility of phospholipids to phospholipase A_2, *J. Biol. Chem.*, 258(19), 11705, 1983.

Gallay, J., De Kruijff, B., and Demel, R. A., Sterol-phospholipid interactions in model membranes. Effect of polar group substitutions in the cholesterol side-chain at C_{20} and C_{22}, *Biochim. Biophys. Acta*, 769, 96, 1984.

Lange, Y., Matthies, H., and Steck, T. L., Cholesterol oxidase susceptibility of the red cell membrane, *Biochim. Biophys. Acta*, 769, 551, 1983.

Rando, R. R., Bangerter, F. W., and Alecio, M. R., The synthesis and properties of a functional fluorescent cholesterol analog, *Biochim. Biophys. Acta*, 684, 12, 1982.

Chapter 4

PHOSPHOLIPID METHYLATION AND MEMBRANE FUNCTION

Fulton T. Crews

TABLE OF CONTENTS

I. INTRODUCTION

It is now widely accepted that the membrane has a dynamic role in the functions of cells. One of the most active and productive areas of biological science in the past decade has been the study of the biochemical and biophysical properties of cell membranes. There is little doubt that membranes are essential components of cellular systems and that each type of membrane has specific functions. The plasma membrane is of particular interest because this membrane constitutes the cell boundary, transports a variety of ions into and out of the cell, contains the cell surface receptors which allow cell to cell communication, and performs a variety of other known and probably unrecognized functions which are essential to cell, tissue, and organism function. In addition, the endoplasmic reticulum contains a large number of specialized enzymes, mitochondria have both inner and outer membranes which are essential for oxidative phosphorylation and cell survival, and numerous other cellular organelles also have specialized membranes.

Phospholipids, one of the major components of membranes, not only provide the membrane with its structural integrity and physical properties, but also play an important role in a number of other cell functions. As the function of each tissue varies in an organism so does its lipid composition. Thus, tissues serving similar functions in different species have similar lipid compositions, whereas tissues within the same organism which have different functions have different lipid compositions.[1,2] A similar generalization can be made on the cellular level where different membranes have different lipid compositions.[1-3] For example, the liver plasma membrane has a cholesterol/phospholipid ratio of 0.83 and approximately 35% of the phospholipid is phosphatidylcholine, while the liver endoplasmic reticulum has a cholesterol/phospholipid ratio of 0.09 and about 60% of the phospholipid is phosphatidylcholine. These differences in cellular organelle membrane composition are likely to reflect differences in organelle function since the lipid composition of rat organelle membranes is typical of the lipid composition for similar membranes from not only mammals, but a variety of other organisms.[1-3] This point is clearly supported by cardiolipin, a phospholipid which is almost exclusively found in the inner mitochondrial membranes of a number of tissues. Two likely explanations for this membrane lipid specificity is the absolute requirement of some enzyme for one or more bound molecules of a particular lipid in order to maintain functional activity, and the requirement of a precisely defined physical state, or fluidity, of each membrane determined in a very sensitive way by interactions among the individual lipids and proteins.

In addition to playing a major role in the biophysical properties of membranes and the requirements of specific enzymes, membrane phospholipids also serve as a reservoir for the storage of certain compounds. The prostaglandins, leukotrienes, platelet-activating factor, and a variety of other active lipids and/or their precursors are released from membrane phospholipids. In addition, choline released from phosphatidylcholine is thought to serve as a precursor for acetylcholine, a major neurotransmitter. Thus, phospholipids appear to modulate the fluid and structural properties of membranes and provide a storage form for the release of bioactive lipids and/or their precursors and other molecules which are important for normal cell, tissue, and organism function.

This review focuses on the phospholipid methylation pathway which involves the sequential methylation of phosphatidylethanolamine to phosphatidylcholine. Phospholipid methylation is a minor route for the synthesis of phosphatidylcholine. For example, in the liver, which has a very active transmethylation pathway, only about 20% of the phosphatidylcholine is made through the methylation pathway;[4] in most tissues, it is probably less than 5%. The CDP-choline pathway is the major pathway for the synthesis of phosphatidylcholine and involves the transfer of phosphocholine from CDP-choline to diacylglycerol. The enzyme which catalyzes this reaction is called "phosphocholine transferase" and in rat liver this enzyme is located in the microsomal fraction.[5] This pathway represents the major pathway for the *de novo* synthesis of phosphatidylcholine. As mentioned above, the transmethylation pathway is a minor route of phosphatidylcholine synthesis. Furthermore, phospholipid methylation does not create additional phospholipid molecules; it simply interconverts phospholipids. Thus, the function of the CDP-choline pathway is clearly the *de novo* synthesis of phosphatidylcholine, whereas the transmethylation pathway appears to have other cell functions.

The functional role(s) of the phospholipid methylation pathway is not completely clear. The activity of this pathway is altered during the growth and differentiation of some cells, by changes in diet, and by receptor activation in other tissues. It has been suggested in different cells to play a role in modifying membrane fluidity, in releasing arachidonic acid, the precursor to prostaglandins and leukotrienes, and in providing a source of choline for acetylcholine. Thus, the phospholipid methylation pathway has been suggested to be involved in almost all of the major roles of phospholipids in general. It is possible that in different tissues and cells phospholipid methylation is involved in all, some, or none of these potential functions. This review presents some general properties of the phospholipid methylation pathway and discusses findings on the actions of this pathway in a variety of specific tissues and cell types. It is possible that the methylation of phospholipids has different functions in different cells and/or different membranes within the same cell.

II. MEMBRANE STRUCTURE

A. Membrane Fluidity

Although several enzymes have specific phospholipid requirements for biological activity, it is becoming increasingly likely that the more significant role of membrane lipids is to provide an environment of proper viscosity and surface ionic milieu for optimal enzyme function.[1] A number of studies have been done during the past decade which implicate membrane fluidity as an important regulator of enzyme activity. The fluid mosaic model of membrane structure[6] has provided a valuable model for protein-lipid interactions within the membrane. Studies of movements within membranes have suggested that phospholipids diffuse in the membrane at rates of approximately 10^{-8} to 10^{-7} cm^2/sec.[1] At these rates, each molecule will exchange with its neighbor approximately 10^6 times per second. However, one of the most important concepts to develop through consideration of the fluid mosaic model is that of coexisting regions of different fluidity in the same membrane. It has been known for many years that artificial bilayers made of pure lipids of the same type found in membranes will undergo a transition from a relatively fluid state (generally termed the liquid crystalline phase) to a more rigid state (the gel phase) when the temperature of the preparation is lowered to a characteristic point. This change is referred to as a "phase transition". In a biological membrane, lipids are present as complex mixtures, with each component possessing its own characteristic phase transition temperature. Chilling such a membrane leads not to a simultaneous cocrystallization of all lipids, but to a lateral migration of the more readily gelled lipid classes into rigid assemblages capable of growing by accretion of similar, but slightly lower freezing lipid species upon further chilling.[1] This process is termed a "phase separation". Thus, a membrane, under certain conditions, contains coexisting domains of lipid crystalline and gel phases in equilibrium. In addition, the membrane may undergo transitions from a bilayer form into a hexagonal phase. Thus, the lipid structure of membranes is likely to be more heterogeneous than that originally envisioned in the fluid mosaic model and rapid changes in lipid composition may alter the fluid properties and structure of the membrane.

B. Membrane Asymmetry

Lipid phase transitions and lateral motions could occur in one or both layers of the bilayer. The composition of the two membrane bilayers appears to be very different. As discussed below, there are differences in composition between the outer and inner layers of membranes despite the fact that there is an active exchange of intact lipid molecules, usually referred to as "flip-flop". Phospholipid flip-flop is never as rapid as lateral motion. Nevertheless, the transbilayer exchange has been reported to be as much as four orders of magnitude faster in biological membranes[8] than it is in artificial bilayers.[9] Since the polar head group of phospholipids must penetrate a seemingly unfavorable energy barrier, it is likely that certain integral membrane proteins regulate transbilayer movement in a manner that maintains the phospholipid asymmetry of the membrane.

The major phospholipids in membranes are phosphatidylcholine, phosphatidylethanolamine, phosphatidylserine, and phosphatidylinositol. As mentioned above, these phospholipids have been found to be asymmetrically arranged in membranes. Phosphatidylcholine and sphingomyelin are preferentially located in the outside layer of the phospholipid bilayer and phosphatidylethanolamine, phosphatidylserine, and phosphatidylinositol are confined primarily to the inner surface of the membrane. These findings are based on several different experimental approaches. The first experiments labeled the amino groups of the phosphatidylethanolamine and phosphatidylserine with the nonpenetrating reagents trinitrobenzenesulfonic acid and formylmethionylsulfone methyl phosphate.[10,11] These phospholipids were found to be localized primarily in the inside layer of the phospholipid bilayer. These studies were confirmed and extended by the use of specific phospholipases.[12] Treatment of intact erythrocytes with phospholipase A under nonlytic conditions hydrolyzed only 20% of the total phospholipid, while hydrolyzing 68% of the phosphatidylcholine. Phosphatidylethanolamine and phosphatidylserine were not hydrolyzed under these conditions. When disrupted ghosts were treated with phospholipase A_2, almost all of the phosphatidylcholine, phosphatidylserine, and phosphatidylethanolamine was hydrolyzed. Another approach to demonstrating membrane lipid asymmetry has been the use of phospholipid exchange proteins. Incubation of phospholipid vesicles containing ^{32}P-phosphatidylcholine with red blood cell ghosts indicates that 60 to 70% of the phosphatidylcholine is available for exchange, suggesting that most of this lipid is exposed on the outside of the membrane.[13] Although most of the studies on membrane asymmetry have been done in erythrocytes, membrane asymmetry has been reported for phospholipids in platelets,[14] the influenza virion,[15] brain synaptosomes,[16] and of other tissues. These studies suggest that the asymmetric arrangement of lipids is likely to be a general membrane property.

III. PHOSPHOLIPID METHYLTRANSFERASE ENZYMES

The methylation of phospholipids was first described in the early 1960s in both microorganisms and vertebrate tissues. These studies indicated that phosphatidylethanolamine could be sequentially methylated three times to form phosphatidylcholine (Figure 1). Studies on choline-requiring mutants of *Neurospora* indicated that there were two intermediates in this pathway of choline synthesis.[17] Subsequent studies have identified these intermediates as phosphatidyl-*N*-monomethylethanolamine and phosphatidyl-*N*-*N*-dimethylethanolamine.[18] *S*-Adenosyl-L-methionine is the methyl donor. In mammalian tissues, Bremer and Greenberg[19-21] were the first to demonstrate that in rat liver phosphatidylcholine could be synthesized by the methylation of phosphatidylethanolamine in addition to the CDP-choline pathway described by Kennedy. These early experiments indicated that the phospholipid methylation pathway was present in almost all tissues examined, with the liver generally having about 10 to 20 times more activity than any other tissue. The methylation pathway of phosphatidylcholine synthesis has since been found in almost every species examined.

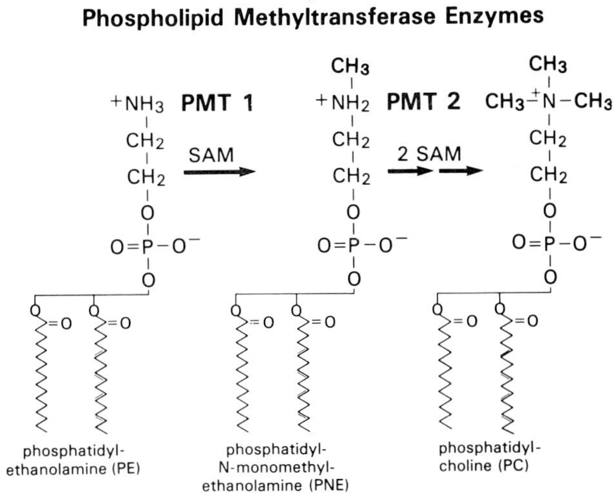

Phospholipid Methyltransferase Enzymes

FIGURE 1. Phosphatidylethanolamine is converted to phosphatidylcholine by three sequential methylations. The rate limiting step in this reaction is the formation of phosphatidyl-*N*-monomethylethanolamine. Two enzymes appear to carry out this transmethylation reaction, phospholipid methyltransferase I (PMT 1) and phospholipid methyltransferase 2 (PMT 2).

These early studies on phospholipid methylation also indicated that the first methylation reaction, e.g., the conversion of phosphatidylethanolamine to phosphatidyl-*N*-monomethylethanolamine, was the rate-limiting step in the synthesis of phosphatidylcholine through the methylation pathway. This finding and others led to the suggestion that there is likely to be more than one enzyme involved in the three sequential methylations. At this writing, the actual number of enzymes in mammalian tissues is not certain, although the data suggest that at least two enzymes are involved in the three methylation reactions. In microorganisms, studies using mutants have indicated that two phospholipid methyltransferases are required to convert phosphatidylethanolamine to phosphatidylcholine.

A. Microorganisms

While phosphatidylcholine is the most abundant phospholipid in many plant and animal tissues, it is a rare component of bacterial lipids. Bacteria of the genus *Agrobacterium* have been shown to synthesize phosphatidylcholine by methylating phosphatidylethanolamine. *Agrobacterium* contains a soluble enzyme which has been purified and is capable of catalyzing the methylation of phosphatidylethanolamine to phosphatidyl-*N*-monomethylethanolamine.[22] This bacterium also has a particulate system which is capable of carrying out all three methylations required to form phosphatidylcholine. These studies support the idea that at least two enzymes are involved in the synthesis of phosphatidylcholine through the methylation pathway. It is likely that this pathway in bacteria is the synthetic route by which bacteria make choline. Studies by Nyc and others,[18,23-25] using choline-requiring mutants of the mold *Neurospora crassa,* strongly suggest that at least two enzymes are involved in the transmethylation pathway of these microorganisms. One strain accumulated phosphatidyl-*N*-monomethylethanolamine and would not methylate it further. A second strain had a defect in the second and possibly third enzymatic *N*-methylations. Assays with mixed microsomes from the two strains showed that there was complementation between the two particulate preparations.[18] These two strains had mutations at two different single genetic loci, suggesting that in molds there are two genes that code for two phospholipid methyltransferases: one which methylates phosphatidylethanolamine to phosphatidyl-*N*-monomethylethanolamine and one which catalyzes the methylation of phosphatidyl-*N*-monomethylethanolamine. Since these mutants are choline-requiring mutants, it is likely that in molds like bacteria, the phospholipid methylation pathway plays an important role in synthesizing choline for other cell functions. This hypothesis is further supported by studies on the fungus *Saccharomyces cervisiae*.[27,28] By growing cells in medium containing various bases, Waechter and Lester[28] were able to show that choline-containing medium suppressed the activity of the entire transmethylation pathway, whereas the presence of monomethylethanolamine suppressed only the first methylation reaction. These authors suggested that in fungi there were at least two phospholipid methyltransferases involved in the synthesis of phosphatidylcholine from phosphatidylethanolamine and that the activity of these enzymes is regulated in some way by the cells' requirements for choline.

Studies on the slime mold *Dictyostelium discoideum* suggest that in this species the phospholipid methylation pathway may be involved in cell signal transduction. In these cells, cyclic AMP (cAMP) acts as a chemotactic stimuli. Extracellular cAMP has been shown to stimulate guanylate cyclase, which increases the cellular content of cGMP.[29] Cell activation also leads to a rapid and transient increase in phospholipid methylation.[29,30] Phospholipid methylation appears to be regulated by cGMP and is activated by an exogenous 8-Br derivative of cGMP.[29] Calmodulin[31] and intracellular calcium[30] also appear to modify the activity of the phospholipid methyltransferases in these cells. Furthermore, recent studies suggest that the cAMP-binding protein, a chemotactic receptor on these cells, contains *S*-adenosyl-L-homocysteine hydrolase activity.[33] *S*-Adenosyl-L-homocysteine is an endogenous inhibitor of methylation reactions. Thus, cAMP might modulate methylation by activating this hydrolyase, reducing the concentration of *S*-adenosyl-L-homocysteine, and thereby increasing methylation reactions.[33] In any case, it appears that phospholipid methylation is involved in the chemotactic response to cAMP in slime molds.

These studies indicate that phospholipid methylation is present in microorganisms and that there are at least two enzymes involved in the enzymatic methylation of phosphatidylethanolamine to phosphatidylcholine. Although phospholipid methylation appears to play a role in cellular chemotaxis to slime molds, it is clear that in a variety of microorganisms phospholipid methylation plays an important role in providing a synthetic pathway for the synthesis of phosphatidylcholine and choline.

B. Mammalian Enzymes

Similar to microorganisms, the phospholipid methylation pathway in mammalian tissues appears to be carried out by at least two enzymes. However, attempts to purify the methyl transferase activity have proven difficult,[34,35] and have not been able to separate two distinct activities. Nonlinear kinetics, different pH optima, ion requirements, and partial differential solubilization have been used to separate two apparent enzymatic activities in bovine adrenal medulla,[36] rat[37-40] and bovine[41] brain, liver,[42,43] erythrocytes,[24,44] mast cells and basophils,[45-48] other leukocytes of all types,[49-51] rat pituitary,[52] and smooth[53] and cardiac[54] muscle. The first methylation reaction, i.e., the methylation of phosphatidylethanolamine to phosphatidyl-*N*-monomethylethanolamine, is generally agreed to be the rate-limiting step in the transmethylation pathway. The studies cited above suggest that this methylation reaction is carried out by a phospholipid methyltransferase 1 (PMT 1). PMT 1 has an optimum pH of about 7.5 and a low apparent K_m for *S*-adenosyl-L-methionine, the methyl donor. In addition, particularly in bovine tissues, this enzyme has a requirement for Mg^{++}.[36,41] This requirement is much less pronounced in rat tissues and makes the resolution of the activities more difficult. It has proved extremely difficult to measure the activity of this enzyme with exogenous phosphatidylethanolamine. It is likely that this enzyme is saturated by endogenous phosphatidylethanolamine and may have a preference for phosphatidylethanolamines containing unsaturated fatty acids.[55] The two additional methylations which convert phosphatidyl-*N*-monomethylethanolamine to phosphatidylcholine are thought to be carried out by a second methyl transferase, phospholipid methyl transferase 2 (PMT 2). PMT 2 has an optimum pH of about 10.5, a high apparent K_m for *S*-adenosyl-L-methionine, and can be differentially solubilized by sonication. The addition of exogenous phosphatidyl-*N*-monomethylethanolamine and/or phosphatidyl-*N,N*-dimethyl-ethanolamine, two trace lipid intermediates in normal membranes, leads to a tremendous increase in the incorporation of methyl groups into phospholipids.

Certain other types of experiments are consistent with the hypothesis that there are two enzymes involved in the three sequential methylations of phosphatidylethanolamine. Prasad and Edwards[52] have separated two apparently different enzymatic activities from rat pituitary extracts. In addition, studies with mutants of a rat basophilic leukemia (RBL) cell line have suggested that there are two different enzymes.[56] RBL cell extracts incorporate methyl groups preferentially into phosphatidyl-*N*-monomethylethanolamine at low concentrations of SAM. This was taken as a measure of PMT 1 activity. In the presence of high concentrations of SAM and exogenous phosphatidyl-*N*-monomethylethanolamine and phosphatidyl-*N,N*-dimethylethanolamine extracts preferentially incorporate methyl groups into phosphatidylcholine. RBL cell mutants were isolated which methylated phosphatidylethanolomine to phosphatidyl-*N*-monomethylethanol, but had an impaired ability to add the two additional methyl groups to form phosphatidylcholine. Similarly, variants were found that could methylate phosphatidyl-*N,N*-dimethylethanolamine, but could not carry out the first methylation. Reconstitution by fusion of the two RBL cell variants results in normal phospholipid methyltransferase activity.[56] These studies strongly suggest that there are at least two phospholipid methyltransferases involved in the conversion of phosphatidylethanolamine to phosphatidylcholine in RBL cells. However, none of these variants has one of the methyltransferase activities exclusively, there always being about 10% residual activity. It is possible that there are two different enzymes which can catalyze all three methylations, but with different pH optima and affinity for SAM and the phospholipid substrates. In vivo, these enzymes could act together with each enzyme acting on its preferred substrates. In vitro purification procedures may not be able to resolve the two activities since the second enzyme appears to be much more readily solubilized[36,37] and can be easily followed by adding exogenous phosphatidyl-*N*-monomethylethanolamine and/or phosphatidyl-*N,N*-dimethylethanolamine to the incubation. In any case, it is generally agreed that the phospholipid methylation pathway is found in a large variety of species and tissues and that the first methylation reaction is the limiting step in the transmethylation pathway.

It is worth mentioning that other lipids are methylated and this has undoubtedly confused some of the literature on phospholipid methylation. Several neutral lipids are formed during incubation of tissue homogenates with SAM or by incubation in intact cells with methionine. Zatz and others[57-59] have identified these lipids as methyl esters of fatty acids, ubiquinone, and 2-(methylthio)benzothiazole.

IV. PHOSPHOLIPID METHYLTRANSFERASE ASYMMETRY

Evidence for the asymmetric localization of phospholipid methyltransferases has been found in rat erythrocyte ghosts,[61] synaptosomal membranes,[38] kidney cortex membranes,[49] and liver microsomal membranes.[43] Hirata and Axelrod,[61] in a particularly well-designed study using rat erythrocyte ghosts, found that the methyltransferases appear to be on opposite sides of the phospholipid bilayer. Erythrocyte ghosts can be prepared as inside- and right-side-out. When *S*-adenosyl-L-methionine was exposed to the outside of right-side-out ghosts, very little phospholipid methylation occurred (Table 1). However, when *S*-adenosyl-L-methionine was inside right-side-out ghosts or ghosts were made inside-out, then there

Table 1
SIDEDNESS OF ENZYMATICALLY METHYLATED PHOSPHOLIPIDS IN ERYTHROCYTE GHOSTS[24]

S-Adenosyl-L-methionine (μM)		[^{3}H]-Methyl incorporation in lipids (pmol/mg protein/hr)	
Inside	Medium	Right-side-out ghosts	Inside-out ghosts
200	0	26.5	2.9
0	200	3.6	29.4

Note: *S*-Adenosyl-L-[methyl-^{3}H]methionine was introduced inside the ghosts or was added to the medium just prior to the start of the reaction. The incubation was carried out at 37°C for 1 hr.

was a considerable incorporation of [^{3}H]-methyl groups into phospholipids. These studies suggested that the methylation of phosphatidylethanolamine takes place on the cytoplasmic side of the membrane. Treatment with trypsin, which does not penetrate the cell membrane and digests only those proteins on the outer surface,[62]destroyed almost all of the activity of PMT 2 while only reducing the activity of PMT 1 approximately 10%, when right-side-out ghosts were used. When inside-out ghosts were used, the activity of PMT 1 was reduced by more than 90%, with only a 10 to 20% decrease in PMT 2 activity. Comparable studies have been done in more complex tissue preparations such as synaptosomes, a pinched-off nerve ending preparation, by comparing intact and lysed preparations.[38] Trypsin treatment caused only a small reduction in PMT 1 activity (approximately 13%) in intact synaptosomes, whereas in lysed synaptosomes trypsin destroyed 83% of the activity (Figure 2). Trypsin treatment destroyed most the PMT 2 activity in both intact and lysed synaptosomes. These results suggest the asymmetric distribution of the two methyltransferases in the erythrocyte and synaptosomal plasma membranes: the first methyltransferase facing mostly the inside of the membrane and the second methyltransferase being mainly localized on the outer surface of the membrane.

The asymmetric distribution of the methylated phospholipids has also been suggested by studies using phospholipase C. This enzyme removes the polar head, which contains radioactive methyl groups, only from phospholipids on the exterior surface of the membrane.[63] Intact membrane preparations can be divided into three fractions. One fraction can be treated with phospholipase C, another was sonicated and then treated with phospholipase, and the third fraction was incubated without phospholipase C. Following these treatments, the phospholipids are extracted and separated. In untreated synaptosomes, monomethyl, dimethyl, and trimethyl (i.e., choline) containing phospholipids are all present. In intact membrane vesicles, a considerable amount of phosphatidylcholine and phosphatidyl-*N,N*-dimethylethanolamine are hydrolyzed by phospholipase C, whereas phosphatidyl-*N*-monomethylethanolamine is not. In sonicated structures where the membrane is disrupted, all three of the methylated phospholipids are hydrolyzed.[38,61] Interestingly, treatment of lysed membrane vesicles does not result in the hydrolysis of phosphatidyl-*N*-monomethylethanolamine, suggesting that this lipid intermediate may be deeply buried within the membrane structure in a configuration which is not accessible to the phospholipase.[38] It could be bound tightly to the enzyme, or in a position not directly in contact with either the inside or outside layers of the membrane. In any case, these data suggest that during the methylation of phospholipids there is a translocation of the methylated lipids from the inside layer of the bilayer to the

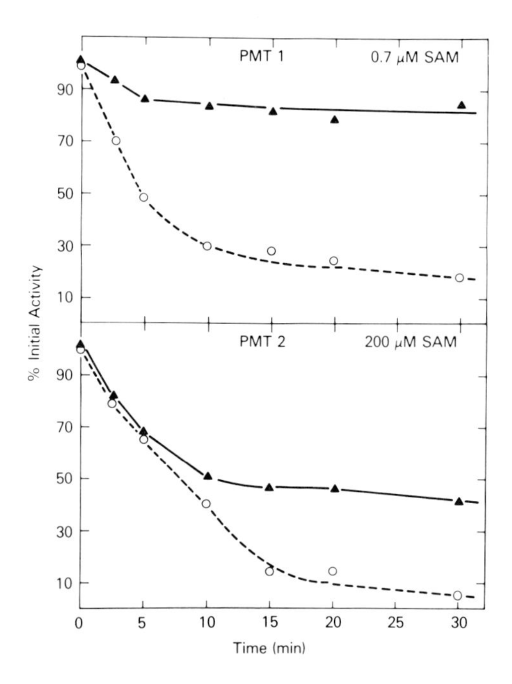

FIGURE 2. Effects of trypsin treatment on the methylation of phospholipids in lysed and intact synaptosomes. Synaptosomes were incubated with trypsin, 1 mg/mℓ, for various periods of time. At the times indicated, the reaction was stopped with buffer containing 5 mg/mℓ pancreatic trypsin inhibitor with 2% nonident and assayed for PMT 1 or PMT 2 activity as described by Crews et al.[38] (▲—▲). Intact synaptosomes show no effect of trypsin treatment on PMT 1 activity and a marked decrease in PMT 2 activity, suggesting that PMT 2 is partially exposed on the exterior surface of synaptosomal membranes. PMT 1 activity is not significantly changed in intact synaptosomes, suggesting that it is inside the synaptosome where trypsin cannot penetrate. Trypsin treatment of lysed synaptosomes (○--○) decrease both PMT 1 and PMT 2 activity.

outside layer (Figure 3). This rapid translocation of the methylated phospholipids could alter membrane structure and could contribute to the asymmetry found in the distribution of phospholipids within the bilayer. However, since the methylation pathway is only a minor pathway for the synthesis of phosphatidylcholine, it is unlikely that the transmethylation pathway alone could maintain phospholipid asymmetry within membranes.

V. PHOSPHOLIPID METHYLATION AND MEMBRANE FLUIDITY

One of the major postulated membrane functions of phospholipid methylation has been that the transmethylation reaction can rapidly and reversibly modify membrane fluidity.[26,64] Studies in erythrocyte ghosts have shown that as phosphatidyl-*N*-monomethylethanolamine is formed there is a corresponding decrease in membrane viscosity, as determined by the fluorescent polarization of diphenylhexatriene. The parallel changes in viscosity and phosphatidyl-*N*-monomethylethanolamine formation are intriguing since this lipid is apparently embedded within the membrane where it is not accessible to phospholipases. It is possible that the formation of phosphatidyl-*N*-monomethylethanolamine triggers a transition change in the membrane structure to hexagonal phase or another structural change in lipids. Studies in liver membranes have shown that the addition of exogenous phosphatidyl-*N*-monome-

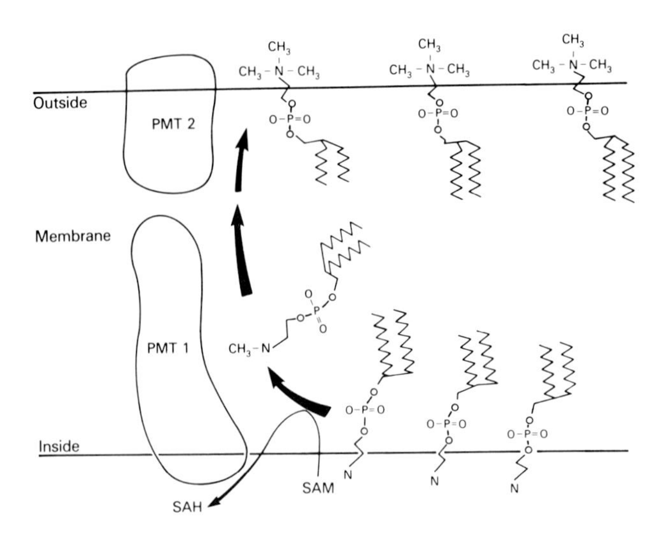

FIGURE 3. Schematic diagram of the methylation and translocation of phospholipids across membranes. PMT 1 appears to be primarily localized on the inside of plasma membranes where the first methylation reaction takes place, and PMT 2 appears to be exposed on the exterior surface of plasma membranes where phosphatidylcholine is primarily localized in the outside layer of the phospholipid bilayer. Phosphatidyl-*N*-monomethylethanolamine appears to be buried within the membrane. SAM is *S*-adenosyl-L-methionine, SAH is *S*-adenosyl-L-homocysteine.

thylethanolamine decreases apparent microviscosity as the proportion of the monomethylated lipid is increased.[42,65] However, larger amounts of exogenous lipid are required than would be predicted from studies on in vivo methylation of lipids. Studies on chicken erythrocytes have shown that the lectin, concanavalin A, can activate phospholipid methylation and thereby decrease membrane fluidity as determined by the ESR spectra of a stearic acid derivative.[66] This change in viscosity could be blocked by methyltransferase inhibitors. Corresponding increases in membrane fluidity and the methylation of phospholipids has also been found in membranes prepared from rat aorta.[67] These studies also suggested that the properties of smooth muscle membranes from hypertensive rats were altered due to an increase in phospholipid methylation and a corresponding decrease in membrane fluidity as determined by fluorescent polarization. Decreases in membrane viscosity which occur during rat liver regeneration[68] and drug-induced enzyme induction[42] have also been shown to correspond to an increase in the content of phosphatidyl-*N*-monomethylethanolamine and may be related to the increased phospholipid methylation found in these cells.[42,68] A change in membrane properties has not been found to correspond to a change in membrane order in kidney brush border membranes[69] and brain membranes.[70] It is possible that the effects of phospholipid methylation on membrane fluidity is specific to certain membranes and/or tissues having a particular membrane composition and/or structure. In any case, one of the postulated functions of the transmethylation pathway is the rapid and transient modification of the fluidity and structure of the membrane.

VI. PHOSPHOLIPID METHYLATION IN MAMMALIAN TISSUES

The phospholipid methyltransferase enzymes have a large tissue and membrane variation. The liver has 10 to 20 times the activity of any other tissue, with the highest specific activity in the microsomal fraction.[5,35] The brain has approximately twice the activity in the kidney with the greatest specific activity in the plasma membrane fraction.[37,41] Leukocytes also have a high specific activity of the phospholipid methyltransferases in the plasma membrane.[49,71,72] These differences in specific activity and membrane localization could represent differences in the function and/or properties of the transmethylation pathway in these various cells. Therefore, studies on phospholipid methylation in different tissues will be discussed separately.

A. Liver

The phospholipid methylation pathway in liver has been extensively studied. This tissue contains by far the highest specific activity of the phospholipid methyltransferases. In the liver, approximately 20% of the phosphatidylcholine made is made through the transmethylation pathway.[4] In rats fed a choline-deficient diet, there is an increase in the turnover of liver phosphatidylcholine as determined using ^{32}P, and a marked increase in the phospholipid methyltransferase enzymes.[73,74,76] There is also a marked change in the molecular species of phosphatidylcholine in the liver with the increase occurring primarily in molecular species containing highly unsaturated fatty acids.[73] Phosphatidylcholine containing arachidonic acid and other highly unsaturated fatty acids in combination with stearic acid is thought to be synthesized primarily through the transmethylation pathway, whereas phosphatidylcholine synthesized through the CDP-choline pathway appears to contain primarily palmitic acid and less unsaturated fatty acids.[7,74,76,77] Thus, there is a good correlation between the methylation of phosphatidylethanolamine in choline-deficient rats and the synthesis of arachidonic acid containing phosphatidylcholine in liver. Since plasma phosphatidylcholines are derived from the liver and choline deficiency induced phospholipid methylation, it is likely that one function of phospholipid methylation in the liver is to provide a source of choline and phosphatidylcholine for various cell, tissue, and organism functions.

During the past several years, evidence has accumulated which suggests that stimulation of a variety of receptors increases phospholipid methylation in hepatocytes. Various studies have suggested that glucagon,[75,78,79] isoproterenol,[80] epinephrine,[79] epidermal growth factor,[81] angiotensin,[82] vasopressin,[82] and the calcium ionophore, A23187,[82,83] can activate phospholipid methylation in hepatocytes, although there are some inconsistent reports (see Reference 83 for review). Glucagon, isoproterenol, and epinephrine all stimulate adenylate cyclase and increase the cellular content of cAMP. Glucagon increases phospholipid methylation approximately twofold and this increase is due to an apparent increase in the V_{max} of the transmethylation pathway without affecting the apparent K_m for *S*-adenosyl-L-methionine.[75] cAMP added exogenously to rat hepatocytes also increases the transmethylation pathway as does an analog of cAMP added to isolated rat liver microsomes in the presence of micromolar concentrations of ATP.[75,82,83] These studies suggest that the activation of a cAMP dependent protein kinase can increase the activity of the phospholipid methyltransferase enzymes (Figure 4).

In addition to receptors which stimulate adenylate cyclase, angiotensin, vasopressin, and the calcium ionophore, A23187, also stimulate phospholipid methylation in hepatocytes. Each of these stimulants is thought to act by mobilizing calcium. In the abscence of external calcium, none of these agents stimulates phospholipid methylation. Furthermore, the addition of calcium to isolated microsomes in the presence of ATP activates phospholipid methylation, probably by acting through calmodulin.[83,132] Thus, stimulation of a variety of cell surface receptors can activate phospholipid methylation in the liver (Figure 4).

As mentioned above, the CDP-choline pathway is the major route of synthesis of phosphatidylcholine. Studies on this pathway in the liver have indicated that increases in cAMP[82,84] or the mobilization of calcium results in a decrease in the synthesis of phosphatidylcholine from choline.[85] This appears to be due to the inhibition of phosphocholine cytidyltransferase, the enzyme that catalyzes the rate-limiting step in the synthesis of phosphatidylcholine through the CDP-choline pathway. This inhibition appears to be due to the phosphorylation of this enzyme by protein kinases.[82,84] These findings have led to the suggestion that the phospholipid methylation pathway and the CDP-choline pathway for phosphatidylcholine synthesis are regulated in a complementary way in the liver such that the inhibition of one pathway results in the stimulation of the alternate pathway (Figure 4). Thus, dietary changes and/or receptor-activated changes result in the activation of one pathway of phosphatidylcholine synthesis and the inhibition of the alternate pathway (see Reference 83 for further discussions of this hypothesis). In any case, it appears likely that the phospholipid methylation pathway in the liver plays an important function in synthesizing phosphatidylcholine and choline and that receptor activation of this pathway is secondary to other second messengers, i.e., both cAMP and calcium appear to activate the transmethylation pathway.

B. Leukocytes

A large number of experiments on phospholipid methylation have been done on leukocytes. These cells offer the advantage that they can be isolated and purified as intact viable cells. Studies on basophils and mast cells have suggested that the phospholipid methylation pathway is involved in IgE-mediated signal transduction. Other studies have concerned neutrophils, lymphocytes, and various types of leukemic cells. The studies on these cell types are reviewed below.

1. Basophils and Mast Cells

Mast cells and basophils secrete histamine, arachidonic acid, and a variety of other substances when stimulated. One type of stimulus for mast cell secretion is the reaction of a specific antigen with antibodies, i.e., IgE, which are bound to the mast cell surface. Mast cells stimulated through antigen IgE accumulate $^{45}Ca^{++}$. Furthermore, antigen-IgE stimulated Ca^{++} accumulation occurs even when the secretory response is inhibited by depriving the cells of antigen-IgE energy.[133] These results support the hypothesis that calcium influx is an early event in the secretory response of mast cells. Studies have shown that IgE receptors need to be bridged to activate calcium influx and histamine release from mast cells.[72,86] Stimulation of mast cells with concanavalin A,[47] a multivalent lectin, or with divalent antibodies against IgE[72] causes a rapid increase in the methylation of phospholipids, which is followed by the release of histamine. After a short time, there is a decrease in the methylated phospholipids, suggesting futher metabolism of these lipids (Figure 5). These studies have been extended by using fragments of anti-IgE antibodies to stimulate mast cells. Monovalent Fab′ fragments fail to increase phospholipid methylation, to induce Ca^{++} uptake, and to release histamine. Divalent F(ab)′2 fragments of anti-IgE cause a rapid increase in phospholipid methylation, Ca^{++} influx, and release histamine.[71,72] These findings suggest that bridging of IgE receptors is necessary to initiate phospholipid methylation, calcium flux, and histamine release.

Several investigators have shown that activation of rat mast cells by anti-IgE or Con A induces changes in intracellular cAMP. There is an initial rapid increase in cAMP after about 15 sec which is followed by a sharp decline at about 1 min (Figure 5).[86] This time course resembles the time course for changes in phospholipid methylation.[72,86] After the initial decline, cAMP slightly increases again after 3 min. Studies using purified mast cell plasma membranes have indicated that monovalent fragments of anti-IgE (i.e., Fab′) do not stimulate either phospholipid methylation or cAMP formation, whereas divalent antibody fragments (i.e., F(ab′)2) do stimulate both cAMP formation and phospholipid methylation in isolated mast cell plasma membranes.[86] These results indicate that both adenylate cyclase and the methyltransferases are closely associated with IgE receptors in the mast cell plasma membrane. Since the methyltransferases could be stimulated in the absence of ATP, whereas cAMP formation required ATP as a substrate, it seems unlikely that the activation of the methyltransferases is secondary to stimulation of a cAMP-dependent protein kinase in these experiments.

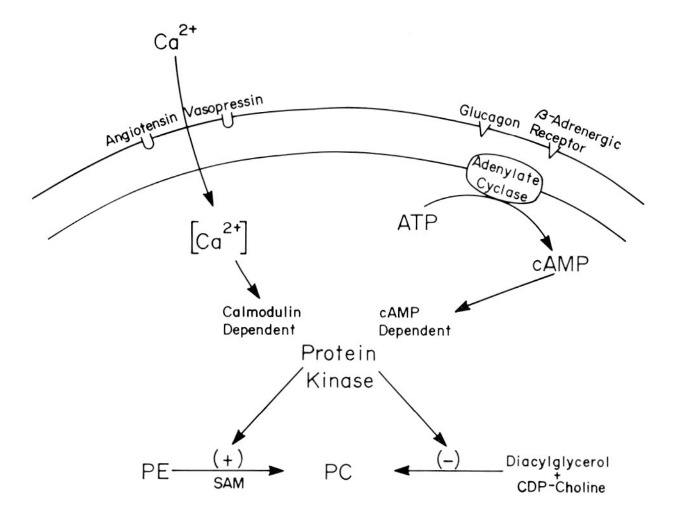

FIGURE 4. A possible model of the effects of calcium and cAMP on the synthesis of phosphatidylcholine by the phospholipid methylation pathway and the CDP-choline pathway in liver microsomes. The stimulation of β-adrenergic receptors and glucagon receptors activates adenylate cyclase to increase cAMP formation. Simulation of angiotensin and vasopressin receptors mobilizes calcium. The elevation of intracellular calcium and cAMP activates protein kinases which inhibit the CDP-choline pathway and activate the transmethylation pathway.[83]

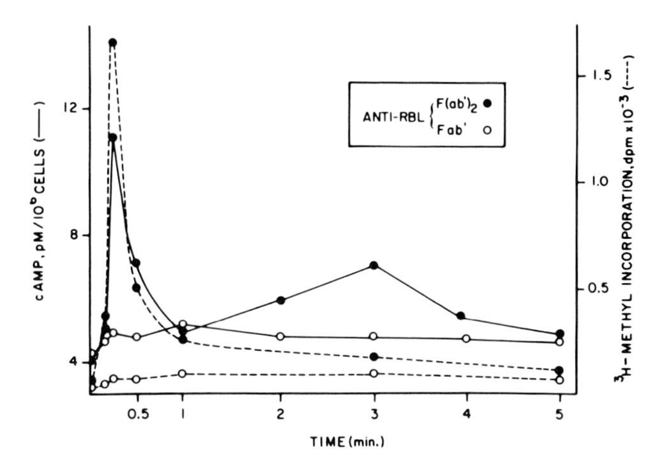

FIGURE 5. Kinetics of changes in phospholipid methylation and cAMP during mast cell activation. Mast cells were treated with either F(ab')2 fragments or Fab' fragments of anti-rat basophilic leukemia cell antibody (anti-RBL). F(ab')2 fragments stimulate calcium influx and histamine release, whereas the monovalent fragments Fab' do not.[71,86,93]

Experiments concerning the interaction of phospholipid methylation and cAMP in mast cell histamine release have used various compounds to inhibit methylation or cAMP metabolism. When intact mast cells are incubated with a methylation inhibitor, 3-deaza-SIBA, phospholipid methylation and histamine release are inhibited, while cyclicAMP is partially inhibited (Figure 6). However, when the same experiments were carried out with a membrane preparation, 3-deaza-SIBA inhibited phospholipid methylation and cAMP in a similar dose-

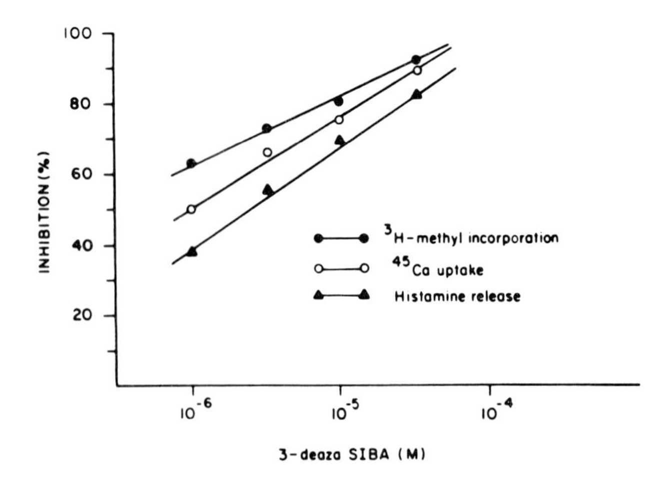

FIGURE 6. Comparison of the inhibition of antibody-stimulated increases in phospholipid methylation, $^{45}Ca^{++}$ uptake and histamine release by the methylation inhibitor 3-deaza-SIBA. Purified rat mast cells were preincubated with 3-deaza-SIBA and then stimulated.[72,86]

response fashion. Preincubation of mast cells with theophylline, which increases mast cell cAMP, inhibited anti-IgE stimulated phospholipid methylation, $^{45}Ca^{++}$ uptake, and subsequent histamine release. Dibutyryl cAMP also inhibited all three reactions with similar dose-response curves. In addition, studies with isoproterenol, which stimulates the accumulation of mast cell cAMP through the β-adrenergic receptor, indicated that β-receptor stimulation causes increases in cAMP, while decreasing phospholipid methylation, $^{45}Ca^{++}$ uptake, and histamine release. cAMP also inhibited phospholipid methylation in purified mast cell membranes. Thus, in mast cells, it appears that an increase in intracellular cAMP results in a suppression of phospholipid methylation and inhibits subsequent calcium flux and histamine release.[86]

Changes in phospholipid methylation have also been found during histamine secretion by rat basophilic leukemia (RBL) cells. RBL cells can be readily grown in tissue culture, have IgE receptors, and contain histamine, serotonin, and other bioactive agents that are found in mast cells.[90] Although not all RBL sublines secrete histamine, there are several sublines in which the IgE mediated secretion of histamine is qualitatively similar to secretion from rat mast cells and human basophils.[87,90] Antigen stimulation of RBL cells results in an initial increase in phospholipid methylation, which is followed by a decline in the methylated phospholipids similar to that found in rat mast cells. The increase in phospholipid methylation precedes the influx of $^{45}Ca^{++}$ and the release of histamine and arachidonic acid, whereas the decline in the methylated phospholipids closely corresponds in time with the release of histamine and arachidonic acid. After antigen stimulation, [methyl-^{3}H]-labeled lysophosphatidylcholine was found suggesting that phosphatidylcholine formed through the methylation pathway is cleaved by phospholipase A_2 to form lysophosphatidylcholine (e.g., lysolecithin) and a free fatty acid such as arachidonic acid. As mentioned above, phosphatidylcholine synthesized through the methylation pathway has been shown to be enriched in arachidonic acid and other unsaturated fatty acids when compared to phosphatidylcholine synthesized through the CDP-choline pathway. Phosphatidylcholine formed through the methylation pathway has been shown to be enriched in arachidonic acid in RBL cells,[88] platelets,[114] and liver.[73,76,77] Stimulation of RBL cells results in the release of arachidonic acid from phospholipids with most of the radioactivity coming from phosphatidylcholine.[87] The release of arachidonic acid parallels the secretion of histamine. The temporal relationships indicated that the initial increase in phospholipid methylation slightly preceded or paralleled the influx of calcium, whereas the decline in the methylated phospholipids corresponded with the time course of the release of histamine and arachidonic acid.

The relationship between phospholipid methylation, secretion of histamine, and arachidonic acid release is further supported by the use of the methylation inhibitor 3-deazaadenosine (3-DZA). 3-DZA inhibited phospholipid methylation and histamine release in almost identical concentration-dependent manners, whereas other methylations were only slightly inhibited.[87] 3-DZA also blocked the release of arachidonic acid. Methylation inhibitors also blocked the antigen-IgE stimulated influx of calcium.[88] It should be pointed out that methylation inhibitors only block IgE-mediated secretion which stimulates phospholipid methylation.[47,89] Other types of mast cell secretagogues, such as compound 48/80, and the calcium ionophore, A23187, are not inhibited by methylation inhibitors and do not activate phospholipid methylation.[47,89] These findings are consistent with the hypothesis that during antigen-IgE mediated activation of mast cells the initial increase in phospholipid methylation alters the membrane permeability to calcium, allowing calcium influx.

Additional evidence for the role of phospholipid methylation in calcium influx and histamine release has been obtained using mutant RBL cells.[56,90] Variant clones of RBL cells were found which did not secrete histamine in response to antigen-IgE, but did secrete when calcium flux was artificially stimulated using the calcium ionophore, A23187. These mutants were found to have IgE receptors, suggesting a defect between the receptor and calcium flux. One cell line was found to have PMT 1 activity and almost no PMT 2 activity. A second variant had PMT 2 activity and no PMT 1 activity. Fusion of each cell line at a ratio of 1:1 resulted in the growth of eight independent hybrids (Table 2). All eight hybrids had normal levels of PMT 1 and PMT 2 activity. In addition, stimulation of the hybrids with antigen-IgE caused calcium influx and histamine release. Thus, by fusing two nonsecreting cell lines that lacked either PMT 1 or PMT 2, the secretory process was reconstituted. These findings suggest that both methyltransferases are required for antigen-IgE stimulated calcium flux.

Recent studies have suggested that the phospholipid methyltransferase enzymes may not be the first enzymes activated following activation of mast cells by antigen-IgE. It has been suggested that a membrane-associated serine esterase plays a role in mast cell secretion.[86] Treatment of mast cells with diisopropyl fluorophosphate (DFP), a potent inhibitor of serine esterases, inhibited antigen-IgE stimulation of phospholipid methylation and histamine release. Other inhibitors of proteases and serine esterases also blocked antigen-IgE stimulated phospholipid methylation in membrane fragments.[86] Thus, it is possible that a membrane-associated proteolytic enzyme is activated by IgE receptor bridging prior to the activation of the phospholipid methyltransferases.

Recent studies have shown that anti-inflammatory steroids block antigen-IgE stimulated histamine release. Only secretagogues which stimulate secretion through IgE are inhibited by steroid treatment. Secretagogues such as compound 48/80, somatostatin, and the calcium ionophore, which do not stimulate phospholipid methylation through IgE, are not inhibited. Recent studies suggest that steroid treatment results in the synthesis of a protein which uncouples the IgE receptor from the phospholipid methyltransferases.[92] Phospholipid methylation is also inhibited in leukemia cells by glucocorticoid treatment.[32] Thus, it is possible that part of the anti-inflammatory action of glucocorticoids is due to the inhibition of phospholipid methylation. Further studies are required to elucidate the steroid methyltransferase interaction.

Studies in mast cells and basophils clearly suggest that the phospholipid methylation is involved in antigen-IgE mediated signal transduction. The correspondence of time courses, the effects of methylation inhibitors, and the reconstitution studies with mutants all strongly suggest that bridging of IgE receptors activates phospholipid methylation, which results in

Table 2
RECONSTITUTION OF TWO PHOSPHOLIPID METHYLTRANSFERASES: Ca^{2+} INFLUX AND HISTAMINE RELEASE BY HYBRIDIZATION OF VARIANTS DEFECTIVE IN ONE PHOSPHOLIPID METHYLTRANSFERASE

	Phospholipid methyltransferase (pmol ^{3}H-methyl group/mg protein)		IgE-Mediated Ca^{2+} influx (pmol/10^6 cells) (30 min)	IgE-Mediated histamine release (%)
Cell line	PMT I	PMT II		
2H3	1.71 ± 0.22	21.7 ± 1.9	69.7	58
Variant/1Cl.1	0.13 ± 0.04	30.2 ± 3.2	0.8	2
2H3.B6	1.04 ± 0.13	4.3 ± 1.3	0.6	2
Hybrid A1	0.82 ± 0.07	20.5 ± 4.7	74.8	28
Hybrid A2	0.70 ± 0.04	17.1 ± 5.0	77.1	33

Note: The two variants were fused and the hybrids grown as described in Reference 59. The phospholipid methyltransferase enzymes were assayed in membrane fractions by incubations with *S*-adenosyl-L-[^{3}H-methyl]-methionine under optimum conditions for each enzyme, respectively. $^{45}Ca^{2+}$ (10 μCi/mℓ) in the presence of 1 m*M* $CaCl_2$ containing buffer was used to measure $^{45}Ca^{2+}$ influx to cells. The supernatant was sampled for histamine release and the cells were assayed for $^{45}Ca^{2+}$ influx. Histamine release is expressed as percent of the total cellular histamine. Control values for unstimulated Ca^{2+} influx and histamine release were approximately 20 pmol/10^6 cells/30 min and 6%, respectively.

the influx of calcium and secretion (Figure 7). Although each of these experiments can be criticized individually, they are rather convincing when considered together. In the mast cell, cAMP and glucocorticoid inhibit phospholipid methylation and histamine and arachidonic acid release. Thus, the transmethylation pathway appears to be a site of receptor response modulation. Recent studies have shown antigen-IgE activation of phospholipid methylation in human lung mast cells.[93] Further studies on the role of phospholipid methylation in immunoglubulin activation of cells could lead to the development of anti-inflammatory compounds equal to or better than the anti-inflammatory steroids. In any case, it seems clear that phospholipid methylation plays a role in IgE-mediated activation of mast cells.

2. Lymphocytes

Studies on lymphocytes are less clear than those on basophils and mast cells. Lymphocytes play a number of important roles in the immune response including secretion of antibodies, secretion of mediators responsible for delayed hypersensitivity reactions, killing of target cells, and other functions regulating the differentiation of certain types of lymphocytes. There are many subpopulations of lymphocytes and this makes detailed biochemical studies difficult, since the cell populations are heterogeneous and frequently differentiating and dividing in vitro such that the population is changing. Early studies suggested that methylation reactions were important for lymphocyte cytolytic, i.e., cell-killing responses.[60] More recent experiments have suggested that the interaction of natural killer cells with susceptible tumor cell targets increased the turnover of the methylated phospholipids.[94] Other studies following the incorporation of methyl groups into phospholipids have suggested that the mitogenic response of T lymphocytes to lectins is related to an increased turnover of the transmethylation

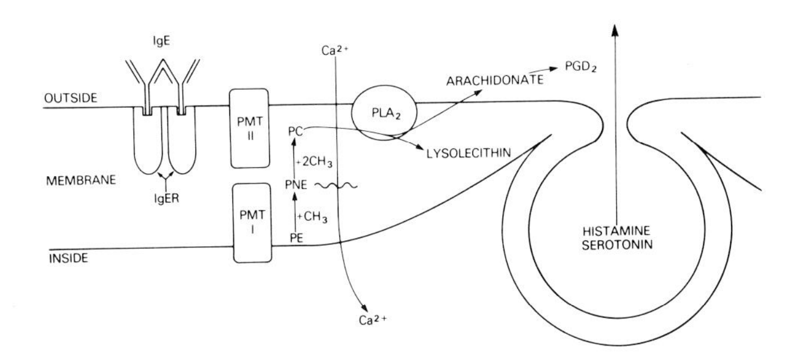

FIGURE 7. Schematic diagram of events during antigen-IgE mediated stimulation of histamine and arachidonic acid release. (IgE is immunoglobulin E and IgER is immunoglobulin E receptor.) The cross-linking of IgER leads to the activation of the phospholipid methyltransferase enzymes. The increase in phospholipid methylation modifies membrane permeability leading to calcium influx which activates exocytotic histamine release and phospholipases which release arachidonic acid.

pathway.[95-97] However, more recent experiments have carefully separated the methylated phospholipids and indicate that increases in the incorporation of methyl groups into lipid extracts may not be related to phospholipid methylation.[98] The phospholipid methylation pathway has also been suggested to be involved in lymphocyte activation by B cell differentiation factor[99] and α and β interferon-induced differentiation of human lymphocytes,[100] but not histocytic lymphoma-derived cell lines.[101] Thus, it is generally agreed that the phospholipid methyltransferase enzymes are present in lymphocytes, but any potential function is blurred by the complexities of lymphocyte responses and heterogeneity.

3. Neutrophils

Neutrophils are the predominant white cell in human blood. It is a highly motile cell and is the first cell to appear at the site of an acute inflammation, providing the first line of defense, usually by phagocytizing the invader. Stimulation of rabbit neutrophils with the chemotactic peptide fMet-Leu-Phe causes calcium influx, the release of arachidonic acid, and the degradation of the methylated phospholipids.[102] Inhibitors of phospholipid methylation block the chemotactic responses as well as calcium flux and arachidonic acid release.[103] This could be related to protein carboxymethylation and/or phospholipid methylation.[102,103] Other inhibitors which block phagocytosis also block phosphatidylcholine synthesis. Although methylation inhibitors suggest a role of methylation reactions in neutrophil responses to stimulation, methylation inhibitors are usually analogs of adenosine and it is not certain that all of their actions can be explained as simply due to an inhibition of methylation.[104] Thus, the methylation of phospholipids takes place in neutrophils, but its function is unknown.

C. Nervous Tissue

The phospholipid methyltransferases have been suggested to have at least two functions in the nervous systems. One suggested function involves receptor-mediated signal transduction. A second possible function is the synthesis of *de novo* choline for acetylcholine. Studies in rat[37-40] and bovine[41] brain have suggested that there are at least two enzymes which carry out the three sequential methylation reactions and that these enzymes are asymmetrically distributed across the synaptosomal membrane.[38] The synaptosomal plasma membrane and microsomal membranes have the highest specific activity, although all membranes appear to have some activity.[37,41] Studies in brain have also shown that ethanolamine plasmalogen, which has a vinyl ether bond in the 1 position instead of an ester bond, can be methylated to form choline plasmalogen in addition to the methylation of the respective phospholipids.[112] Thus, nervous tissue can synthesize phosphatidylcholine through the phospholipid methylation pathway.

Phospholipid methyltransferase activity in brain is highest in young rats.[113,120] The two methyltransferases appear to develop at different rates in rat brain. PMT 1 activity is greatest in neonates, whereas PMT 2 activity is highest approximately 2 weeks after birth.[120] The activity of PMT 1 and 2 remains relatively constant in rats 1 to 5 months old. However, as rats approach senescence there is an increase in PMT 1 activity while PMT 2 activity remains relatively constant.[114] The activity of PMT 1 is approximately 30% higher in the brains of rats 21 months old compared to the brains of rats 1 month old. The significance of these changes in phospholipid methylation during development and aging is not clear. It is possible that these changes reflect differences in actual amounts of the enzymes and/or they are secondary to the stimulation of the methyltransferases by certain growth factors and/or neutrotransmitters.

Nerve growth factor has been shown to stimulate phospholipid methylation in rat superior cervical ganglion[106] and in suspensions of chick sympathetic ganglion.[107] The stimulation of phospholipid methylation by nerve growth factor appears to be independent and separate from the activation of the Na^+, K^+-ATPase, and adenylate cyclase by nerve growth factor.[107] Thus, changes in development and/or aging could be secondary to changes in growth factor stimulation of the methyltransferases.

Phospholipid methylation appears to interact with a variety of neurotransmitter receptors in nervous tissue. Studies with C6-astrocytoma cells have demonstrated that both β-adrenergic receptor agonists and benzodiazepines stimulate phospholipid methylation in a dose-dependent manner. When β-adrenergic agonists and benzodiazepines are added together, phospholipid methyltransferase activity was increased in an additive manner.[108] This suggests that different receptors may be located in separate areas of the membrane and are associated with their own component of methyltransferase enzymes. Viral infections of these cells uncouples the β-adrenergic stimulation of phospholipid methylation in these cells.[109] Similar studies in rat brain synaptosomes have indicated that both norepinephrine and dopamine can increase the methylation of phospholipids.[117] It is interesting that a receptor coupled to adenylate cyclase, e.g., the β-adrenergic receptor, and a receptor coupled to the chloride channel-GABA receptor would both activate phospholipid methylation.

Other studies in nervous tissue have correlated changes in receptors and other processes with changes in phospholipid methylation. Increases in phospholipid methylation in synaptosomes decrease norepinephrine uptake and the K^+-stimulated release of norepinephrine.[110] Inhibitors of methylation actually enhance release of norephinephrine from neuronal cultures, e.g., PC-12 cells.[125] Thus, changes in the transmethylation pathway may modulate neurotransmitter release. In other studies, the binding of diazepam and β-carboline-3-carboxylic acid ethyl ester (BCCE) to synaptosomal membranes increases almost twofold by incubating membranes with *S*-adenosyl-L-methionine. The increase in the density of benzodiazepine receptors (i.e., diazepam binding) and GABA receptors (i.e., BCCE binding) is correlated with the methylation of phospholipids.[121] Thus, the methylation of phospholipids may be important for neurotransmitter release as well as the regulation of receptor function in the brain.

The transmethylation pathway may also play a role in providing a source of choline for acetylcholine. Choline availability for acetylcholine synthesis has been extensively studied and phospholipids have been suggested to provide a source of choline for acetylcholine synthesis. Phospholipid methylation is the only known metabolic route by which *de novo* choline is formed. Studies in rat brain synaptosomes have shown that phospholipid methylation can rapidly generate free choline.[131] Although the transmethylation pathway has relatively little activity in brain compared to the liver, the amount of choline generated suggested that the methylation pathway in the brain was turning over and generating choline much more rapidly than would be expected from experiments following the incorporation of [^{3}H]-methyl groups into phosphatidylcholine.[131] Experiments have not been done to determine if choline-deficient diets increase phospholipid methylation in the brain as they do in liver. Studies have shown that intercranial and systemic administration of [^{3}H]-methionine or [^{14}C]-ethanolamine into rats is converted to phosphatidylcholine in the brain.[120] Time course studies following systemic administration indicate that the liver converts these precursors to phosphatidylcholine before significant amounts appear in the brain.[120] It is possible that phosphatidylcholine synthesized through the transmethylation pathway in the liver is secreted into the plasma, transported to the brain, and then used as a source of choline for acetylcholine. Thus, one function of phospholipid methylation is likely to be to provide a source of choline for acetylcholine synthesis.

VII. OTHER TISSUES

Phospholipid methylation has been studied in a large variety of tissues. Numerous studies have been done in liver, leukocytes, and nervous tissue as described above. Experiments on other tissues have been less extensive. In this section, these studies on phospholipid methylation will be briefly reviewed.

Early studies on phospholipid methylation in rat erythrocytes suggested that the enzymes were asymmetrically distributed and that they may play a role in membrane fluidity. Hirata et al.[44] used rat reticulocytes, i.e., immature erythrocytes which contain β-adrenergic receptors linked to adenylate cyclase, to study the interaction between β-adrenergic receptors and phospholipid methylation. These studies were the first experiments to suggest that the phospholipid methylation pathway may be linked to receptor activation. Stimulation of β-adrenergic receptors on reticulocytes increased the incorporation of [^{3}H]-methyl groups from [^{3}H]-methyl-SAM into phospholipids in a dose-dependent manner. The increased methylation of phospholipids did not occur in leaky reticulocyte ghosts, suggesting that the structural integrity of the membrane was important for the response. Propranolol, a β-receptor antagonist blocked the response. To determine if the activation of phospholipid methylation was secondary to activation of adenylate cyclase, reticulocytes were incubated with cholera-toxin and NaF, which stimulate adenylate cyclase directly, bypassing the receptor. The addition of these compounds has no effect on phospholipid methylation.[26,44] Thus, stimulation of phospholipid methylation by β-adrenergic receptors in reticulocytes, unlike that in liver, appears to be due to the binding of β-receptors agonists to the receptor and not secondary to the activation of adenylate cyclase. By introducing various concentrations of SAM into reticulocyte ghosts, it was demonstrated that an increase in the methylation of phospholipids increased the coupling of the β-adrenergic receptor to adenylate cyclase. Increasing membrane fluidity by the addition of vaccenic acid to turkey erythrocytes also facilitated receptor adenylate cyclase coupling, while decreasing fluidity with cholesterol reduces hormone cyclase coupling. These studies have supported the hypothesis that phospholipid methylation increases the fluidity of reticulocyte membranes. This increased fluidity enhances the lateral mobility and rotation of the β-adrenergic receptor, which facilitates coupling of the receptor to adenylate cyclase.[26,44] This hypothesis is supported by the finding that the greatest increases in β receptor cyclase coupling occur at the same concentration of SAM as those which cause the greatest change in membrane viscosity, i.e., 1 to 10 μM SAM. Complementary studies demonstrating an interaction between β-adrenergic receptor stimulation of adenylate cyclase and phospholipid methylation have been done in HeLa cells[136] and C-6 astocytoma cells.[135] In addition to enhancing β-adrenergic receptor coupling in reticulocytes, studies have shown that the formation of phosphatidylcholine through the transmethylation pathway increases the number of β-adrenergic receptor-binding sites. Phospholipid methylation also modifies calcium transport in erythrocytes by increasing the calcium ATPase activity.[134] These studies suggest that changes in phospholipid methylation do modify erythrocyte membranes and thereby alter a variety of processes including β-receptor adenylate cyclase coupling.

Platelets have been studied by a number of groups since they are known to undergo large changes in lipid metabolism during stimulation of aggregation and serotonin release. The phospholipid methyltransferase enzymes have been found in rat,[137] rabbit,[115,116] and human platelets.[118,119,122] Stimulation of platelets results in the release of large amounts of arachidonic acid, with phosphatidylcholine and phosphatidylinositol being the primary sources. Studies in both rabbit[115] and human platelets[118] indicate that phosphatidylcholine synthesized through both the transmethylation and the CDP-choline pathway appear to be metabolized during platelet activation. A number of studies have shown that the methylated phospholipids are decreased during platelet activation;[118,119,122] however, synthesis of phosphatidylcholine through the CDP-choline pathway is also inhibited.[122] Furthermore, several studies have shown that the inhibition of phospholipid methylation has no effect on platelet aggregation and serotonin release stimulated by a variety of agonists including thrombin, collagen, and the calcium ionophore A23187.[115,118,137] Recent studies have suggested that the decreased methylation in activated platelets is secondary to calcium entry and calmodulin-mediated inhibition of the phospholipid methyltransferase enzymes.[115,116] The significance of this calmodulin-mediated inhibition is not clear; however, the data clearly suggests that in platelets phospholipid methylation does not have a major functional role in receptor-stimulated serotonin release and platelet aggregation. Thus, the transmethylation pathway in red cells appears to be coupled to receptor activation, while in platelets any potential functional role is clearly secondary to cell activation.

A few studies have looked at phospholipid methylation in several types of muscle including rat aorta,[138] myogenic cell lines,[123] rat diaphragm,[55] and cardiac muscle.[54,124,126,127] When intact rat aortic strips are incubated with methionine, there is an increase in the fluidity of the microsomal membranes which is correlated to the methylation of phospholipids. Methylation inhibitors decrease microsomal fluidity. Phospholipid methylation was found to be substantially greater in spontaneous hypertensive rat (SHR) aortic microsomes than in Wistar-Kyoto controls and the membrane fluidity was also greater in SHR microsomes.[138] Experiments on the rat hemidiaphragm have shown that stimulation of the muscle directly or through the nerve increases the methylation of lipids. The methylation of lipids and the contractile response of the muscle are increased by exogenous methionine, particularly in young rat hemidiaphragm.[139] These studies have led to the suggestion that the methylation pathway may play a role in regulating the fluidity of sarcoplasmic reticulum membranes[139] and thereby alter the transport of calcium and other ions (as shown in erythrocytes[134]). Thus, changes in methylation appear to alter the contractility of muscle and may play a role in the changes in muscle which occur during aging[139] and hypertension.[138] In contrast to other cell types, experiments on the coupling of the β-adrenergic receptor in myogenic cell lines have suggested that β-receptor activation does not stimulate the transmethylation pathway in smooth muscle cells.[123] It is possible that the primary function of phospholipid methylation in muscle cells is to regulate the lipid composition of the sarcoplasmic reticulum and thereby modify ion transport. This is clearly an area which deserves more research.

The phospholipid methyltransferases are unevenly distributed in the heart with the sinoatrial node having about twice the activity of ventricular muscle.[140] Stimulation of β-adrenergic receptors in dog cardiac membranes enhances phospholipid methylation.[124] Furthermore, increases in phospholipid methylation increase the number of β-adrenergic receptor binding sites in rat[127] and dog heart.[124] These studies suggest that phospholipid methylation in cardiac muscle may play a functional role in β-receptor adenylate cyclase coupling similar to that found in red cells. Chronic alcohol treatment,[140] hyperthyroidism,[126] and hypertrophy secondary to aortic constriction[127] all increase phospholipid methylation in rat heart. The functional significance of these changes is not clear although there is an increase in the number of β-adrenergic receptors in hypertrophied hearts which may be related to the increased phospholipid methylation.[127] Thus, studies in cardiac muscle are consistent with phospholipid methylation playing a role in receptor function.

Phospholipid methylation has also been studied in cells of the male reproductive system, e.g., Leydig cells[128] and sperm.[129,130] Human chorionic gonadotropin (HCG) stimulation of Leydig cells causes a concentration-dependent increase in phospholipid methylation. The 8-bromo-derivative of cAMP also increased phospholipid methylation in Leydig cells. Both treatments increased the maximum velocity of the methylation reaction without changing the K_m for *S*-adenosyl-L-methionine. The largest increase in methyl incorporation occurred in phosphatidyl-*N*-monomethylethanolamine, suggesting that HCG may activate PMT 1 through cAMP. The phospholipid methyltransferases are also found in human[130] and hamster sperm.[129] Methyltransferase inhibitors reduce sperm phospholipid methylation, capacitation,[129] and motility[130] with similar dose-response curves. However, protein carboxymethylation is known to play a role in sperm motility and is also inhibited. The functions of the transmethylation pathway in these responses is still uncertain.

VIII. SUMMARY AND CONCLUSIONS

The studies described clearly indicate that the phospholipid methylation pathway is present in a wide variety of cell types and organisms. In several microorganisms, there are clearly at least two enzymes involved in the three sequential methylations of phosphatidylethanolamine to phosphatidylcholine. In mammalian systems there is considerable data suggesting two enzymes; however, the clear purification of two distinct enzymes has proved difficult and it is possible that there is only one or perhaps three enzymes. In any case, the first methylation reaction which converts phosphatidylethanolamine to phosphatidyl-*N*-monomethylethanolamine is the rate-limiting reaction. Studies on asymmetry suggest that this reaction takes place in the cytoplasmic side of the plasma membrane and that during the methylation of phospholipids there is a flip-flop of the methylated phospholipid from the inside layer of the bilayer to the outside layer. Thus, one possible function of the transmethylation pathway could be related to maintaining the asymmetric distribution of phospholipids. Since phospholipid methylation is a minor pathway of phosphatidylcholine synthesis, it is likely that other factors are more important in maintaining membrane phospholipid asymmetry.

One function of phospholipid methylation is the synthesis of phosphatidylcholine, which involves the *de novo* synthesis of choline. In both microorganisms and mammalian liver, the transmethylation pathway is induced during choline deficiency. Since the liver has a very high activity and secretes phosphatidylcholine into the plasma, it is possible that the transmethylation pathway in liver provides a source of choline and phosphatidylcholine for a number of tissues. Phosphatidylcholine synthesized through the transmethylation pathway contains a much higher percentage of unsaturated fatty acids than the phosphatidylcholine synthesized through the CDP-choline pathway. Thus, the transmethylation pathway synthesizes a particular molecular species of phosphatidylcholine which could function as a mechanism for altering membrane viscosity and/or in providing a source of phosphatidylcholine which contains arachidonic acid.

The phospholipid methylation pathway is clearly activated by a large variety of receptors. In the liver, the transmethylation pathway appears to be regulated by receptors which stimulate both adenylate cyclase and calcium flux. The liver microsomes contain very high methyltransferase activity which appears to be regulated by protein kinases. These are activated by cAMP or by calcium. The CDP-choline pathway appears to be regulated in a reciprocal manner. In contrast, phospholipid methylation is clearly directly linked to receptor activation in mast cells and erythrocytes. Studies have shown that the activation of the transmethylation pathway in these cells is independent of cAMP and a variety of other second messengers. In mast cells, the transmethylation pathway appears to play a primary role in modifying the permeability of the membrane to calcium. The fact that methylation inhibitors, glucocorticoids, and cAMP all inhibit immunoglobulin stimulation of phospholipid methylation, calcium flux, and histamine and arachidonic acid release suggests that the transmethylation pathway may be regulated in such a way that the transduction of signals by certain receptors may be regulated depending upon the activity of the transmethylation pathway. In the case of immunoglobin receptors on mast cells, the transmethylation pathway appears to be essential for the receptor activation of calcium flux. In the case of the β-adrenergic receptor, the transmethylation pathway appears to facilitate coupling in many tissues and to modulate receptor number, but it is not essential for adenylate cyclase activation. These studies concern phospholipid methyltransferases which are clearly localized in the plasma membrane. Thus, receptor stimulation activates phospholipid methylation in a wide variety of tissues. This activation is a primary response to agonist occupancy of the receptor in some cases and is secondary to a second messenger in other cases. Plasma membrane phospholipid methyltransferases may play a key role in transducing and modulating the responses to receptor stimulation, while microsomal methyltransferases may represent an important aspect of the response in other cases. Clearly, the transmethylation pathway functions as a transducer, a modifier, and a response to receptor activation.

The modification of membrane fluid properties by the phospholipid methyltransferases has been suggested by a number of authors to be the mechanism by which the transmethylation pathway modulates receptor responses. Membrane viscosity is a difficult parameter to clearly measure. The synthesis of phosphatidylcholine containing a high degree of unsaturated fatty acids as well as the unknown location of phosphatidyl-*N*-monomethylethanolamine could alter membrane properties. Changes in membrane viscosity can alter membrane permeability, ion transport, and receptor responses. It is possible that phospholipid methylation could alter membrane properties in the localized domain of certain proteins and thereby alter their function. Further experiments and new methodologies to investigate membrane properties are required before this hypothesis can be clearly tested.

In summary, the phospholipid methylation pathway appears to have several related functions. It provides a synthetic route of phosphatidylcholine enriched in unsaturated fatty acids and synthesizes *de novo* choline. In addition, the transmethylation pathway is involved in the cellular responses of hormones and transmitters at several levels. Phospholipid methylation appears to function as a transducer, a modifier, and a response to receptor activation.

REFERENCES

1. **Thompson, G. A., Jr., Ed.,** *The Regulation of Membrane Lipid Metabolism,* CRC Press, Boca Raton, Fla., 1980, chap. 15.
2. **Ansell, G. B., Hawthorne, J. N., and Dawson, R. M. C.,** *Form and Function of Phospholipids,* Elsevier, Amsterdam, 1973.
3. **Quinn, P. J.,** *The Molecular Biology of Cell Membranes,* Macmillan, London, 1976, 33.
4. **Sundler, R. and Akesson, B.,** Regulation of phospholipid biosynthesis in isolated rat hepatocytes: effect of different substrates, *J. Biol. Chem.,* 256, 3359, 1975.
5. **Skurdal, D. N. and Cornatzer, W. E.,** Choline phosphotransferase and phosphatidyl ethanolamine methyltransferase activities, *Int. J. Biochem,* 6, 579, 1975.
6. **Singer, S. J. and Nicholson, G. L.,** The fluid mosaic model of the structure of cell membranes, *Science,* 175, 720, 1972.

7. **Kimelbert, H. K.,** The influence of membrane fluidity on the activity of membrane-bound enzymes, in *Cell Surface Review,* Vol. 3, Poste, G. and Nicholson, G. L., Eds., North-Holland, Amsterdam, 1977, 205.
8. **Rothman, J. E. and Kennedy, E. P.,** Rapid transmembrane movement of newly synthesized phospholipids during membrane assembly, *Proc. Natl. Acad. Sci. U.S.A.,* 74, 1821, 1977.
9. **Roseman, M., Litman, B. J., and Thompson, T. E.,** Transbilayer exchange of phosphatidylethanolamine for phosphatidylcholine and *N*-acetimidoyl-phosphatidylethanolamine in single-walled bilayer vesicles, *Biochemistry,* 14, 4826, 1975.
10. **Bretscher, M. S.,** Phosphatidylethanolamine: differential labelling in intact cells and cell ghosts of human erythrocytes by a membrane-impermeable reagent, *J. Mol. Biol.,* 71, 523, 1972.
11. **Gordesky, S. E. and Marinetti, G. V.,** The asymetric arrangement of phospholipids in the human erythrocyte membrane, *Biochem. Biophys. Res. Commun.,* 50, 1027, 1973.
12. **Zwaal, R. F., Roelofsen, B., and Colley, C. M.,** Localization of red cell membrane constituents, *Biochim. Biophys. Acta,* 300, 159, 1973.
13. **Bloj, B. and Zilversmit, D. B.,** Asymetry and transposition rates of phosphatidylcholine in rat erythrocyte ghosts, *Biochemistry,* 15, 1277, 1976.
14. **Chap, H. J., Zwaal, R. F., and Van Deenen, L. L. M.,** Action of highly purified phospholipases on blood platelets. Evidence for an asymetric distribution of phospholipids in the surface membrane, *Biochim. Biophys. Acta,* 467, 146, 1977.
15. **Tsai, K.-H. and Lenard, J.,** Asymmetry of influenza virus membrane bilayer demonstrated with phospholipase C, *Nature (London),* 253, 554, 1975.
16. **Fontaine, R. N., Harris, R. A., and Schroeder, F.,** Aminophospholipid asymmetry in murine synaptosomal plasma membrane, *J. Neurochem.,* 34, 269, 1980.
17. **Horowitz, N. H.,** The isolation and identification of a natural precursor of choline, *J. Biol. Chem.,* 162, 413, 1946.
18. **Scarborough, G. A. and Nyc, J. F.,** Methylation of ethanolamine phosphatides by microsomes from normal and mutant strains of *Neurospora crassa, J. Biol. Chem.,* 242(2), 238, 1967.
19. **Bremer, J. and Greenberg, D. M.,** Mono and dimethylethanolamine isolated from rat liver phospholipids, *Biochim. Biophys. Acta,* 35, 287, 1959.
20. **Bremer, J. and Greenberg, D. M.,** Biosynthesis of choline *in vitro, Biochim. Biophys. Acta,* 37, 173, 1960.
21. **Bremer, J. and Greenberg, D. M.,** Methyl transfering enzyme system of microsomes in the biosynthesis of lecithin (phosphatidylcholine), *Biochim. Biophys. Acta,* 46, 205, 1961.
22. **Kaneshiro, T. and Law, J. H.,** Phosphatidylcholine synthesis in agrobacterium tumefaciens. I. Purification and properties of a phosphatidylethanolamine *N*-methyltransferase, *J. Biol. Chem.,* 239, 1705, 1964.
23. **Scarborough, G. A. and Nyc, J. F.,** Properties of a phosphatidylmonomethylethanolamine *N*-methyltransferase from *Neurospora crassa, Biochim. Biophys. Acta,* 146, 11, 1967.
24. **Hirata, F. and Axelrod, J.,** Enzymatic synthesis and rapid translocation of phosphatidylcholine by two methyltransferases in erythrocyte membranes, *Proc. Natl. Acad. Sci. U.S.A.,* 75(5), 2348, 1978.
25. **Crocken, B. J. and Nyc, J. F.,** Phospholipid variations in mutant strains of *Neurospora crassa, J. Biol. Chem.,* 239, 1727, 1964.
26. **Hirata, F. and Axelrod, J.,** Phospholipid methylation and biological signal transmission, *Biochem. Biophys. Res. Commun.,* 94(4), 1325, 1980.
27. **Carson, M. A., Atkinson, K. D., and Waechter, C. J.,** Properties of particulate and solubilized phosphatidylserine synthase activity from *Saccharomyces cerevisiae, J. Biol. Chem.,* 257(14), 8115, 1982.
28. **Waechter, C. J. and Lester, R. L.,** Differential regulation of the N-methyl transferases responsible for phosphatidylcholine synthesis in *Saccharomyces cerevisiae, Arch. Biochem. Biophys.,* 158, 401, 1973.
29. **Alemany, S., Gil, M. G., and Mato, J. M.,** Regulation by guanosine 3′:5′-cyclic monophosphate of phospholipid methylation during chemotaxis in *Dictyostelium discoideum, Proc. Natl. Acad. Sci. U.S.A.,* 77(12), 6996, 1980.
30. **Mato, J. M. and Marin-Cao, D.,** Protein and phospholipid methylation during chemotaxis in *Dictyostelium discoideum* and its relationship to calcium movements, *Proc. Natl. Acad. Sci. U.S.A.,* 76(12), 6109, 1979.
31. **Gil, M. G., Alemany, S., Cao, D. M., Castano, J. G., and Mato, J. M.,** Calmodulin modulates phospholipid methylation in *Dictyostelium discoideum, Biochem. Biophys. Res. Commun.,* 94(4), 1325, 1980.
32. **Ramachandran, C. K. and Melnykovych, G.,** Transient changes in phospholipid methylation induced by dexamethasone in lymphoid cells, *Cancer Res.,* 5725, 1983.
33. **de Gunzburg, J., Hohman, R., Part, D., and Veron, M.,** Evidence that a cAMP binding protein from *Dictyostelium discoideum* carries *S*-adenosyl-L-homocysteine hydrolase activity, *Biochimie,* 65(1), 33, 1983.

34. **Rehbinder, D. and Greenberg, D. M.,** Studies of the methylation of ethanolamine phosphatides by liver preparations, *Arch. Biochem. Biophys.,* 108, 110, 1965.
35. **Schneider, W. J. and Vance, D. E.,** Conversion of phosphotidylethanolamine to phosphotidylcholine in rat liver (partial purification and characterization of the enzymatic activities), *J. Biol. Chem.,* 254(10), 3886, 1979.
36. **Hirata, F., Viveros, O. H., Diliberto, E. J., Jr., and Axelrod, J.,** Identification and properties of two methyltransferases in conversion of phosphatidylethanolamine to phosphatidylcholine, *Proc. Natl. Acad. Sci. U.S.A.,* 75(4), 1718, 1978.
37. **Crews, F. T., Hirata, F., and Axelrod, J.,** Identification and properties of methyltransferases that synthesize phosphatidylcholine in rat brain synaptosomes, *J. Neurochem.,* 34(6), 1491, 1980.
38. **Crews, F. T., Hirata, F., and Axelrod, J.,** Phospholipid methyltransferase asymmetry in synaptosomal membranes, *Neurochem. Res.,* 5(9), 983, 1980.
39. **Mozzi, R. and Porcellati, G.,** Conversion of phosphatidylcholine in rat brain by the methylation pathway, *FEBS Lett.,* 100(2), 363, 1979.
40. **Fonlupt, P., Rey, C., and Pacheco, H.,** Phosphatidylethanolamine methylation in membranes from rat cerebral cortex: effect of exogenous phospholipids and S-adenosylhomocysteine, *Biochem. Biophys. Res. Commun.,* 100(4), 1720, 1981.
41. **Blusztajn, J. K., Zeizel, S. H., and Wurtman, R. J.,** Synthesis of lecithin (phosphatidylcholine) from phosphatidylethanolamine in bovine brain, *Brain Res.,* 179, 319, 1979.
42. **Sastry, B. V., Statham, C. N., Meeks, R. G., and Axelrod, J.,** Changes in phospholipid methyltransferases and membrane microviscosity during induction of rat liver microsomal cytochrome P-450 by phenobarbitol and 3-methylcholanthrene, *Pharmacology,* 23(4), 211, 1981.
43. **Hutson, J. L. and Higgins, J. L.,** Asymmetric synthesis followed by transmembrane movement of phosphatidylethanolamine in rat liver endoplasmic reticulum, *Biochim. Biophys. Acta,* 687, 247, 1982.
44. **Hirata, F., Strittmatter, W. J., and Axelrod, J.,** Beta-adrenergic receptor agonists increase phospholipid methylation, membrane fluidity, and beta-adrenergic receptor-adenylate cyclase coupling, *Proc. Natl. Acad. Sci. U.S.A.,* 76(1), 368, 1979.
45. **Siraganian, R. P., McGivney, A., Barsumian, E. L., Crews, F. T., Hirata, F., and Axelrod, J.,** Variants of the rat basophilic leukemia cell line for the study of histamine release, *Fed. Proc. Fed. Am. Soc. Exp. Biol.,* 41(1), 30, 1982.
46. **McGivney, A., Crews, F. T., Hirata, F., Axelrod, J., and Siraganian, R. P.,** Rat basophilic leukemia cell lines defective in phospholipid methyltransferase enzymes, Ca^{2+} influx, and histamine release: reconstitution by hybridization, *Proc. Natl. Acad. Sci. U.S.A.,* 78(10), 6176, 1981.
47. **Hirata, F., Axelrod, J., and Crews, F. T.,** Concanavalin A stimulates phospholipid methylation and phosphatidylserine decarboxylation in rat mast cells, *Proc. Natl. Acad. Sci. U.S.A.,* 76(10), 4813, 1979.
48. **Crews, F. T., Morita, Y., Hirata, F., Axelrod, J., and Siraganian, R. P.,** Phospholipid methylation affects immunoglobulin E mediated histamine and arachidonic-acid release in rat leukemic basophils, *Biochem. Biophys. Res. Commun.,* 93(1), 42, 1980.
49. **Garg, L. C. and Brown, J. C.,** Surface membrane-associated phosphatidylethanolamine N-methyltransferase activity in L-929 cells, *Arch. Biochem. Biophys.,* 220(1), 22, 1983.
50. **Bareis, D. L., Hirata, F., Schiffmann, E., and Axelrod, J.,** Phospholipid metabolism, calcium flux, and the receptor-mediated induction of chemotaxis in rabbit neutrophils, *J. Cell Biol.,* 93(3), 690, 1982.
51. **Hirata, F., Toyoshima, S., Axelrod, J., and Waxdal, M. J.** Phospholipid methylation a biochemical signal modulating lymphocyte mitogenesis, *Proc. Natl. Acad. Sci. U.S.A.,* 77(2), 862, 1980.
52. **Prasad, C. and Edwards, R. M.,** Synthesis of phosphatidylcholine from phosphatidylethanolamine by at least two methyltransferases in rat pituitary extracts, *J. Biol. Chem.,* 256, 13000, 1981.
53. **Sastry, B. V., Owens, L. K., and Janson, V. E.,** Enhancement of contraction of the rat diaphragm by L-methionine and phospholipid methylation and their relationships to aging, *J. Pharmacol. Exp. Ther.,* 221(3), 629, 1982.
54. **Prasad, C. and Edwards, R. M.,** Increased phospholipid methylation in the myocardium of alcoholic rats, *Biochim. Biophys. Res. Commun.,* 111(2), 710, 1983.
55. **Crews, F. T., Morita, Y., McGivney, A., Hirata, F., Siragnian, R. P., and Axelrod, J.,** IgE-mediated histamine release in rat basophilic leukemia cells: receptor activation, phospholipid methylation, Ca^{2+} flux, and release of arachidonic acid, *Arch. Biochem. Biophys.,* 121(2), 561, 1981.
56. **McGivney, A., Crews, F. T., Hirata, F., Axelrod, J., and Siragnian, R. P.,** Rat basophilic leukemia cell lines defective in phospholipid methyltransferase enzymes, Ca^{2+} influx, and histamine release: reconstitution by hybridization, *Proc. Natl. Acad. Sci. U.S.A.,* 78(10), 6176, 1981.
57. **Zatz, M.,** Nonpolar lipid methylation: biosynthesis of fatty acid methyl esters by rat lung membranes using *S*-adenosylmethionine, *J. Biol. Chem.,* 256, 10028, 1981.

58. **Kloog, Y., Zatz, M., Rivay, B., Dudley, P. A., and Markey, S. P.,** Nonpolar lipid methylation-identification of nonpolar methylated products synthesized by rat basophilic leukemia cells, retina and parotid, *Biochem. Pharmacol.*, 31(5), 753, 1982.
59. **Zatz, M.,** Biosynthesis of *S*-methyl-*N*-oleolymercaptoethylamide from oleoyl coenzyme A and *S*-adenosymethionine, *J. Biol. Chem.*, 257, 13673, 1982.
60. **Zimmerman, T. P., Wolberg, G., and Duncan, G. S.,** Inhibition of lymphocyte-mediated cytolysis by 3-diazaadenosine: evidence for a methylation reaction essential to cytolysis, *Proc. Natl. Acad. Sci. U.S.A.*, 75, 6220, 1978.
61. **Hirata, F. and Axelrod, J.,** Enzymatic synthesis and rapid translocation of phosphatidylcholine by two methyltransferases in erythrocyte membranes, *Proc. Natl. Acad. Sci. U.S.A.*, 75(5), 2348, 1978.
62. **Vance, D. E., Choy, P. C., Farren, S. B., Lim, P. H., and Schneider, W. J.,** Asymetry of phospholipid biosynthesis, *Nature (London)*, 270, 268, 1977.
63. **Kahlenberg, A., Walker, C., and Rohrlick, R.,** Evidence for asymetric distribution of phospholipids in the human erythrocyte membrane, *Can. J. Biochem.*, 52, 803, 1974.
64. **Hirata, F. and Axelrod, J.,** Enzymatic methylation of phosphatidylethanolamine increases erythrocyte membrane fluidity, *Nature (London)*, 275, 219, 1978.
65. **Sastry, B. V., Statham, C. N., Meeks, R. G., and Axelrod, J.,** Changes in phospholipid methyltransferases and membrane microviscosity during induction of rat liver microsomal cytochrome P-450 by phenobarbitol and 3-methylcholanthrene, *Pharmacology*, 23(4), 211, 1981.
66. **Nakajima, M., Tamura, E., Irimura, T., Toyoshima, S., Hirano, H., and Osawa, T.,** Mechanism of the concavalin A-induced change of membrane fluidity of chicken erythrocytes, *J. Biochem.*, 89(2), 665, 1981.
67. **Jaiswal, R. K., Landon, E. J., and Sastry, B. V.,** Methylation of phospholipids in microsomes of the rat aorta, *Biochem. Biophys. Acta*, 735(3), 367, 1983.
68. **Jaiswal, R. K., Rama Sastry, B. V., and Landon, E. J.,** Changes in microsomal membrane microviscosity and phospholipid methyltransferases during rat liver regeneration, *Pharmacology*, 24(6), 355, 1982.
69. **Chauhan, V. P., Sikka, S. C., and Kalra, V. K.,** Phospholipid methylation of kidney cortex brush border membranes. Effect on fluidity and transport, *Biochim. Biophys. Acta*, 688(2), 357, 1982.
70. **Goldstein, D. B. and Brett, P. B.,** Role of membrane fluidity in events mediated by phospholipid methylatransferases, *Neurochemistry*, 41, 551A, 1983.
71. **Ishizaka, T., Hirata, F., Sterk, A. R., Ishizaka, K., and Axelrod, J. A.,** Bridging of IgE receptors activates phospholipid methylation and adenylate cyclase in mast cell plasma membranes, *Proc. Natl. Acad. Sci. U.S.A.*, 78(11), 6812, 1981.
72. **Ishizaka, T., Hirata, F., Ishizaka, K., and Axelrod, J.,** Stimulation of phospholipid methylation, Ca^{2+} influx, and histamine release by bridging of IgE receptors on rat mast cells, *Proc. Natl. Acad. Sci. U.S.A.*, 77(4), 1903, 1980.
73. **Glenn, J. L. and Austin, W.,** The conversion of phosphatidyl ethanolamines to lecithins in normal and choline deficient rats, *Biochim. Biophys. Acta*, 231, 153, 1971.
74. **Lombardi, B., Pani, F., Schlunk, F. F., and Shi-hau, C.,** Labeling of liver and plasma lecithins after injection of 1-2-(^{14})C-2-dimethylaminoethanol and (^{14})C-L-methionine-methyl to choline deficient rats, *Lipids*, 4(1), 67, 1969.
75. **Castano, J. G., Alemany, S., Nieto, A., and Mato, J. M.,** Activation of phospholipid methyltransferase by glucagon in rat hepatocytes, *J. Biol. Chem.*, 255, 9041, 1980.
76. **Lyman, R. L., Tinoca, J., Bouchard, P., Sheehan, G., Ostwald, R., and Miljanich, P.,** Sex differences in the metabolism of phosphatidylcholines in rat liver, *Biochim. Biophys. Acta*, 137, 107, 1967.
77. **Lyman, R. L., Hopkins, S. M., Sheehan, G., and Tinoca, J.,** Effects of estradial and testosterone on the incorporation and distribution of [methyl^{14}C]methionine in rat liver lecithins, *Biochim. Biophys. Acta*, 152, 197, 1968.
78. **Kraus-Friedmann, N. and Zimniak, P.,** $^{45}Ca^{2+}$ uptake and phospholipid methylation in isolated rat liver microsomes, *Cell Calcium*, 4(3), 139, 1983.
79. **Kraus-Friedmann, N. and Zimniak, P.,** Glucagon and epinephrine-stimulated phospholipid methylation in hepatic microsomes, *Life Sci.*, 28(13), 1483, 1981.
80. **Marin-Cao, D., Alvarez Chiva, V., and Mato, J. M.,** Beta-adrenergic control of phosphatidylcholine synthesis by transmethylation in hepatocytes from juvenile, adult and adrenalectomized rats, *Biochem. J.*, 216(3), 675, 1983.
81. **Chiva, V. A., Cao, D. M., and Mato, J. M.,** Stimulation by epidermal growth factor of phospholipid methyltransferase in isolated rat hepatocytes, *FEBS Lett.*, 160(1-2), 101, 1983.
82. **Prelach, S. L. and Vance, D. E.,** Regulation of rat liver cytosolic CTP: phosphocholine cytidylyltransferase by phosphorylation and dephosphorylation, *J. Biol. Chem.*, 257, 14198, 1982.
83. **Mato, J. M. and Alemany, S.,** What is the function of phospholipid N-methylation?, *Biochem. J.*, 213, 1, 1983.

84. **Perlach, S. L., Pritchard, P. H., and Vance, D. E.,** cAMP analogues inhibit phosphatidylcholine biosynthesis in cultured rat hepatocytes, *J. Biol. Chem.*, 256(16), 8283, 1981.
85. **Alemany, S., Varela, I., and Mato, J. M.,** Inhibition of phosphatidylcholine synthesis by vasopressin and angiotensin in rat hepatocytes, *Biochem. J.*, 208, 453, 1982.
86. **Ishizaka, T.,** Biochemical analysis of triggering signals induced by bridging of IgE receptors, *Fed. Proc. Fed. Am. Soc. Exp. Biol.*, 41(1), 17, 1982.
87. **Crews, F. T., Morita, Y., Hirata, F., Axelrod, J., and Siraganian, R. P.,** Phospholipid methylation affects immunoglubulin E-mediated histamine and arachidonic acid release in rat leukemia basophils, *Biochem. Biophys. Res. Commun.*, 93(1), 42, 1980.
88. **Crews, F. T., Morita, Y., McGivney, A., Hirata, F., Siraganian, R. P., and Axelrod, J.,** IgE-mediated histamine release in rat basophilic leukemia cells: receptor activation, phospholipid methylation, Ca^{2+} flux, and release of arachidonic acid, *Arch. Biochem. Biophys.*, 212(2), 561, 1981.
89. **Morita, Y. and Siraganian, R. P.,** Inhibition of IgE-mediated histamine release from rat basophilic leukemia cells and rat mast cells by inhibitors of transmethylation, *J. Immunol.*, 127(4), 1339, 1981.
90. **Siraganian, R. P., McGivney, A., Barsumian, E. L., Crews, F. T., Hirata, F., and Axelrod, J.,** Variants of the rat basophilic leukemia cell line for the study of histamine release, *Fed. Proc. Fed. Am. Soc. Exp. Biol.*, 41(1), 30, 1982.
91. **Schnieder, W. J. and Vance, D. E.,** Effect of choline deficiency on the enzymes that synthesize phosphatidylcholine and phosphatidylethanolamine in rat liver, *Eur. J. Biochem.*, 85, 181, 1978.
92. **Daeron, M., Sterk, A. R., Hirata, F., and Ishizaka, T.,** Biochemical analysis of glucocorticoid-induced inhibition of IgE-mediated histamine release from mouse mast cells, *J. Immunol.*, 129(3), 1212, 1982.
93. **Ishizaka, T., Conrad, D. H., Schulman, E. S., Sterk, A., and Ishizaka, K.,** Biochemical analysis of initial triggering events of IgE-mediated histamine release from human lung mast cells, *J. Immunol.*, 130(5), 2357, 1983.
94. **Hoffman, T., Hirata, F., Bougnoux, P., Fraser, B. A., Goldfarb, R. H., Herberman, R. B., and Axelrod, J.,** Phospholipid methylation and phospholipase A2 activation in cytotoxicity by human natural killer cells, *Proc. Natl. Acad. Sci. U.S.A.*, 78(6), 3839, 1981.
95. **Toyoshima, S., Hirata, F., Iwata, M., Axelrod, J., Osawa, T., and Waxdal, M. J.,** Lectin-induced mitosis and phospholipid methylation, *Mol. Immunol.*, 19(3), 467, 1982.
96. **Toyoshima, S., Nakajima, S., Oguchi, Y., and Osawa, T.,** Effect of p-aminobenzoic acid *N*-xyloside sodium salt (K-247) on metabolism and functions of normal lymphocytes and leukemic cells, *J. Pharmacobiodyn.*, 5(6), 430, 1982.
97. **Hirata, F., Toyoshima, S., Axelrod, J., and Waxdal, M. J.,** Phospholipid methylation: a biochemical signal modulating lymphocyte mitogenesis, *Proc. Natl. Acad. Sci. U.S.A.*, 77(2), 862, 1980.
98. **Moore, J. P., Smith, G. A., Hesketh, T. R., and Metcalfe, J. C.,** Early increases in phospholipid methylation are not necessary for the mitogenic stimulation of lymphocytes, *J. Biol. Chem.*, 257(14), 8183, 1982.
99. **Kishi, H., Miki, Y., Kikutani, H., Yamamura, Y., and Kishimoto, T.,** Sequential induction of phospholipid methylation and serine esterase activation in a B cell differentiation factor (BDCF)-stimulated human B cell line, *J. Immunol.*, 131(4), 1961, 1983.
100. **Bougnoux, P., Bonvini, E., Chang, Z. L., and Hoffman, T.,** Effect of interferon on phospholipid methylation by peripheral blood mononuclear cells, *J. Cell Biochem.*, 20(3), 215, 1982.
101. **Hattori, T., Pack, M., Bougnoux, P., Chang, Z. L., and Hoffman, T.,** Interferon-induced differentiation of U937 cells. Comparison with other agents that promote differentiation of human myeloid or monocyte like cell lines, *J. Clin. Invest.*, 73(1), 237, 1983.
102. **Hirata, F., Corcoran, B. A., Venkatasubramanian, K., Schiffman, E., and Axelrod, J.,** Chemoattractants stimulate degradation of methylated phospholipids and release of arachidonic acid in rabbit leukocytes, *Proc. Natl. Acad. Sci. U.S.A.*, 77, 2533, 1979.
103. **Bareis, D. L., Hirata, F., Schiffmann, E., and Axelrod, J.,** Phospholipid metabolism, calcium flux, and the receptor-mediated induction of chemotaxis in rabbit neutrophils, *J. Cell Biol.*, 93(3), 690, 1982.
104. **Garcia, C. I., Mato, J. M., Vasanthakumar, G., Wiesmann, W. P., Schiffmann, E., and Chiang, P. K.,** Paradoxical effects of adenosine on neutrophil chemotaxis, *J. Biol. Chem.*, 258(7), 4345, 1983.
105. **Kahlenberg, A., Walker, C., and Rohrlick, R.,** Evidence for asymmetric distribution of phospholipids in the human erythrocyte membrane, *Can. J. Biochem.*, 52, 803, 1974.
106. **Pfenninger, K. H. and Johnson, M. P.,** Nerve growth factor stimulates phospholipid methylation in growing neurites, *Proc. Natl. Acad. Sci. U.S.A.*, 78(12), 7797, 1981.
107. **Skaper, S. D. and Varon, S.,** Nerve growth factor stimulates phospholipid methylation in target ganglionic neurons independently of the cyclic AMP and sodium pump responses, *J. Neurochem.*, 42(1), 116, 1984.
108. **Strittmatter, W. J., Hirata, F., Axelrod, J., Mallorga, P., Tallman, J. F., and Henneberry, R. C.,** Benzodiazepine and beta-adrenergic receptor ligands independently stimulate phospholipid methylation, *Nature, (London)*, 282(5741), 857, 1979.

109. **Munzel, P. and Koschel, K.** Alteration in phospholipid methylation and impairment of signal transmission in persistently paramyxovirus-infected C65 rat glioma cells, *Proc. Natl. Acad. Sci. U.S.A.*, 79(12), 3692, 1982.
110. **Fonlupt, P., Rey, C., and Pacheco, H.** Phospholipid methylation and noradrenaline exchanges in a synaptosomal preparation from the rat brain, *J. Neurochem.*, 38(6), 1615, 1982.
111. **Vance, D. E., Choy, P. C., Farren, S. B., Lim, P. H., and Schneider, W. J.**, Asymetry of phospholipid biosynthesis, *Nature (London)*, 270, 268, 1977.
112. **Mozzi, R., Siepa, D., Andreoli, V., and Porcellati, G.**, The synthesis of choline plasmalogen by the methylation pathway in rat brain, *FEBS Lett.*, 131(1), 115, 1981.
113. **Hitzemann, R.**, Developmental regulation of phospholipid methylation in rat brain synaptosomes, *Life Sci.*, 30(15), 1297, 1982.
114. **Crews, F. T., Calderini, G., Battistella, A., and Toffano, G.**, Age dependent changes in the methylation of rat brain phospholipids, *Brain Res.*, 229, 256, 1981.
115. **Taniguchi, S., Mori, K., Hayashi, H., Fujiwara, M., and Fujiwara, M.**, Calcium-dependent regulation of phospholipid methylation of rabbit platelets, *Thromb. Res.*, 32(5), 495, 1983.
116. **Mori, K., Taniguchi, S., Kumada, K. Nakazawa, K., Fujiwara, M., and Fujiwara, M.**, Enzymatic properties of phospholipid methylation in rabbit platelets, *Thromb. Res.*, 29(2), 215, 1983.
117. **Leprohon, C. E., Blusztajn, J. K., and Wurtman, R. J.**, Dopamine stimulation of phosphatidylcholine (lecithin) biosynthesis in rat brain neurons, *Proc. Natl. Acad. Sci. U.S.A.*, 80, 2063, 1983.
118. **Shattil, S. J., Montgomery, J. A., and Chiang, P. K.**, The effect of pharmacologic inhibition of phospholipid methylation on human platelet function, *Blood*, 59(5), 906, 1982.
119. **Cordasco, D. M., Segarnick, D. J., and Rotrosen, J.**, Human platelet phospholipid methylation, *Life Sci.*, 29(22), 2299, 1981.
120. **Morganstern, R. D. and Abdel-Latif, A. A.**, Incorporation of [^{14}C] ethanolamine and [^{3}H] methionine into phospholipids of rat brain and liver in *in vivo* and *in vitro*, *J. Neurobiol.*, 5(5), 393, 1974.
121. **DePerri, B., Calderini, G., Battistella, A., and Toffano, G.**, Phospholipid methylation increases [^{3}H]diazepam and [^{3}H]GABA binding in membrane preparations of rat cerebellum, *J. Neurochem.*, 41, 302, 1983.
122. **Shattil, S. J., McDonough, M., and Burch, J. W.**, Inhibition of platelet phospholipid methylation during platelet secretion, *Blood*, 57(3), 537, 1981.
123. **Koch, T. K., Gordon, A. S., and Diamond, I.**, Phospholipid methylation in myogenic cells, *Biochem. Biophys. Res. Commun.*, 114(1), 339, 1983.
124. **Okumura, K., Ogawa, K., and Satake, T.**, Phospholipid methylation in canine cardiac membranes. Relations to beta-adrenergic receptors and digitalis receptors, *Jpn. Heart J.*, 24(2), 215, 1983.
125. **Rabe, C. S., Williams, T. P., and McGee, R.**, Enhancement of depolarization-dependent release of norepinephrine by an inhibitor of *S*-adenosylmethionine-dependent transmethylations *Life Sci.*, 27, 1753, 1980.
126. **Limas, C. J.**, Increased phospholipid methylation in the myocardium of hyperthyroid rats, *Biochim. Biophys. Acta*, 632(2), 254, 1980.
127. **Limas, C. J.**, Effect of phospholipid methylation on beta-adrenergic receptors in the normal and hypertrophied rat myocardium, *Circ. Res.*, 47(4), 536, 1980.
128. **Nieto, A. and Catt, K. J.**, Hormonal activation of phospholipid methyltransferase in the Leydig cell, *Endocrinology*, 113(2), 758, 1983.
129. **Llanos, M. N. and Meizel, S.**, Phospholipid methylation increased during capacitation of golden hamster sperm *in vitro*, *Biol. Reprod.*, 28(5), 1043, 1983.
130. **Sastry, B. V. and Janson, V. E.**, Depression of human sperm motility by inhibition of enzymatic methylation, *Biochem. Pharmacol.*, 32(8), 1423, 1983.
131. **Blusztajn, J. K. and Wurtman, R. M.**, Choline biosynthesis by a preparation enriched in synaptosomes from rat brain, *Nature (London)*, 290, 417, 1981.
132. **Alemany, S., Varela, I., Harper, J. F., and Mato, J. M.**, Calmodulin regulation of phospholipid and fatty acid methylation by rat liver microsomes, *J. Biol. Chem.*, 257(16), 9249, 1982.
133. **Foreman, J., Hallett, M., and Monger, J.**, Relationship between histamine secretion and ^{45}calcium uptake by mast cells, *J. Physiol. (London)*, 271, 193, 1977.
134. **Strittmatter, W., Hirata, F., and Axelrod, J.**, Increased calcium ATPase activity associated with methylation of phospholipids in human erythrocytes, *Biochem. Biophys. Res. Commun.*, 88(1), 147, 1979.
135. **Hirata, F., Tallman, J. F., Jr., Henneberry, R. C., Mallorga, P., Strittmatter, W. J., and Axelrod, J.**, Regulation of beta-adrenergic receptors by phospholipid methylation, *Adv. Biochem. Psychopharmacol.*, 21, 91, 1980.
136. **Hirata, F., Tallman, J. F., Henneberry, R. C., Mallorga, P., Strittmatter, W. J., and Axelrod, J.**, Phospholipid methylation: a possible mechanism of signal transduction across biomembranes, *Prog. Clin. Biol. Res.*, 63, 383, 1981.

137. **Hotchkiss, A., Jordan, J. V., Hirata, F., Shulman, N. R., and Axelrod, J.,** Phospholipid methylation and human platelet function, *Biochem. Pharmacol.*, 30(15), 2089, 1981.
138. **Jaiswal, R. K., Landon, E. J., and Sastry, B. V.,** Methylation of phospholipids in microsomes of the rat aorta, *Biochem. Biophys. Acta*, 735(3), 367, 1983.
139. **Sastry, B. V., Owens, L. K., and Janson, V. E.,** Enhancement of the rat diaphragm by L-methionine and phospholipid methylation and their relationships to aging, *J. Pharmacol. Exp. Ther.*, 221(3), 629, 1982.
140. **Prasad, C. and Edwards, R. M.,** Increased phospholipid methylation in the myocardium of alcoholic rats, *Biochem. Biophys. Res. Commun.*, 111(2), 710, 1983.

Chapter 5

PHOSPHATIDYLINOSITOL TURNOVER AND Ca^{2+} GATING

Irene Litosch and John N. Fain

TABLE OF CONTENTS

I. OVERVIEW

Hokin and Hokin[1] were the first to demonstrate that cholinergic stimulation of pigeon pancreases increased ^{32}P incorporation into phosphatidylinositol and its precursor phosphatidic acid. The increased incorporation of ^{32}P into phosphatidylinositol (the phosphatidylinositol effect) was independent of the secretory response and not secondary to the elevation of cytosolic Ca^{2+}. Hokin-Neaverson[2,3] found that levels of phosphatidylinositol decreased during cholinergic stimulation and that there was an increase in diacylglycerol and inositol. Similar results were reported by Jones and Michell[4] using rat parotid glands, suggesting that agonist actually stimulated breakdown of phosphatidylinositol by an apparent activation of a phosphatidylinositol-specific phospholipase C. Pharmacological studies have shown that stimulated phosphatidylinositol turnover is associated primarily with muscarinic cholinergic and α_1-adrenergic amine stimulation.[5] Activation of this class of receptors results in an increase in the levels of intracellular Ca^{2+} which acts as a second messenger to initiate cellular responses such as secretion, contraction, or glycogenolysis.

Durell et al.[6] suggested that stimulated phosphatidylinositol turnover might be linked to receptor activation. Michell[5] elaborated on this view and proposed that phosphatidylinositol breakdown was part of the receptor mechanism linked to the entry of extracellular Ca^{2+}. Breakdown of a small pool of plasma membrane phosphatidylinositol was postulated to result in activation of hormone-regulated Ca^{2+} channels. Stimulated turnover of phosphatidylinositol thus became a metabolic consequence of an initial hormone-mediated breakdown of phosphatidylinositol at the plasma membrane.

Our understanding of the phosphatidylinositol response has increased considerably since its original observation by Hokin and Hokin.[1] The number of agonists reported to stimulate phosphatidylinositol turnover has increased to include 5-hydroxytryptamine,[7] f-MetLeuPhe,[8] vasopressin,[9] angiotensin,[9] substance P,[10] thrombin,[11] thyrotrophin-releasing hormone,[12] and epidermal growth factor.[13] An examination of the phosphatidylinositol response in different tissues suggests that several types of phosphatidylinositol responses exist. In systems such as rat hepatocytes and blowfly salivary glands, where phosphatidylinositol breakdown appears to be linked to Ca^{2+} entry, there is a decrease of less than 10% of the cellular phosphatidylinositol in response to hormone stimulation.[14] This is in contrast to tissues such as platelets, parotid, and acinar pancreas where agonists stimulate breakdown of up to 50% of the cellular phosphatidylinositol. Breakdown of such a large fraction of cellular phosphatidylinositol is unlikely to be directly linked to receptor activation and Ca^{2+} gating. Phosphatidylinositol breakdown of this magnitude may be secondary to an initial breakdown of a small pool of plasma membrane phosphatidylinositol linked to the entry of extracellular Ca^{2+}. The subsequent elevation of intracellular Ca^{2+} and diacylglycerol may activate the soluble Ca^{2+}-activated phospholipase C.[14]

Several groups have shown that phosphatidylinositol breakdown occurs within seconds of hormone addition.[11,12,15] Stimulated resynthesis of phosphatidylinositol is observed, however, after a lag of approximately 2.5 min.[15] One of the most significant observations has been that the direct addition of hormone to cell-free systems increases the degradation of phosphatidylinositol.[16-19] These studies indicate that breakdown of phosphatidylinositol is a direct consequence of receptor activation.

Recent studies have added a new dimension to the phosphatidylinositol story. It has been shown that the polyphosphoinositides, phosphatidylinositol-4,5bisphosphate (PtdIns(4,5)P_2) and phosphatidylinositol-4phosphate (PtdIns4P), also undergo rapid metabolic changes upon hormone addition, indicating that this class of phosphoinositides is also under receptor control.[20] Streb et al.[21] have suggested that inositol-1,4,5-trisphosphate, produced as a consequence of phospholipase C-mediated breakdown of PtdIns(4,5)P_2, is a second messenger which releases Ca^{2+} from intracellular storage sites. The diacylglycerol generated from phospholipase C-mediated breakdown of phosphoinositides activates a C-kinase[22] and results in phosphorylation of cellular proteins; this implicates another role for phosphoinositides in hormone action. At present, the precise mechanism by which hormones stimulate phosphoinositide breakdown, the metabolic interrelationships between phosphoinositides, and the link between phosphoinositide breakdown and Ca^{2+} gating is not known.

II. PHOSPHOINOSITIDE SYNTHESIS

The phosphoinositides comprise approximately 8 to 10% of cellular phospholipids. Phosphatidylinositol is the major phosphoinositide and represents at least 90% of the total phosphoinositide pool. PtdIns(4,5)P_2 and PtdIns4P together comprise no more than 10% of phosphoinositide content.[5] The major site for phosphatidylinositol biosynthesis is the endoplasmic reticulum,[23] although appreciable synthesis also occurs in the Golgi.[24] *De novo* synthesis of phosphatidylinositol occurs from phosphatidic acid which is derived from diacylglycerol via the action of diacylglycerol kinase (Figure 1). The enzyme CTP-phosphatidic acid cytidyltransferase, in the presence of CTP, converts phosphatidic acid to CDP-diacyglycerol. In the presence of inositol, CDP-diacylglycerol is converted to phosphatidylinositol by the action of CDP-diacylglycerol:inositol transferase. The activity of this enzyme is markedly inhibited by micromolar calcium concentrations.[25]

Inositol may be incorporated into phosphatidylinositol by CDP-diacylglycerol:inositol transferase. Another enzyme involved in inositol incorporation is the Mn^{2+}-stimulated exchange enzyme[26] which catalyzes a simple exchange of inositol. Tolbert et al.[27] showed that in rat hepatocytes there was appreciable Mn^{2+}-exchange activity. The inositol incorporated into phosphatidylinositol via the exchange enzyme did not equilibriate with the hormone-sensitive pool of phosphatidylinositol. Thus, while ^{32}P uptake into phosphatidylinositol was stimulated by vasopressin or α_1-adrenergic amines, inositol uptake was also not affected in the absence or presence of Mn^{2+}.[27]

The polyphosphoinositides are synthesized from phosphatidylinositol by phosphorylation of phosphatidylinositol as indicated below:

$$\text{Phosphatidylinositol} + \text{ATP} \rightarrow \text{PtdIns4P}$$
$$\text{(phosphatidylinositol kinase)}$$
$$\text{PtdIns4P} + \text{ATP} \rightarrow \text{PtdIns(4,5)P}_2$$
$$\text{(PtdIns4P kinase)}$$

Phosphatidylinositol kinase is found in plasma membrane,[28] microsomal membrane fractions,[29] and lysosomes.[30] Phosphatidylinositol kinase requires Mg^{2+} for activity.[31] Ca^{2+} does not activate phosphatidylinositol kinase in liver homogenate.[28] However, in the presence of Mg^{2+}, Ca^{2+} inhibits phosphatidylinositol kinase possibly by competing for a cation binding site.[28]

PtdIns4P kinase has been found in cytosol,[32] Golgi,[33] and plasma membrane.[34] In brain supernatant, the enzyme is activated by Ca^{2+} and Mg^{2+}. Ca^{2+} inhibits enzymatic activity in the presence of Mg^{2+}.[32]

A sequential series of reactions involving phosphatidylinositol and PtdIns4P kinase appears to exist in brain tissue for formation of PtdIns(4,5)P_2.[35] In rabbit erythrocyte membranes, however, there were differences in the salt requirement, detergent sensitivity, and optimum temperature for formation of labeled PtdIns4P and PtdIns(4,5)P_2.[36,37] It was found that ^{32}P-PtdIns4P was not converted to ^{32}P-PtdIns(4,5)P_2 under conditions in which the membranes should convert PtdIns4P to PtdIns(4,5)P_2. Thus, once ^{32}P became incorporated into PtdIns4P and PtdIns(4,5)P_2 these ^{32}P-labeled lipids did not interconvert even in the presence of added nonradioactive ATP. It was proposed that PtdIns4P and PtdIns(4,5)P_2 are not exclusively

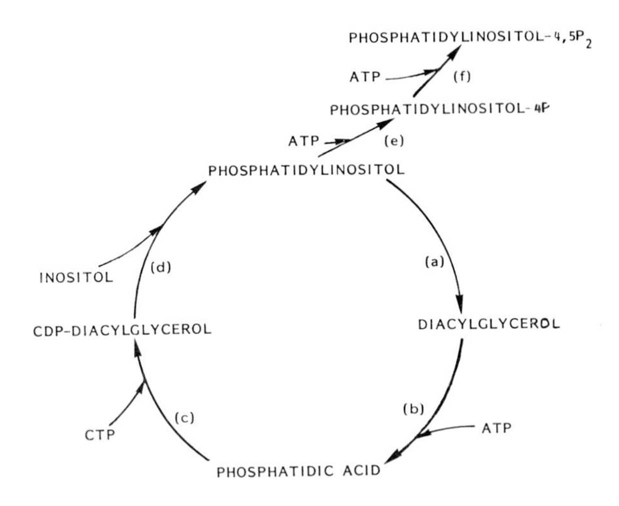

FIGURE 1. Enzymes involved in phosphoinositide synthesis. (a) Receptor activation results in the formation of diacylglycerol which is derived from phospholipase C-mediated breakdown of phosphatidylinositol, phosphatidylinositol 4-phosphate and phosphatidylinositol 4,5-bisphosphate. (b) Diacylglycerol is rapidly utilized for the synthesis of phosphatidic acid via diacylglycerol kinase. (c) Phosphatidic acid is converted to CDP-diacylglycerol via CTP-phosphatidic acid cytidyltransferase. (d) The enzyme CDP-diacylglycerol:inositol transferase converts CDP-diacylglycerol to phosphatidylinositol in the presence of inositol (e) Phosphatidylinositol is converted to phosphatidylinositol-4P by phosphatidylinositol kinase. (f) Phosphatidylinositol-4P is converted to phosphatidylinositol-4,5P_2 via phosphatidylinositol-4P kinase.

synthesized by a sequential phosphorylation of phosphatidylinositol. PtdIns(4,5)P_2 may be synthesized through a separate PtdIns4P kinase pathway. Under these conditions changes in isotopic content of PtdIns(4,5)P_2 might not correlate with changes in total content.

Phosphorylation of phosphatidylinositol may occur through enzymes other than PtdIns4P and phosphatidylinositol kinase. In lymphocyte membranes, the addition of catalytic subunit of the cyclic AMP (cAMP)-dependent protein kinase increased the formation of polyphosphoinositides from endogenous phosphatidylinositol.[38]

Gispen and co-workers have shown that rat brain plasma membrane contains a Ca^{2+}-dependent protein kinase which is similar to protein kinase C.[39,40] This kinase mediates the phosphorylation of a B-50 protein present in synaptic membranes. Direct addition of ACTH to synaptic membranes inhibited B-50 phosphorylation and this was associated with an increased conversion of exogenously added PtdIns4P to PtdIns(4,5)P_2. These membranes did not convert exogenous phosphatidylinositol to PtdIns4P.[41] It is not clear whether PtdIns4P kinase per se mediates the formation of PtdIns(4,5)P_2 from PtdIns4P. Possibly other kinases are activated during ACTH action that can phosphorylate PtdIns4P. This remains a crucial point to address.

These studies on polyphosphoinositide synthesis in vitro indicate that multiple pathways for PtdIns(4,5)P_2 formation exist. The physiological significance of these processes remains to be established.

III. CATABOLISM OF PHOSPHOINOSITIDES

Phosphatidylinositol breakdown occurs through a phosphatidylinositol-specific phospholipase C with production of diacylglycerol and inositol-1,2-cyclic phosphate. Cellular phosphatases rapidly degrade inositol-1,2-cyclic phosphate to inositol-1-phosphate and inositol. A phosphatidylinositol-specific phospholipase C requiring high concentrations of Ca^{2+} for activity is present in the cytosol of most cells.[42-46] Metabolism of phosphatidylinositol can also occur through lysosomal enzymes which contain a Ca^{2+}-inhibited phospholipase C as well as a phospholipase A_1 which deacylates phosphatidylinositol to lysophosphatidylinositol and eventually glycerolphosphoinositol.[47] The lysosomal phospholipase C has been shown to hydrolyze all phospholipids Matsuzawa and Hostetter.[48] A membrane-bound phospholipase C activity distinguishable from cytosolic activity was described in cerebral cortex,[49] but appears to have reflected trapped cytosolic enzyme.[43] To date, there has been no conclusive demonstration of a membrane-bound phosphatidylinositol-specific phospholipase C. The majority of the cited studies, however, would have failed to localize the enzyme to the plasma membrane if the enzyme were maintained in a latent form requiring hormone activation for expression of activity. In this regard it is interesting to note that Bormann et al.[50] reported that neutrophil plasma membranes contain a phospholipase A_2 activity activated by the addition of agonists. There is little activity in the absence of agonist.

The Ca^{2+} requirement for phospholipase C activity appears to be at variance with its proposed role in hormone action. Activation of phospholipase C by Ca^{2+} in vitro may reflect Ca^{2+} modulation of a catalytic or regulatory site on the enzyme. Alternatively, Ca^{2+} may simply activate the phosphatidylinositol substrate by neutralizing the negative charges on the lipid headgroups. It has been shown that phospholipase activity depends on the physiochemical state of the substrate.[51] Multiple forms of phospholipase C complicate the situation. Hoffman and Majerus[45] suggested that there were at least two forms of phospholipase C in seminal vesicles, while Low and Weglicki[46] isolated four forms of phospholipase C in myocardium. It is not clear if the multiple forms of phospholipase C represent isozymes or whether they are generated during the purification procedure by protease action.

The contribution of substrate in modulating expression of enzymatic activity must also be considered. Phospholipase C activity against phosphatidylinositol present in intact membranes is lower than against pure substrate. This low activity of phospholipase C against membrane-associated phosphatidylinositol may reflect inhibition by positively charged proteins and phosphatidylcholine.[51] In contrast, unsaturated fatty acids, such as arachidonic acid and diacylglycerol stimulated enzymatic activity.[51] The membrane environment of the hormone-sensitive pool of phosphatidylinositol may differ from the majority of the cellular phosphatidylinositol. Less than 10% of the cellular phosphatidylinositol is degraded during hormone stimulation in hepatocytes[9] and blowfly glands.[7]

Platelets appear to have multiple pools of phosphatidylinositol. Thrombin stimulates an initial breakdown of phosphatidylinositol followed by a secondary breakdown of phosphatidylinositol that may be a consequence of the elevation in cytosolic Ca^{2+}.[52] In blowfly salivary glands, the hormone-sensitive pool of phosphatidylinositol is not in metabolic equilibrium with cellular phosphatidylinositol. Depletion of the hormone-sensitive pool of phosphatidylinositol in blowfly salivary glands by prolonged exposure to hormone does not result in recruitment of phosphatidylinositol from other cellular compartments to replenish the hormone sensitive pool of phosphatidylinositol. The hormone-sensitive pool of phosphatidylinositol is restored only during incubation of depleted glands in inositol-containing medium.[53,54] In hepatocytes, the hormone-sensitive pool of phosphatidylinositol has a higher turnover rate than the rest of the cellular phosphatidylinositol.[15] These observations indicate that conclusions about the regulation of phospholipase C will depend upon the particular assay conditions employed.

Polyphosphoinositide breakdown also occurs by the phospholipase C route with generation of diacylglycerol and the corresponding inositol phosphates. Most studies on phosphoinositide breakdown have been done with crude cellular fractions which contain other phospholipases and phosphatases. In addition the presence of soluble inhibitors and activators may greatly complicate interpretation of these studies.[55] All the available evidence suggests that the same phospholipase C hydrolyzes both phosphatidylinositol and PtdIns(4,5)P_2. Low and Weglicki isolated four fractions containing phospholipase C activity from beef heart ventricles by ion exchange chromatography. All forms of phospholipase C required Ca^{2+} for activity and hydrolyzed phosphatidylinositol as well as polyphosphoinositides. Recent studies by Irvine et al.[56] on rat brain supernatant have shown that rat brain supernatant contains a soluble Ca^{2+}-activated phospholipase C which degrades PtdIns(4,5)P_2, PtdIns4P, and phosphatidylinositol. The lipid environment has pronounced effects on the hydrolysis of the phosphoinositides. Phosphatidylcholine has little effect on PtdIns(4,5)P_2 hydrolysis,[56] but it markedly inhibits phosphatidylinositol hydrolysis.[51] Phosphatidylethanolamine stimulates hydrolysis of both phosphatidylinositol and PtdIns(4,5)P_2. However, it appears that the degree of stimulation afforded by phosphatidylethanolamine differs for phosphatidylinositol and PtdIns(4,5)P_2. A phosphatidylethanolamine:phosphatidylinositol ratio of 2:7 produced maximal stimulation of phosphatidylinositol hydrolysis. A 10:1 and 50:1 ratio was employed by Irvine et al.[56] for stimulating PtdIns(4,5)P_2 hydrolysis. Under these conditions, there was a preferential hydrolysis of PtdIns(4,5)P_2 over phosphatidylinositol. These type of studies emphasize the necessity to consider the role of the native substrate environment on enzymatic activity. It is obvious that in vitro assay conditions may bias the activity of the enzyme under investigation. This may generate misleading data concerning activity of the enzyme on all its substrates.

Loss of PtdIns(4,5)P_2 can also occur through activation of a phosphomonoesterase which converts PtdIns(4,5)P_2 to phosphatidylinositol. Both Ca^{2+} and Mg^{2+} stimulate phosphomonoesterase activity.[57,58] In tissues which contain appreciable phosphomonoesterase activity, the agonist-induced increase in cytosolic Ca^{2+} might mask the breakdown of labeled phosphatidylinositol by stimulating conversion of PtdIns(4,5)P_2 to phosphatidylinositol. Enzymes which degrade the polyphosphoinositides through a deacylase route have not yet been described. As more effort is directed at elucidating the mechanism of polyphosphoinositide breakdown, deacylase activity against the polyphosphoinositides will no doubt be found.

IV. KINETICS OF AGONIST-INDUCED PHOSPHOINOSITIDE BREAKDOWN

The mechanism by which receptor activation stimulates phosphoinositide breakdown remains elusive and consequently controversial. Studies on intact cells have shown that agonists induce a rapid breakdown of phosphoinositides through an apparent activation of phospholipase C activity. Diacylglycerol[59-61] and inositol phosphates[62-66] are formed during hormone stimulation. Lin and Fain[67] as well as Bennett et al.[68] have shown that agonists induce a preferential loss of phosphatidylinositol from the plasma membrane, linking phosphatidylinositol breakdown to a primary change at the plasma membrane.

When hormone effects are monitored by measuring changes in lipid content, it is apparent that there is a rapid decrease in the levels of labeled phosphatidylinositol and labeled PtdIns(4,5)P_2.[11,12,15,61] A consistent effect of hormones on breakdown of labeled PtdIns4P has been more difficult to demonstrate. In some studies, labeled PtdIns4P decreased during hormone stimulation[64,69,70] while in others, the amount of labeled PtdIns4P did not change significantly.[15,71,72] This variability in the PtdIns4P response may be a consequence of its intermediate position in the phosphoinositide synthetic pathway.

The initial rate of agonist-induced phosphatidylinositol and PtdIns(4,5)P_2 disappearance is similar.[15,61,64,65] In hepatocytes, half-maximal loss of phosphoinositides is observed within 15 sec.[15,61] Breakdown of ^{3}H-inositol-labeled phosphoinositides is followed by a rapid resynthesis of all phosphoinositides to control unstimulated levels by 1 min. It is not known whether the Mn^{2+}-stimulated exchange enzyme or the *de novo* pathway is involved in phosphatidylinositol resynthesis. In contrast with ^{32}P-labeled phosphoinositides, PtdIns(4,5)P_2 and PtdIns4P levels recover to control levels by 5 min. Stimulated resynthesis of phosphatidylinositol to levels greater than control is observed after 2.5 min.[15]

It is not clear why there is such a rapid recovery of labeled lipid even in the presence of agonist which stimulates lipid breakdown. Depletion of the hormone-sensitive pool of phosphoinositides could initiate a rapid compensatory resynthesis mechanism, allowing the cell to replenish its phosphoinositide pool and respond to a second challenge of hormone. Cellular kinases may become activated as a consequence of the increase in cytosolic Ca^{2+} and diacylglycerol. These may stimulate diacylglycerol kinase and phosphoinositide kinase activity resulting in formation of phosphatidic acid, phosphatidylinositol, and eventually polyphosphoinositides. Alternatively, recovery of the phosphoinositides may be a consequence of a gradual desensitization of the receptor. Studies on A-431 cells have shown that C-kinase mediates phosphorylation of the EGF receptor resulting in desensitization.[73] A diacylglycerol-mediated increase in C-kinase activity may initiate a feedback inhibition of phosphoinositide breakdown. Resynthesis of phosphoinositides in presence of agonist is characteristic of many mammalian systems but not in blowfly salivary glands. In the salivary gland, 5-hydroxytryptamine stimulates net breakdown of ^{32}P or ^{3}H-labeled phosphoinositides. Resynthesis of phosphoinositides during hormone stimulation is inhibited by a mechanism which appears to involve Ca^{2+}.[54]

The amount of phosphatidylinositol degraded during hormone stimulation is approximately five- to tenfold greater than the amount of PtdIns(4,5)P_2 degraded based on both isotopic and chemical analysis.[11,15] In rat hepatocytes, both phosphatidylinositol and PtdIns(4,5)P_2 disappear at the same rate during hormone stimulation indicating that rapid degradation of both lipids is initiated upon receptor activation.[15,61,65] It is not obvious, however, if the phosphoinositides are degraded through the same pathway or whether different metabolic pathways are involved. Chemical analysis of the phosphoinositide content has shown that phosphatidylinositol decreases within 30 sec of hormone addition.[11,12,15] Changes in PtdIns(4,5)P_2 content have, in most cases, been estimated from changes in radioactivity. In horse platelets, thrombin caused a 15% decrease in the amount of PtdIns(4,5)P_2.[72] In rat hepatocytes[15] and human platelets,[74] the amount of PtdIns(4,5)P_2 increased during hormone stimulation. These observations suggest that part of the decrease in phosphatidylinositol is due to its conversion to PtdIns(4,5)P_2. Resynthesis of PtdIns(4,5)P_2 can occur from phosphatidylinositol via phosphatidylinositol kinase and PtdIns4P kinase[35] or from PtdIns4P via PtdIns4P kinase.[37] Studies in rat parotid support the concept that some loss of phosphatidylinositol occurs through its conversion to PtdIns(4,5)P_2. The amount of labeled inositol-trisphosphate generated during hormone stimulation appeared to be greater than the amount of labeled PtdIns(4,5)P_2 lost.[70] Interpretation of these results is complicated since the specific activity of the labeled hormone-sensitive pool of PtdIns(4,5)P_2 is not known.

It is possible that stimulation of phosphoinositide kinase activity is a feature of the phosphoinositide cycle. Phosphoinositide kinase activity may be stimulated as a direct consequence of receptor activation. The insulin-stimulated synthesis of PtdIns(4,5)P_2 in isolated adipocytes[75] may reflect such a receptor-mediated increase in phosphatidylinositol phosphorylation since the insulin receptor appears to have intrinsic kinase activity.[76] It is possible that similar mechanisms also exist for α_1-adrenergic amines and other Ca^{2+} mobilizing hormones. Alternatively, loss of PtdIns(4,5)P_2 from critical membrane sites may initiate rapid compensatory resynthesis of PtdIns(4,5)P_2. In blowfly salivary glands, depletion of phosphoinositides by prior exposure to hormone resulted in a preferential resynthesis of PtdIns(4,5)P_2 that occurred at the expense of phosphatidylinositol resynthesis.[77] Clearly, studies in more systems are needed to determine whether stimulated PtdIns(4,5)P_2 synthesis is a universal feature of the phosphoinositide cycle.

It has recently been suggested that hormones only stimulate PtdIns(4,5)P_2 breakdown and that all loss of phosphatidylinositol is secondary to its conversion to PtdIns(4,5)P_2.[70,79] Evidence in support of this mechanism comes primarily from studies on the kinetics of inositol phosphate formation. Using Dowex-formate columns to separate the inositol phosphates generated from phosphoinositide breakdown it has been shown that inositol-1,4,5-trisphosphate formation preceded that of inositol-1-phosphate formation.[66,70,79,80] Increases in inositol-1,4,5-trisphosphate were detected within 5 sec of hormone stimulation, while significant increases in the amount of inositol-1-phosphate were not detected until 30 sec later.[79] It has been suggested that inositol-1-phosphate is derived from dephosphorylation of inositol-1,4,5-trisphosphate rather than from phosphatidylinositol breakdown. According to this scheme loss of phosphatidylinositol does not occur through phospholipase C to generate inositol-1-phosphate, but only through its phosphorylation to regenerate PtdIns(4,5)P_2. It has thus been suggested that PtdIns(4,5)P_2 is the sole substrate in agonist action.

There are several problems with this proposed mechanism. (1) Levels of inositol-1,4-bisphosphate increase at the same rate as inositol-1,4,5-trisphosphate, indicating that at least two phosphoinositides are degraded simultaneously during receptor activation.[66,70,79,80] PtdIns(4,5)P_2 therefore cannot be regarded as the primary substrate. (2) Despite the use of heroic amounts of LiCl to inhibit breakdown of inositol-1,4,5-trisphosphate by cellular phosphatases, there is considerable formation of inositol-1-phosphate.[66,70,80] This suggests that inositol-1-phosphate may be derived from phospholipase C-mediated breakdown of phosphatidylinositol.

An increase in the amount of ^{3}H-inositol or ^{32}P-labeled PtdIns(4,5)P_2 is not observed during hormone stimulation as would be predicted if labeled phosphatidylinositol were converted to PtdIns(4,5)P_2.[15] In hepatocytes prelabeled for 5 min in medium containing ^{32}P, vasopressin stimulated a net loss of ^{32}P-labeled PtdIns(4,5)P_2. A transient increase in the amount of ^{32}P-labeled PtdIns(4,5)P_2 was not observed.[15]

A consideration of the actual change in lipid mass that must occur according to this scheme indicates another difficulty. Approximately 10 to 50% of the cellular phosphatidylinositol is degraded within 30 sec of hormone stimulation.[11,12,15,61] PtdIns(4,5)P_2 constitutes approximately 10% of the phosphoinositide content and possibly only a small fraction of the total PtdIns(4,5)P_2 is part of the hormone-sensitive pool. A considerable stimulation of the synthetic pathway would have to occur in order to convert so much phosphatidylinositol to PtdIns(4,5)P_2 within 30 sec. It is not clear that phosphoinositide kinases have such a high capacity. Furthermore, conversion of each molecule of phosphatidylinositol to PtdIns(4,5)P_2 would require the utilization of at least 2 molecules of ATP. This would appear to be a costly scheme to generate a lipid for the sole purpose of degrading it to diacylglycerol and inositol-1,4,5-trisphosphate.

Since the major evidence for an indirect role of phosphatidylinositol in agonist-induced hydrolysis of the phosphoinositides comes from data using ion-exchange columns, it would be useful to examine this procedure. In the past, identification of inositol phosphates has been accomplished primarily by high voltage paper ionophoresis. This is a time-consuming procedure and it is difficult to process a large number of samples. Berridge et al.[62] employed Dowex-formate columns to characterize inositol phosphates produced during agonist stimulation. Since nonspecific adsorption of charged molecules to the resin may occur, it is essential to obtain optimum and reliable recovery of each inositol phosphate. Berridge et al.[62] compared the recovery of the individual inositol phosphates using Dowex-formate columns and high voltage paper ionophoresis. There was a discrepancy between these two methods regarding recovery of the different inositol phosphates. As compared to ionophoresis, the Dowex-formate method resulted in only a 46% recovery of inositol and a 14% recovery of inositol-1-phosphate generated during hormone stimulation. There was a 246% increase in the recovery of inositol-1,4-bisphosphate and a 172% increase in the recovery of glycerolphosphoinositol. Since these studies were done in the absence of LiCl there was little inositol-1,4,5-trisphosphate formed and thus no index of its recovery on the Dowex-formate columns. The marked decrease in inositol-1-phosphate recovery would obviously underestimate the contribution of phosphatidylinositol to breakdown. It would seem essential that inositol phosphate recovery standards be included in all analysis of inositol phosphates. This is routinely done in all cAMP assays utilizing ion exchange columns to isolate cAMP.

Fain[80] has suggested that hormones stimulate breakdown of the hormone-sensitive pool of phosphoinositides through an apparent activation of phospholipase C (Figure 2). Breakdown of the phosphoinositides may be initiated simultaneously upon receptor activation due to an increase in phospholipase C activity or through substrate activation. This mechanism is consistent with studies that have been done on the kinetics of lipid loss. Alternatively, hormones may stimulate a preferential breakdown of PtdIns(4,5)P_2 and PtdIns4P followed by a loss of phosphatidylinositol through both a phospholipase C-mediated breakdown of the lipid as well as through its conversion to PtdIns(4,5)P_2. This would be consistent with the data obtained from studies on the inositol phosphates. The phospholipase C must recognize both PtdIns(4,5)P_2 and PtdIns4P as the preferred substrate. In the absence of specific inhibitors of the phosphatidylinositol cycle, it is impossible to distinguish unequivocally between these possibilities.

The agonist-induced breakdown of phosphoinositides generates at least two intracellular messengers. Diacylglycerol is derived primarily from phosphatidylinositol breakdown and activates C-kinase resulting in phosphorylation of key cellular proteins.[22] Since the predominant fatty acid in the sn-2 position of phosphatidylinositol is arachidonic acid, phosphatidylinositol breakdown provides a source of free arachidonic acid through the concerted action of phospholipase C and diacylglycerol lipase. The arachidonic acid is utilized for synthesis of prostaglandins, an important class of regulatory effectors.[14]

Another potential second messenger generated during phosphoinositide breakdown is inositol-1,4,5-trisphosphate derived from PtdIns(4,5)P_2 breakdown. It appears to trigger the release of Ca^{2+} from endoplasmic reticulum, thus initiating a rise in intracellular Ca^{2+} levels.[21]

V. STUDIES ON CELL-FREE SYSTEMS

Durell and Garland[16] found that the addition of acetylcholine to a crude mitochondria fraction stimulated hydrolysis of ^{3}H-inositol-labeled phosphoinositides with accumulation of inositol-1-phosphate and inositol-1,4-bisphosphate consistent with activation of a phospholipase C. This effect of acetylcholine was greater at 3 than at 10 min, suggesting that acetylcholine increased the initial rate of phosphoinositide breakdown. Atropine blocked the effect of acetylcholine.

Activation of phosphoinositide breakdown by 5-hydroxytryptamine was observed in homogenates and cytosolic-depleted membranes from blowfly salivary glands.[17] As in the studies of Durell and Garland, there was a considerable basal rate of phosphoinositide breakdown. Addition of 5-hydroxytryptamine produced a dose-dependent stimulation of the initial rate of phosphoinositide breakdown. This effect of hormone was observed in medium containing EGTA to chelate Ca^{2+}, indicating that added Ca^{2+} was not required for hormone-activation of phosphoinositide breakdown.

Stimulation of phosphatidylinositol breakdown in isolated rat liver plasma membrane due to vasopressin[18] and norepinephrine[19,82] has been reported. Wallace et al.[18] demonstrated that approximately 10% of plasma membrane phosphatidylinositol was degraded in the presence of vasopressin after 30 min. Deoxycholate-stimulated breakdown of 20% of the phosphatidylinositol. These effects were observed in medium containing EGTA. A requirement for cytosol was reported in the studies of Harrington and Eichberg.[19] They observed that norepinephrine stimulated a dose-dependent breakdown of phosphatidylinositol. Ap-

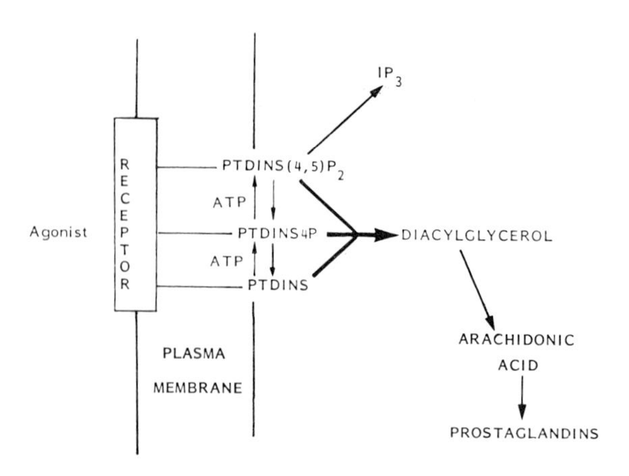

FIGURE 2. Agonist-induced phosphoinositide breakdown results in formation of two intracellular messengers. The interaction of Ca^{2+} mobilizing hormones with receptors activates a membrane-bound phospholipase C which degrades a small pool of cellular phosphoinositides. As a consequence of phosphoinositide breakdown, there is an increase in diacylglycerol and inositol phosphates. Diacylglycerol is derived primarily from phosphatidylinositol breakdown and results in the activation of C-kinase. Diacylglycerol is also a source of arachidonic acid which is utilized for formation of prostaglandins. Breakdown of phosphatidylinositol-4,5P_2 generates inositol-1,4,5-trisphosphate which may be an intracellular messenger for the release of Ca^{2+} from intracellular stores. During breakdown of the phosphoinositides there is some conversion of phosphatidylinositol to phosphatidylinositol-4,5P_2 that is dependent on ATP levels. The amount of inositol-1,4,5-trisphosphate formed will be determined, in part, by the degree of conversion of phosphatidylinositol to phosphatidylinositol-4,5P_2.

proximately 50% of the plasma membrane phosphatidylinositol was degraded after a 30 min incubation with hormone. Breakdown of phosphatidylinositol was not accompanied by a detectable formation of lysophosphatidylinositol or PtdIns(4,5)P_2. This effect of hormone was also observed in the presence of EGTA.

In the studies on isolated rat liver plasma membranes, loss of phosphatidylinositol was quantitated by chemical methods. Breakdown products of phosphatidylinositol were not identified and thus the mechanism by which hormones stimulate phosphatidylinositol breakdown is not known.

Breakdown of phosphatidylinositol and possibly the polyphosphoinositides may be mediated through the activation of a membrane-bound or membrane-associated phospholipase C with formation of diacylglycerol and inositol phosphates. So far, a membrane-bound phospholipase C active on all phosphoinositides has not been identified. Possibly the enzyme is maintained in an inactive form and requires receptor activation for expression of activity.

Part of the loss of phosphatidylinositol may occur as a consequence of its conversion to PtdIns(4,5)P_2. Since an increase in PtdIns(4,5)P_2 was not detected,[19] this indicates that the newly formed PtdIns(4,5)P_2 is rapidly degraded. This would again necessitate activation of a membrane-associated phospholipase C. Clearly, future studies need to be directed at establishing the mechanism of hormone-activated phosphoinositide breakdown and the metabolic interrelationships among the phosphoinositides.

VI. PHOSPHOINOSITIDE BREAKDOWN AND ELEVATION IN CYTOSOLIC Ca^{2+}

In resting cells, the free intracellular Ca^{2+} is maintained at approximately 0.1 μM. The addition of hormones, such as the α_1-adrenergic amines, induces a rapid increase in cytosolic Ca^{2+} to values approaching 1 μM.[83-86] Increases in cytosolic Ca^{2+} can be detected within 1 sec of hormone addition.[86] It appears that the initial action of many hormones is to trigger the release of Ca^{2+} from intracellular sites such as the endoplasmic reticulum, mitochondria, or plasma membrane. Hormones also increase the influx of extracellular Ca^{2+}.[85,86] The increased influx of Ca^{2+} (Ca^{2+} gating) appears to be essential to augment and sustain the initial rise in cytosolic Ca^{2+}. The relationship between mobilization of intracellular Ca^{2+} and Ca^{2+} influx is not known. The increase in intracellular Ca^{2+} might trigger the influx of Ca^{2+}. Alternatively, a change in plasma membrane properties due to phosphoinositide breakdown may result in Ca^{2+} influx. It remains to be established if the same mechanism mediates both the release of intracellular Ca^{2+} and the influx of extracellular Ca^{2+}. Whether Ca^{2+} influx occurs through distinct receptor-regulated Ca^{2+} channels or through a generalized increase in permeability is not known.

Support for a role of phosphoinositides in the elevation of cytosolic Ca^{2+} due to hormones comes from the studies of Fain and Berridge[7,53] who demonstrated that in salivary glands of the blowfly (*Calliphora erythrocephala*), the maintenance of a critical pool of phosphatidylinositol was essential to obtain a Ca^{2+} gating response. There are several unique features of the salivary gland which have made it a valuable model system to study the relationship between phosphoinositide breakdown and Ca^{2+} gating. The intracellular pool of inositol appears to be quite small in these glands. When glands are incubated in medium containing ^{3}H-inositol, the labeled inositol is selectively incorporated into a small pool of phosphoinositides that constitutes less than 5% of the cellular phosphoinositide content. 5-Hydroxytryptamine stimulates breakdown of the labeled pool of phosphoinositides and this is associated with an increase in plasma membrane Ca^{2+} permeability and activation of secretion. Breakdown of the phosphoinositides is rapid with generation of diacylglycerol within 30 sec[60] and formation of inositol phosphates derived from phosphoinositide breakdown.[62,64] Breakdown of the phosphoinositides is mediated through a phosphoinositide-specific phospholipase C and is not secondary to the entry of extracellular Ca^{2+}.[7,64] Electrophysiological studies have shown that one major effect of hormone is to elevate intracellular Ca^{2+} by increasing the influx of extracellular Ca^{2+} across the plasma membrane.[87] There is also release of Ca^{2+} from intracellular stores. Salivary gland secretion exhibits a temporary independence from extracellular Ca^{2+}, however, influx of Ca^{2+} is clearly needed to sustain the secretory response. The relationship, if any, between release of intracellular Ca^{2+} and Ca^{2+} gating is not known. Thus, the parameters which affect Ca^{2+} gating cannot *a priori* be taken to affect Ca^{2+} mobilization.

Glands depleted of the labeled pool of phosphoinositides by a prolonged stimulation with 5-hydroxytryptamine were unable to gate Ca^{2+} upon a second stimulation with hormone. If the depleted glands were allowed to replenish their phosphatidylinositol levels by recovering in medium containing inositol, their Ca^{2+} gating response was restored.[53,54] These results suggest that the ability of hormones to gate Ca^{2+} was linked to phosphatidylinositol breakdown.

Recent studies by Sadler et al.[77] have characterized in more detail the changes in phosphatidylinositol and polyphosphoinositide content that occur during recovery of the Ca^{2+} gating response. Litosch et al.[64] showed that a 30-min stimulation with 5-hydroxytryptamine was sufficient to maximally deplete the levels of phosphatidylinositol and polyphosphoinositides due to phospholipase C-mediated breakdown as well as inhibition of phosphoinositide resynthesis. Sadler et al.[77] correlated the resynthesis of phosphoinositides with the recovery of the Ca^{2+} gating response. Restoration of Ca^{2+} gating was controlled by limiting the amount of inositol added to the recovery medium. It was found that at low medium inositol concentrations (3 to 5 μM), there was a preferential resynthesis of PtdIns(4,5)P_2 that occurred at the expense of phosphatidylinositol synthesis. There was little change in the amount of labeled PtdIns4P formed. In glands allowed to recover in medium containing

3 to 5 μM ^{3}H-inositol, labeled PtdIns(4,5)P_2 constituted approximately 70% of the labeled phosphoinositide pool. As the medium inositol concentration was increased to 30 and 300 μM inositol, there was an increase in the synthesis of phosphoinositides and a proportionately greater amount of labeled phosphatidylinositol formed. At 300 μM inositol, phosphatidylinositol constituted 70% of the phosphoinositide pool. Maximal recovery of Ca^{2+} gating was observed in glands in which phosphatidylinositol comprised the major labeled phosphoinositide. Glands allowed to recover at 3 μM inositol exhibited a poor Ca^{2+} gating response. The ability of 5-hydroxytryptamine to stimulate phosphoinositide breakdown in desensitized glands was similar to control glands indicating that the inability to gate Ca^{2+} was not at the level of hormone-induced phosphoinositide breakdown.

These findings indicate that both phosphatidylinositol and PtdIns(4,5)P_2 levels must be restored to control levels to obtain an optimal Ca^{2+} gating response. It is not clear if mobilization of intracellular Ca^{2+} was also inhibited by the desensitization procedure since secretion was not measured under Ca^{2+}-free conditions. In the absence of medium Ca^{2+}, the increase in secretion due to hormone should reflect primarily Ca^{2+} release from intracellular stores. The data suggest that desensitization results from the depletion of phosphoinositides from a cellular site essential for hormone-sensitive Ca^{2+} gating. Breakdown of a critical amount of phosphoinositides may be necessary to open a receptor-regulated Ca^{2+} channel. Breakdown of salivary gland phosphoinositides results in the formation of diacylglycerol and the corresponding inositol phosphates which may act as second messengers to release Ca^{2+} from intracellular sites and increase the influx of extracellular Ca^{2+}. In glands depleted of phosphoinositides the production of these putative messengers would be reduced and there would be a corresponding reduction in the generated signal. Alternatively, resynthesis of a critical amount of phosphoinositides may be essential to maintain the Ca^{2+} channel in a hormone-responsive state. Depletion of the hormone sensitive pool of phosphoinositides during hormone action might render the Ca^{2+} channel insensitive to further hormone-stimulated Ca^{2+} entry. In the latter case, depletion of phosphoinositides would result in closure of the Ca^{2+} channel. Critical experiments have yet to be designed to distinguish between these possibilities.

A. Role of Ca^{2+} ATPase

Reinhart et al.[83] demonstrated that there was a 7-sec lag period between the elevation in cytosolic Ca^{2+} due to phenylephrine and the onset of Ca^{2+} efflux from perfused livers. Korchak et al.[86] found a 10-sec lag for f-MetLeuPhe-induced Ca^{2+} efflux from neutrophils. Similar delays in Ca^{2+} efflux have been reported by Joseph and Williamson[88] for phenylephrine- and vasopressin-induced Ca^{2+} efflux from hepatocytes. These observations suggested that an early effect of Ca^{2+} mobilizing hormones was to inhibit the cellular mechanisms involved in restoring low intracellular Ca^{2+} levels. The lag in Ca^{2+} efflux may be necessary to maintain the agonist-mediated increase in intracellular Ca^{2+} levels elevated sufficiently long to produce activation of cellular response.

The cell has highly efficient mechanisms for maintaining low intracellular Ca^{2+} levels. A low plasma membrane Ca^{2+} permeability prevents continual leakage of Ca^{2+} into the cells from medium. A Na^+-Ca^{2+} exchange mechanism is operative in some cells and utilizes the Na^+-K^+ gradient to drive efflux of Ca^{2+}. In addition, there is a high affinity plasma membrane Ca^{2+} pump which maintains a 10,000-fold Ca^{2+} concentration gradient between extracellular and intracellular Ca^{2+}. The Ca^{2+} pump is expressed enzymatically as a Ca^{2+}-ATPase.[89,90]

Lotersztajn et al.[91] demonstrated that rat liver plasma membranes contain a high affinity Ca^{2+}-ATPase activity whose activity is not modulated by calmodulin.[92] Lin et al.[93] confirmed these findings and demonstrated that an early action of vasopressin and phenylephrine was to inhibit the plasma membrane Ca^{2+}-ATPase activity. This effect was maximal at 15 sec and transitory. Within 12 min, the inhibitory action of hormone was no longer apparent. Inhibition of Ca^{2+}-ATPase by vasopressin and phenylephrine is not secondary to the elevation of intracellular Ca^{2+} since this effect was seen even in hepatocytes which had been incubated in medium containing 1 m*M* EGTA. Inhibition of Ca^{2+}-ATPase was not induced by insulin or glucagon. Neither of these hormones stimulate phosphoinositide breakdown or elevate Ca^{2+} in hepatocytes.

It is obvious that an inhibition of the Ca^{2+} efflux mechanism helps sustain an elevation of intracellular Ca^{2+} due to release from endoplasmic reticulum. An inhibition of Ca^{2+} efflux may also result in an increased influx of extracellular Ca^{2+} as extracellular Ca^{2+} enters the cell along its electrochemical gradient. This might represent one mechanism by which hormones increase the influx of extracellular Ca^{2+} and elevate intracellular Ca^{2+}.

The mechanism whereby vasopressin and α_1-adrenergic amines inhibit plasma membrane Ca^{2+}-ATPase has not been established. Inhibition of the hepatocyte Ca^{2+}-ATPase may be related to the rapid breakdown of phosphoinositides that occurs upon addition of these hormones to hepatocytes. Buckley and Hawthorne[95] found that increasing the PtdIns(4,5)P_2 levels in red cell membranes resulted in an increased Ca^{2+}-Mg^{2+}ATPase activity. Possibly the agonist induced decrease in PtdIns(4,5)P_2 levels results in a decrease in the Ca^{2+}-transport mechanism. The initial breakdown of PtdIns(4,5)P_2 is followed by resynthesis of PtdIns(4,5)P_2 even in the presence of agonist.[15] Resynthesis of PtdIns(4,5)P_2 may restore Ca^{2+}-ATPase activity and thus account for the transitory effect of phenylephrine or vasopressin on inhibition of Ca^{2+}-ATPase.[93] Lin and Fain[96] purified Ca^{2+}-ATPase from rat liver plasma membrane. The activity of the purified Ca^{2+}-ATPase was stimulated by PtdIns(4,5)P_2.[96] Burgess et al.[97] reported that α_1-adrenergic amines increased membrane fluidity. An increase in membrane fluidity may inhibit Ca^{2+}-ATPase activity. Breakdown products of the phosphoinositides such as diacylglycerol and inositol phosphates may influence Ca^{2+}-ATPase activity. Finally, receptor activation might cause the redistribution of an inhibitor or activator of the enzyme. Further studies are needed to establish the mechanism of the hormone inhibition.

B. Role of Inositol-1,4,5-Trisphosphate in Ca^{2+} Mobilization

Inositol phosphates are generated during phospholipase C-stimulated phosphoinositide breakdown. Streb et al.[21] demonstrated that the addition of inositol-1,4,5-trisphosphate to permeabilized pancreatic acinar cells resulted in a transient release of Ca^{2+} from an oligomycin-insensitive intracellular storage site. Carbachol also stimulated Ca^{2+} release from these cells. However, neither inositol, inositol-1-phosphate, inositol-1,4-bisphosphate, nor inositol-1,2-cyclic phosphate could release Ca^{2+} from these cells under the conditions tested. The effects of carbachol and inositol-1,4,5-trisphosphate were not additive, suggesting that both agents were stimulating Ca^{2+} release from the same pool, presumably endoplasmic reticulum. Similar effects of inositol-1,4,5-trisphosphate on the release of Ca^{2+} from saponin-permeabilized hepatocytes have been reported by Burgess et al.[98] and Thomas et al.[99]

These studies suggest that inositol-1,4,5-trisphosphate might be the chemical messenger which mediates the effect of hormones on mobilizing Ca^{2+} from intracellular storage sites. Inositol-1,4,5-trisphosphate appears to have all the desired properties of a second messenger. It is produced rapidly upon hormone addition, it is derived from breakdown of phosphoinositides which have already been linked to the Ca^{2+} mobilizing mechanism, it has rapid physiological effects when added to permeabilized cells, and it is rapidly metabolized by cellular phosphatases. This later point is essential since it provides a mechanism for the termination of the signal. In most systems, appreciable amounts of inositol-1,4,5-trisphosphate cannot be detected unless LiCl is added to inhibit inositol phosphate phosphatases. A notable exception to this has been the studies of Berridge[79] on the blowfly salivary gland where appreciable amounts of inositol-1,4,5-trisphosphate accumulate even in the absence of LiCl.

In postulating a role for inositol-1,4,5-trisphosphate as a second messenger, it is important to obtain a clear perspective on the actions of Ca^{2+} mobilizing hormones in intact cells and the effects of inositol-1,4,5-trisphosphate in permeabilized cells. A comparison of the kinetics for vasopressin-stimulated increase in inositol-1,4,5-trisphosphate and Ca^{2+} elevation in intact hepatocytes showed that the maximal increase in cytosolic Ca^{2+} occurred before peak levels of inositol-1,4,5-trisphosphate were attained in the cell.[99] However, only a fraction of the total inositol trisphosphate generated during vasopressin action appears to be the active inositol-1,4,5-trisphosphate compound.[99] Addition of LiCl inhibits breakdown of inositol-trisphosphate, resulting in a marked increase in the accumulation of this compound but the degree of Ca^{2+} mobilization is not affected. Phenylephrine produces a barely detectable increase in inositol-1,4,5-trisphosphate, although it produced an increase in cytosolic Ca^{2+} similar to that of vasopressin.[100] It is not known whether phenylephrine, like vasopressin, stimulates formation of inactive inositol-trisphosphate in addition to inositol-1,4,5-trisphosphate. The apparent discrepancy between the amount of inositol trisphosphate generated and Ca^{2+} mobilization may simply reflect the amount of active compound formed in response to either hormone.

The data suggest that only a small increase in inositol-1,4,5-trisphosphate is needed to evoke maximal Ca^{2+} release. This is reminiscent of the cAMP receptor system where small increases in cAMP trigger the maximal physiological response. In the latter case, cAMP initiates a series of phosphorylation-dephosphorylation reactions which greatly amplify the original signal generated by the rise in cAMP. If the analogy of inositol-1,4,5-trisphosphate to cAMP is valid it is probable that the activity of key regulatory enzymes or the conformation of target proteins is modulated by inositol-1,4,5-trisphosphate. Such intracellular receptors for inositol-1,4,5-trisphosphate have yet to be identified.

An alternative explanation for the differential effect of vasopressin and phenylephrine on inositol-1,4,5-trisphosphate formation, despite having similar effects on Ca^{2+} elevation, is that phenylephrine does not utilize inositol-1,4,5-trisphosphate as its sole second messenger. An examination of the kinetics for release of Ca^{2+} from permeabilized cells and the effects of agonists on the elevation of intracellular Ca^{2+} in whole cells shows some interesting differences. Addition of increasing doses of vasopressin to hepatocytes primarily decreases the lag period for Ca^{2+} release. On the other hand, inositol-1,4,5-trisphosphate addition to permeabilized cells increases the amount of Ca^{2+} released and delays re-uptake of Ca^{2+} into storage sites.[100] This latter point is important since all studies thus far have shown that the inositol-1,4,5-trisphosphate-stimulated release of Ca^{2+} is transient. Rapid release of Ca^{2+} is followed by rapid reuptake of Ca^{2+} by the cells. Unfortunately, these studies were not done in the presence of LiCl to prevent enzymatic degradation of inositol-1,4,5-trisphosphate. It is therefore unclear whether the transient nature of the response reflects inositol-1,4,5-trisphosphate metabolism, binding to intracellular proteins, or other as yet unknown effects.

There is some evidence which indicates that phenylephrine and vasopressin release Ca^{2+} from both mitochondrial and endoplasmic reticulum stores. Inositol-1,4,5-trisphosphate apparently does not release Ca^{2+} from mitochondria. This suggests that other cellular messengers might be involved in the release of Ca^{2+} from mitochondria. Joseph and Williamson[88] have observed different kinetics for Ca^{2+} release from mitochondrial and endoplasmic reticulum stores, suggesting that separate release mechanisms may be operative. Furthermore, the role of inositol-1,4,5-trisphosphate in Ca^{2+} gating is not clear.

Whether inositol-1,4,5-trisphosphate is indeed a second messenger in mobilizing Ca^{2+} from the endoplasmic reticulum will require additional documentation. In particular, it will be necessary to reconstitute the effect in a cell-free system by showing that inositol-1,4,5-trisphosphate causes Ca^{2+} release when added directly to isolated organelles.

C. Role of Guanine Nucleotide Binding Protein

The binding of labeled vasopressin[101] and angiotensin[102] to isolated rat liver plasma membrane is inhibited by GTP. GTP interacts with a guanine-nucleotide binding protein to decrease the affinity of agonists for binding to receptors linked to adenylate cyclase.[103] In hepatocytes, vasopressin and angiotensin do not stimulate cAMP production, but rather increase phosphoinositide breakdown and elevate Ca^{2+}. Similarly, ligand binding at muscarinic receptors is modulated by GTP.[104] Snavely and Insel[105] reported that GTP regulates the binding of both α_1- and α_2-antagonists. α_1-Receptors are linked to phosphoinositide breakdown and Ca^{2+} elevation while α_2-receptors are linked to the inhibition of adenylate cyclase.[106] The modulatory role of GTP on ligand binding suggests that a guanine nucleotide binding protein may mediate the effects of hormones on phosphoinositide breakdown and Ca^{2+} mobilization.

There are some effects of vasopressin, angiotensin, and α_1-adrenergic agonists that are linked to the inhibition of adenylate cyclase. Vasopressin[102] and angiotensin[102,107] inhibit cAMP formation due to stimulatory agents in intact cells and isolated plasma membrane. Norepinephrine, in the presence of propranolol, also inhibited agonist-stimulated cAMP production through an α_1-receptor.[107] In platelets, the vasopressin-mediated inhibition of adenylate cyclase appears to be mediated through the same class of receptors which stimulate phosphoinositide breakdown.[108]

The data suggest that there may be separate subtypes of receptors linked to phosphoinositide breakdown and adenylate cyclase inhibition. Modulation of agonist binding by GTP may affect only that subtype of receptors associated with the inhibition of adenylate cyclase. Alternatively the inhibition of adenylate cyclase may be a consequence of phosphoinositide breakdown and consequent alteration in membrane fluidity. The apparent requirement for GTP to inhibit adenylate cyclase suggests that GTP may be necessary for phosphoinositide breakdown, thus implicating a role for guanine nucleotide binding proteins in phosphoinositide breakdown. Alternatively, GTP may be required to link phosphoinositide breakdown to inhibition of adenylate cyclase.

A recent report that pertussis toxin treatment inhibited epinephrine-stimulated phosphatidylinositol synthesis in adipocytes provides more evidence for a possible link between guanine nucleotide binding proteins and phosphoinositide breakdown.[109] Pertussis toxin ADP-ribosylates a 41K guanine nucleotide binding protein and, as a consequence, inhibitory agonists such as the α_2-adrenergic agonists can no longer inhibit adenylate cyclase.[108] The data suggest that a guanine nucleotide-binding protein is involved in some way with phosphatidylinositol turnover. However, phosphoinositide breakdown was not measured in these studies and thus the effect of pertussis toxin on phosphatidylinositol turnover may be exerted at a site secondary to the initial breakdown of phosphoinositide.

Gomperts[110] has shown that introduction of GTP analogues to permeabilized resealed mast cells caused them to secrete in response to extracellular Ca^{2+}. It was proposed that guanine nucleotide-binding proteins might be involved in the gating of Ca^{2+}. Unfortunately, all measurements were made at 10 min and therefore there is no indication as to the rapidity of the effect. Further studies need to be done to establish whether the GTP effects on agonist binding and exocytosis are related to a primary receptor event involved in Ca^{2+} gating and phosphoinositide breakdown.

D. Phosphatidic Acid as an Ionophore

One consequence of agonist-stimulated phosphoinositide breakdown is an increased formation of phosphatidic acid derived from diacylglycerol (Figure 1). In a Pressman cell, phosphatidic acid translocates Ca^{2+} across an organic phase separating two aqueous compartments.[111] Addition of phosphatidic acid to rat parotid slices stimulated a Ca^{2+}-dependent efflux of ^{86}Rb. Maximal effects were obtained with 10^{-4} *M* phosphatidic acid.[112] In isolated smooth muscle cells, 10^{-8} *M* phosphatidic acid induced cellular contraction within 10 sec.[113] Phosphatidic acid causes a rapid translocation of Ca^{2+} in liposomes.[114] It was suggested that phosphatidic acid functions as an ionophore to increase intracellular Ca^{2+} levels.[113-115]

Holmes and Yoss[116] compared the ability of phosphatidic acid derived from several commercial sources to translocate Ca^{2+} across liposome membranes. They found that pure phosphatidic acid was ineffective in translocating Ca^{2+}. If the phosphatidic acid had undergone appreciable fatty acid oxidation, it became an effective ionophore. Since oxidized fatty acids act as ionophores in liposomes,[114] these findings suggest that the reported effects of phosphatidic acid actually reflect the ability of oxidized fatty acids to facilitate the movement of Ca^{2+} across membranes.

In blowfly salivary glands, 5-hydroxytryptamine stimulates a rapid increase in Ca^{2+} gating.[7,86] Diacylglycerol is generated within 30 sec, but is not converted to phosphatidic acid.[60] Stimulated phosphatidic acid formation occurs after 30 min of hormone stimulation. The latency between the onset of Ca^{2+} gating and increased phosphatidic acid formation makes it unlikely that phosphatidic acid is an ionophore in blowfly salivary glands. The observations of Holmes and Yoss[116] indicate that it is unlikely that phosphatidic acid acts as a cellular ionophore.

VII. CONCLUSION

The aim of this review has been to summarize our current concepts concerning the relationship between hormone-activated phosphatidylinositol turnover and elevation of intracellular Ca^{2+}. A summary scheme for cellular activation due to Ca^{2+} mobilizing hormones is shown in Figure 3. Activation of the appropriate receptor stimulates breakdown of the phosphoinositides through an apparent activation of phospholipase C. There is approximately a five- to tenfold greater loss of phosphatidylinositol than PtdIns(4,5)P_2 during hormone stimulation. Loss of phosphatidylinositol occurs both through direct phospholipase C-mediated hydrolysis and through conversion of phosphatidylinositol to PtdIns(4,5)P_2 via phosphoinositide kinases. Some PtdIns(4,5)P_2 may also be converted to phosphatidylinositol through activation of a phosphomonoesterase.

Inositol-1,4,5-trisphosphate derived from PtdIns(4,5)P_2 breakdown triggers the release of Ca^{2+} from endoplasmic reticulum, presumably through interaction with a target protein involved in regulating Ca^{2+} movement from the endoplasmic reticulum. There may be a concurrent influx of Ca^{2+} through receptor-regulated Ca^{2+} channels that is initiated through an unknown mechanism. The plasma membrane Ca^{2+} pump is inhibited, resulting in a sustained elevation of the intracellular Ca^{2+} and augmentation of Ca^{2+}-dependent responses. Ca^{2+} and diacylglycerol activate calmodulin-dependent kinases and C-kinase, respectively, resulting in phosphorylation of cellular proteins. One consequence of increased C-Kinase activity may be an increased conversion of phosphatidylinositol to PtdIns(4,5)P_2. In addition, C-kinase or other intermediates may induce receptor desensitization. This would result in termination of phosphoinositide breakdown and stimulated resynthesis of the phosphoinositides. Inositol-1,4,5-trisphosphate and diacylglycerol would be metabolized to inactive compounds by cellular phosphatases and diacylglycerol kinase. Release of Ca^{2+} from endoplasmic reticulum and influx of extracellular Ca^{2+} would be terminated. The plasma membrane Ca^{2+} pump would be reactivated and low intracellular Ca^{2+} levels restored. The cell would be primed for another response.

This scheme is consistent with our current knowledge of events occurring during receptor activation. The data discussed in this review illustrate both the unity and controversy that exists within the field. The validity of the many proposed models will be determined through further experimentation and a critical evaluation of all the available data. The ultimate result will be an understanding of the receptor mechanism linked to the elevation of cytosolic Ca^{2+}.

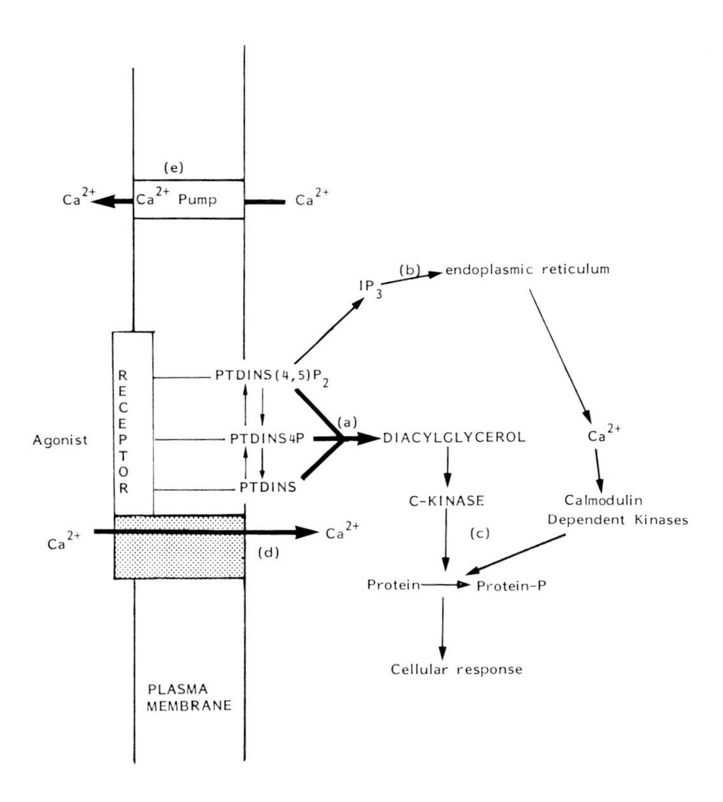

FIGURE 3. Generalized scheme for hormone action. (a) Receptor activation results in phosphoinositide breakdown via an apparent activation of phospholipase C activity. (b) The inositol-1,4,5-trisphosphate (IP_3) acts as a second messenger to release Ca^{2+} from endoplasmic reticulum. (c) Diglyceride, in conjunction with Ca^{2+}, activates C-kinase which mediates the phosphorylation of regulatory cell proteins. (d) Ca^{2+} gating is activated and there is an influx of extracellular Ca^{2+} which may augment and sustain the initial elevation in cytosolic Ca^{2+}. (e) Extrusion of Ca^{2+} is transiently reduced due to the inhibition of the Ca^{2+} pump and this results in a further augmentation of the Ca^{2+} signal.

REFERENCES

1. **Hokin, L. E. and Hokin, M. R.,** Effects of acetylcholine on the turnover of phosphoryl units in individual phospholipids of pancreas slices and brain cortex slices, *Biochim. Biophys. Acta*, 18, 102, 1955.
2. **Hokin-Neaverson, M.,** Acetylcholine causes a net decrease in phosphatidylinositol and a net increase in phosphatidic acid in mouse pancreas, *Biochem. Biophys. Res. Commun.*, 58, 763, 1974.
3. **Banschbach, M. W., Geison, R. L., and Hokin-Neaverson M.,** Acetylcholine increases the level of diglyceride in mouse pancreas, *Biochem. Biophys. Res. Commun.*, 58, 714, 1974.
4. **Jones, L. M. and Michell, R. H.,** Breakdown of phosphatidylinositol provoked by muscarinic cholineric stimulation of rat parotid-gland fragments, *Biochem. J.*, 142, 583, 1974.
5. **Michell, R. H.,** Inositol phospholipids and cell surface receptor function, *Biochim. Biophys. Acta*, 415, 81, 1975.
6. **Durell, J., Garland, J. T., and Friedel, R. O.,** Acetylcholine action: biochemical aspects, *Science*, 165, 862, 1969.
7. **Fain, J. N. and Berridge, M. J.,** Relationship between hormonal activation of phosphatidylinositol hydrolysis, fluid secretion and calcium flux in the blowfly salivary gland, *Biochem. J.*, 178, 45, 1979.
8. **Cockroft, S. and Gomperts, B. D.,** Evidence for a role of phosphatidylinositol turnover in stimulus-secretion coupling, *Biochem. J.*, 178, 681, 1979.

9. **Billah, M. M. and Michell, R. H.,** Phosphatidylinositol metabolism in rat hepatocytes stimulated by glycogenolytic hormones, *Biochem. J.*, 182, 661, 1979.
10. **Weiss, S. J. and Putney, J. W., Jr.,** The relationship of phosphatidylinositol turnover to receptors and calcium ion channels in rat parotid acinar cells, *Biochem. J.*, 194, 463, 1981.
11. **Bell, R. L. and Majerus, P. W.,** Thrombin-induced hydrolysis of phosphatidylinositol in human platelets, *J. Biol. Chem.*, 255, 1790, 1980.
12. **Rebecchi, M. F., Kolesnick, R. N., and Gershengorn, M. C.,** Thyrotropin-releasing hormone stimulates rapid loss of phosphatidylinositol and its conversion to 1,2-diacylglycerol and phosphatidic acid in rat mammotropic pituitary cells, *J. Biol. Chem.*, 258, 227, 1983.
13. **Sawyer, S. T. and Cohen, S.,** Enhancement of calcium uptake and phosphatidylinositol turnover by epidermal growth factor in A-431 cells, *Biochemistry*, 20, 6280, 1981.
14. **Fain, J. N.,** Involvement of phosphatidylinositol breakdown in elevation of cytosol Ca^{2+} by hormones and relationship to prostaglandin formation, in *Hormone Receptors*, Kohn, L. D., Ed., John Wiley & Sons, New York, 1982, chap. 11.
15. **Litosch, I., Lin, S.-H., and Fain, J. N.,** Rapid changes in hepatocyte phosphoinositides induced by vasopressin, *J. Biol. Chem.*, 258, 13727, 1983.
16. **Durell, J. and Garland, J. T.,** Acetylcholine-stimulated phosphodiesteratic cleavage of phosphoinositides: hypothetical role in membrane depolarization, *Ann. N. Y. Acad. Sci.*, 165, 743, 1969.
17. **Fain, J. N., Lin, S.-H., Litosch, I., and Wallace, M.,** Hormonal regulation of phosphatidylinositol breakdown, *Life Sci.*, 32, 2055, 1983.
18. **Wallace, M., Randazzo, P., Li, S.-Y., and Fain, J. N.,** Direct stimulation of phosphatidylinositol degradation by addition of vasopressin to purified rat liver plasma membranes, *Endocrinology*, 111, 341, 1982.
19. **Harrington, C. A. and Eichberg, J.,** Norepinephrine causes α_1-adrenergic receptor mediated decrease of phosphatidylinositol in isolated rat liver plasma membranes supplemented with cytosol, *J. Biol. Chem.*, 258, 2087, 1983.
20. **Kirk, C. J., Creba, A. A., Downes, C. P., and Michell, R. H.,** Hormone-stimulated metabolism of inositol lipids and its relationship to hepatic receptor function, *Biochem. Soc. Trans.*, 9, 377, 1981.
21. **Streb, H., Irvine, R. F., Berridge, M. J., and Schulz, I.,** Release of Ca^{2+} from a non-mitochondrial store in pancreatic acinar cells by inositol-1,4,5-trisphosphate, *Nature (London)*, 306, 67, 1983.
22. **Takai, Y., Kikkawa, U., Kaibuchi, K., and Nishizuka, Y.,** Membrane phospholipid metabolism and signal transduction for protein phosphorylation, *Adv. Cyclic Nucleotide Res.*, in press.
23. **Takenawa, T., Saito, M., Nagai, Y., and Egawa, K.,** Solubilization of the enzyme catalyzing CDP-diglyceride-independent incorporation of myo-inositol into phosphatidylinositol and its comparison to CDP diglyceride:inositol transferase, *Arch. Biochem. Biophys.*, 182, 244, 1977.
24. **Jelsema, C. L. and Morre, D. J.,** Distribution of phospholipid biosynthetic enzymes among cell components of the rat liver, *J. Biol. Chem.*, 253, 7960, 1978.
25. **Bleasdale, J. E., Wallis, P., MacDonald, P. C., and Johnston, J. M.,** Characterization of the forward and reverse reactions catalyzed by CDP-diacylglycerol:inositol transferase in rabbit lung tissue, *Biochim. Biophys. Acta*, 575, 135, 1979.
26. **Takenawa, T. and Egawa, K.,** CDP-diglyceride: inositol transferase from rat liver, *J. Biol. Chem.*, 252, 5419, 1977.
27. **Tolbert, M. E. M., White, A. C., Aspry, K., Cutts, J., and Fain, J. N.,** Stimulation by vasopressin and α-catecholamines of phosphatidylinositol formation in isolated rat liver parenchymal cells, *J. Biol. Chem.*, 255, 1938, 1980.
28. **Michell, R. H., Harwood, J. L., Coleman, R., and Hawthorne, J. N.,** Characteristics of rat liver phosphatidylinositol kinase and its presence in the plasma membrane, *Biochim. Biophys. Acta*, 144, 649, 1967.
29. **Colodzin, M. and Kennedy, E. P.,** Biosynthesis of diphosphoinositide in brain, *J. Biol. Chem.*, 240, 3771, 1965.
30. **Collins, C. A. and Wells, W. W.,** Identification of phosphatidylinositol kinase in rat liver lysosomal membranes, *J. Biol. Chem.*, 258, 2130, 1983.
31. **Harwood, J. L. and Hawthorne, J. N.,** The properties and subcellular distribution of phosphatidylinositol kinase in mammalian tissues, *Biochim. Biophys. Acta*, 171, 75, 1969.
32. **Kai, M., Salway, J. G., and Hawthorne, J. N.,** The diphosphoinositide kinase of rat brain, *Biochem. J.*, 106, 791, 1968.
33. **Jergel, B. and Sundler, R.,** Phosphorylation of phosphatidylinositol in rat liver golgi, *J. Biol. Chem.*, 258, 7968, 1983.
34. **Tou, J.-S., Hurst, M. W., and Huggins, C. G.,** Phosphatidylinositol kinase in rat kidney cortex, *Arch. Biochem. Biophys.*, 131, 596, 1968.
35. **Hawthorne, J. N. and Kemp, P.,** The brain phosphoinositides, *Adv. Lipid. Res.*, 2, 127, 1964.

36. **Quist, E. E.,** Polyphosphoinositide synthesis in rabbit erythrocyte membranes, *Arch. Biochem. Biophys.*, 219, 58, 1982.
37. **Quist, E. E. and Barker, R. C.,** Properties of phosphatidylinostol kinase activities in rabbit erythrocyte membranes, *Arch. Biochem. Biophys.*, 222, 170, 1983.
38. **Sarkadi, B., Enyedi, A., Farago, A., Meszaros, G., Kremmer, T., and Gardos, G.,** Cyclic AMP-dependent protein kinase stimulates the formation of polyphosphoinositides in lymphocyte plasma membranes, *FEBS Lett.*, 152, 95, 1983.
39. **Oestreicher, A. B., Zwiers, H., Gispen, W. H., and Roberts, S.,** Characterization of infant rat cerebral cortical membrane proteins phosphorylated *in vivo*. Identification of the ACTH sensitive phosphoprotein B-50, *J. Neurochem.*, 39, 683, 1982.
40. **Aloyo, V. J., Zwiers, H., and Gispen, W. H.,** Phosphorylation of B-50 protein by calcium activated phospholipid-dependent protein kinase and B-50 protein kinase, *J. Neurochem.*, 41, 649, 1983.
41. **Jolles, J., Zwiers, H., Dongen, C. J., Schotman, P., Wirtz, K. W. A., and Gispen, W. H.,** Modulation of brain polyphosphoinositide metabolism by ACTH-sensitive protein phosphorylation, *Nature (London)*, 286, 623, 1980.
42. **Thompson, W.,** The hydrolysis of monophosphoinositide by extracts of brain, *Can. J. Biochem.*, 45, 853, 1967.
43. **Irvine, R. F. and Dawson, R. M. C.,** The distribution of calcium-dependent phosphatidylinositol-specific phosphodiesterase in rat brain, *J. Neurochem.*, 31, 1427, 1978.
44. **Takenawa, T. and Nagai, Y.,** Purification of phosphatidylinositol-specific phospholipase C from rat liver, *J. Biol. Chem.*, 256, 6769, 1981.
45. **Hoffman, S. L. and Majerus, P. W.,** Identification and properties of two distinct phosphatidylinositol-specific phospholipase C enzymes from sheep seminal vesicular glands, *J. Biol. Chem.*, 257, 6461, 1982.
46. **Low, M. and Weglicki, W. B.,** Resolution of myocardial phospholipase C into several forms with distinct properties, *Biochem. J.*, 215, 325, 1983.
47. **Irvine, R. F., Hemington, N., and Dawson, R. M. C.,** Phosphatidylinositol-degrading enzymes in liver lysosomes, *Biochem. J.*, 164, 177, 1977.
48. **Matsuzawa, Y. and Hostetter, K. Y.,** Properties of phospholipase C isolated from rat liver lysosomes, *J. Biol. Chem.*, 255, 646, 1980.
49. **Lapetina, E. G. and Michell, R. H.,** A membrane-bound activity catalysing phosphatidylinositol breakdown to 1,2-diacylglycerol, D-myoinositol 1:2-cyclic phosphate and D-myoinositol-1-phosphate, *Biochem. J.*, 131, 433, 1973.
50. **Bormann, B. J., Huang, C.-K., Mackin, W. M., and Becker, E. L.,** Receptor-mediated activation of a phospholipase A_2 in rabbit neutrophil plasma membrane, *Proc. Natl. Acad. Sci. U.S.A.*, 81, 767, 1984.
51. **Irvine, R. F., Hemington, N., and Dawson, R. M. C.,** The Ca^{2+} dependent phosphatidylinositol-phosphodiesterase of rat brain, *Eur. J. Biochem.*, 99, 525, 1979.
52. **Billah, M. and Lapetina, E. G.,** Evidence for multiple metabolic pools of phosphatidylinositol in stimulated platelets, *J. Biol. Chem.*, 257, 11856, 1982.
53. **Fain, J. N. and Berridge, M. J.,** Relationship between phosphatidylinositol synthesis and recovery of 5-hydroxytryptamine responsive Ca^{2+} flux in blowfly salivary glands, *Biochem. J.*, 180, 655, 1979.
54. **Berridge, M. J. and Fain, J. N.,** Inhibition of phosphatidylinositol synthesis and the inactivation of calcium entry after prolonged exposure of the blowfly salivary gland to 5-hydroxytryptamine, *Biochem. J.*, 178, 59, 1979.
55. **Irvine, R. F. and Dawson, R. M. C.,** Phosphatidylinositol phosphodiesterase of rat brain: Ca^{2+}-dependency, pH optima and heterogeneity, *Biochem. J.*, 215, 431, 1983.
56. **Irvine, R. F., Letcher, A. J., and Dawson, R. M. C.,** Phosphatidylinositol-4,5-bisphosphate phosphodiesterase and phosphomonesterase activities of rat brain, *Biochem. J.*, 218, 177, 1984.
57. **Dawson, R. M. C. and Thompson, W.,** The triphosphoinositide phosphomonoesterase of brain tissue, *Biochem. J.*, 91, 244, 1964.
58. **Koutouzov, S. and Marche, P.,** The Mg^{2+}-activated phosphatidylinositol 4,5-bisphosphate specific phosphomonoesterase of erythrocyte membrane, *FEBS Lett.*, 144, 344, 1982.
59. **Rittenhouse-Simmons, S.,** Production of diglyceride from phosphatidylinositol in activated human platelets, *J. Clin. Invest.*, 63, 5805, 1979.
60. **Litosch, I., Saito, Y., and Fain, J. N.,** 5HT-stimulated arachidonic acid release from labeled phosphatidylinositol in blowfly salivary glands, *Am. J. Physiol.*, 243, C222, 1982.
61. **Thomas, A. P., Mark, J. S., Coll, K. E., and Williamson, J. R.,** Quantitation and early kinetics of inositol lipid changes induced by vasopressin in isolated and cultured hepatocytes, *J. Biol. Chem.*, 258, 5716, 1983.
62. **Berridge, M. J., Dawson, R. M. C., Downes, C. P., Heslop, J. P., and Irvine, R. F.,** Changes in the levels of inositol phosphates after agonist-dependent hydrolysis of membrane phosphoinositides, *Biochem. J.*, 212, 473, 1983.

63. **Agranoff, B. W., Pushpalatha, M., and Seguin, E. B.,** Thrombin-induced phosphodiesteratic cleavage of phosphatidylinositol bisphosphate in human platelets, *J. Biol. Chem.*, 258, 2076, 1983.
64. **Litosch, I., Lee, H. S., and Fain, J. N.,** Phosphoinositide breakdown in blowfly salivary glands, *Am. J. Physiol.*, 246, C141, 1984.
65. **Creba, J. A., Downes, C. P., Hawkins, P. T., Brewster, G., Michell, R. H., and Kirk, C. J.,** Rapid breakdown of phosphatidylinositol 4,5-bisphosphate in rat hepatocytes stimulated by vasopressin and other Ca^{2+}-mobilizing hormones, *Biochem. J.*, 212, 733, 1983.
66. **Aub, D. and Putney, J. W.,** Metabolism of inositol phosphates in parotid cells, *Life Sci.*, 34, 1347, 1984.
67. **Lin, S.-H. and Fain, J. N.,** Vasopressin and epinephrine stimulation of phosphatidylinositol breakdown in the plasma membrane of rat hepatocytes, *Life Sci.*, 18, 1905, 1981.
68. **Bennett, J. P., Cockcroft, S., Caswell, A. H., and Gomperts, B. D.,** Plasma membrane location of phosphatidylinositol hydrolysis in rabbit neutrophils stimulated with formylmethionyl leucyl phenylalanine, *Biochem. J.*, 208, 801, 1982.
69. **Godfrey, P. P. and Putney, J. W.,** Receptor-mediated metabolism of the phosphoinositides and phosphatidic acid in rat lacrimal acinar cells, *Biochem. J.* 218, 187, 1984.
70. **Downes, C. P. and Wusterman, M. M.,** Breakdown of polyphosphoinositides and not phosphatidylinositol accounts for muscarinic agonist-stimulated inositol phospholipid metabolism in rat parotid glands, *Biochem. J.*, 216, 633, 1983.
71. **Putney, J. W., Burgess, G. M., Halenda, S. P., McKinney, J. S., and Rubin, R.,** Effects of secretagogues on [^{32}P]phosphatidylinositol-4,5-bisphosphate metabolism in the exocrine pancreas, *Biochem. J.*, 212, 483, 1983.
72. **Billah, M. M. and Lapetina, E. G.,** Rapid decrease of phosphatidylinositol-4,5-bisphosphate in thrombin-stimulated platelets, *J. Biol. Chem.*, 257, 12705, 1982.
73. **Cochet, C., Gill, G. N., Meisenhelder, J., Cooper, J. A., and Hunter, T.,** C-kinase phosphorylates the epidermal growth factor receptor and reduces its epidermal growth factor stimulated tyrosine protein kinase activity, *J. Biol. Chem.*, 259, 2553, 1984.
74. **Bertrand, P. P., Plantavid, M., Chap, H., and Douste-Blazy, L.,** Are polyphosphoinositides involved in platelet activation?, *Biochem. Biophys. Res. Commun.*, 110, 660, 1983.
75. **Farese, R. V., Larson, R. E., and Sabir, M. A.,** Insulin acutely increases phospholipids in the phosphatidate-inositide cycle in rat adipose tissue, *J. Biol. Chem.*, 257, 4042, 1982.
76. **Häring, H.-U., Kasuga, M., and Kahn, C. R.,** Insulin receptor phosphorylation in intact adipocytes and in a cell-free system, *Biochem. Biophys. Res. Commun.*, 108, 1538, 1982.
77. **Sadler, K., Litosch, I., and Fain, J. N.,** Stimulation of phosphatidylinositol-4,5-bisphosphate formation by prior incubation of blowfly salivary glands with 5-HT, *Biochem. J.*, 222, 327, 1984.
78. **Michell, R. H., Kirk, C. J., Jones, L. M., Downes, C. P., and Creba, J. A.,** The stimulation of inositol lipid metabolism that accompanies calcium mobilization in stimulated cells: defined characteristics and unanswered questions, *Philos. Trans. R. Soc. London, on Ser. B:* 296, 123, 1981.
79. **Berridge, M. J.,** Rapid accumulation of inositol trisphosphate reveals that agonists hydrolyze polyphosphoinositides instead of phosphatidylinositol, *Biochem. J.*, 212, 849, 1983.
80. **Fain, J. N.,** Activation of plasma membrane phosphatidylinositol turnover by hormones, *Vitamins Hormones,* 41, 117, 1984.
81. **Thomas, A. P., Alexander, J., and Williamson, J. R.,** Relationship between inositol polyphosphate production and the increase of cytosolic free Ca^{2+} induced by vasopressin in isolated hepatocytes, *J. Biol. Chem.*, 259, 5574, 1984.
82. **Wallace, M. A., Giraud, F., Poggioli, J., and Claret, M.,** Norepinephrine-induced loss of phosphatidylinositol from isolated rat liver plasma membrane, *FEBS Lett.*, 156, 239, 1983.
83. **Reinhart, P. H., Taylor, W. M., and Bygrave, F. L.,** Calcium ion fluxes induced by the action of α-adrenergic agonists in perfused rat liver, *Biochem. J.*, 208, 619, 1982.
84. **Charest, R., Blackmore, P. F., Berthon, B., and Exton, J. H.,** Changes in free cytosolic Ca^{2+} in hepatocytes following α_1-adrenergic stimulation, *J. Biol. Chem.*, 258, 8769, 1983.
85. **Berthon, B., Binet, A., Mauger, J. P., and Claret, M.,** Cytosolic free Ca^{2+} in isolated rat hepatocytes as measured by Quin 2, *FEBS Lett.*, 167, 19, 1984.
86. **Korchak, H. M., Vienne, K., Rutherford, L. E., Wilkenfeld, C., Finkelstein, M. C., and Weissman, G.,** Stimulus response coupling in the human neutrophil. II. Temporal analysis of changes in cytosolic calcium and calcium efflux, *J. Biol. Chem.*, 259, 4076, 1984.
87. **Berridge, M. J.,** Preliminary measurements of intracellular calcium in a insect salivary gland using a calcium sensitive-microelectrode, *Cell Calcium,* 1, 217, 1980.
88. **Joseph, S. K. and Williamson, J. R.,** The origin, quantitation and kinetics of intracellular calcium mobilization by vasopressin and phenylephrine in hepatocytes, *J. Biol. Chem.*, 258, 10425, 1983.
89. **Gill, D. L.,** Receptor-mediated modulation of plasma membrane calcium transport, in *Horizons in Biochemistry and Biophysics,* Kohn, L. D., Ed., John Wiley & Sons, New York, 1982, chap. 10.

90. **Penniston, J. T.,** Plasma membrane Ca^{2+}-ATPases as active Ca^{2+} pumps, in *Calcium and Cell Function,* Cheung, W. Y., Ed., Academic Press, New York, 1983, chap. 3.
91. **Lotersztajn, S., Hanoune, J., and Pecker, F.,** A high affinity calcium stimulated magnesium-dependent ATPase in rat liver plasma membranes, *J. Biol. Chem.,* 256, 11209, 1981.
92. **Lotersztajn, S. and Pecker, F.,** A membrane-bound protein inhibitor of the high affinity Ca ATPase in rat liver plasma membranes, *J. Biol. Chem.,* 257, 6638, 1982.
93. **Lin, S.-H., Wallace, M. A., and Fain, J. N.,** Regulation of Ca^{2+}-Mg^{2+}-ATPase activity in hepatocyte plasma membranes by vasopressin and phenylephrine, *Endocrinology,* 113, 2268, 1983.
94. **Lin, S.-H. and Fain, J. N.,** Purification of Ca^{2+}-Mg^{2+}-ATPase from rat liver plasma membranes, *J. Biol. Chem.,* 259, 3016, 1984.
95. **Buckley, J. T. and Hawthorne, J. N.,** Erythrocyte membrane polyphosphoinositide metabolism and the regulation of calcium binding, *J. Biol. Chem.,* 247, 7218, 1972.
96. **Lin, S.-H. and Fain, J. N.,** Ca^{2+}-Mg^{2+}-ATPase in rat hepatocyte membranes: inhibition by vasopressin and purification of the enzyme, in *Epithelial Calcium and Phosphate Transport,* Bronner, F. and Peterlik, M., Eds., Alan R. Liss, New York, 1984, 25.
97. **Burgess, G. M., Giraud, F., Poggioli, J., and Claret, M.,** α-Adrenergically mediated changes in membrane lipid fluidity and Ca^{2+} binding in isolated rat liver plasma membranes, *Biochim. Biophys. Acta,* 731, 387, 1983.
98. **Burgess, G. M., Godfrey, P. P., McKinney, J. S., Berridge, M. F., Irvine, R. F., and Putney, J. W.,** The second messenger linking receptor activation to internal Ca^{2+} release in liver, *Nature (London),* 309, 63, 1984.
99. **Suresh, K., Thomas, A. P., Williams, R. J., Irvine, R. F., and Williamson, J. R.,** myo-Inositol 1,4,5-Trisphosphate: a second messenger for hormonal mobilization of intracellular Ca^{2+} in liver, *J. Biol. Chem.,* 259, 3077, 1984.
100. **Williamson, J. R., Thomas, A. P., and Joseph, S. K.,** Second messenger role of inositol trisphosphate for mobilization of intrallular calcium in liver, *Proc. Chilton Conf.,* in press.
101. **Cantau, B., Keppens, S., deWulf, H., and Jard, S.,** (^{3}H)Vasopressin binding to isolated rat hepatocytes and liver membranes: regulation by GTP and relation to glycogen phosphorylase activation, *J. Receptor Res.,* 1, 137, 1980.
102. **Crane, J. K., Campanile, C. P., and Garrison, J. C.,** The hepatic angiotensin II receptor. II. Effect of guanine nucleotides and interaction with cyclic AMP production, *J. Biol. Chem.,* 257, 4959, 1982.
103. **Gilman, A. G.,** G proteins and dual control of adenylate cyclase, *Cell,* 36, 577, 1984.
104. **Hulme, E. C., Berrie, C. P., Birdsall, N. J. M., and Burgen, A. S. V.,** Interactions of muscarinic receptors with guanine nucleotides and adenylate cyclase in *Drug Receptors and Their Effectors,* Birdsall, N. J. M., Ed., MacMillian, London, 1981, chap. 3.
105. **Snavely, M. D. and Insel, P. A.,** Characterization of α-adrenergic subtypes in the rat renal cortex. Differentiated regulation of α_1 and α_2 adrenergic receptors by guanine nucleotides and Na^+, *Mol. Pharmacol.,* 22, 532, 1982.
106. **Fain, J. N. and Garcia-Sainz, A.,** Role of phosphatidylinositol turnover in α_1 and of adenylate cyclase inhibition in α_2 effects of catacholamines, *Life Sci.,* 26, 1183, 1980.
107. **Jard, S., Cantau, B., and Jacobs, K. H.,** Angiotensin II and α-adrenergic agonists inhibit rat liver adenylate cyclase, *J. Biol. Chem.,* 256, 2603, 1981.
108. **Vanderwel, M., Lum, D. S., and Haslam, R. J.,** Vasopressin inhibits the adenylate cyclase activity of human platelet particulate fraction through V_1-receptors, *FEBS Lett.,* 164, 340, 1983.
109. **Moreno, F. J., Mills, I., Garcia-Sâinz, J. A., and Fain, J. N.,** Effects of pertussis toxin treatment on the metabolism of rat adipocytes, *J. Biol. Chem.,* 258, 10938, 1983.
110. **Gomperts, B. D.,** Involvement of guanine nucleotide-binding protein in the gating of Ca^{2+} by receptors, *Nature (London),* 306, 64, 1983.
111. **Tyson, C. A., Zande, H. V., and Green, D. E.,** Phospholipids as ionophores, *J. Biol. Chem.,* 251, 1326, 1976.
112. **Putney, J. W., Weiss, S. J., Van de Walle, C. M., and Haddas, R. A.,** Is phosphatidic acid a calcium ionophore under neurohumoral control? *Nature (London),* 284, 345, 1980.
113. **Salmon, D. M. and Honeyman, T. W.,** Proposed mechanism of cholinergic action in smooth muscle, *Nature (London),* 184, 344, 1980.
114. **Serhan, C., Anderson, P., Goodman, E., Dunham, P., and Weissman, G.,** Phosphatidate and oxidized fatty acids are calcium ionophores, *J. Biol. Chem.,* 256, 2736, 1981.
115. **Putney, J. W.,** Recent hypothesis regarding the phosphatidylinositol effect, *Life Sci.,* 29, 1183, 1931.
116. **Holmes, R. P. and Yoss, N. L.,** Failure of phosphatidic acid to translocate Ca^{2+} across phosphatidylcholine membrane, *Nature (London),* 305, 637, 1983.

Chapter 6

THE ROLE OF PHOSPHOLIPASES IN HUMAN DISEASES

Karl Y. Hostetler

TABLE OF CONTENTS

I. INTRODUCTION

Phospholipases are found in all intracellular membranes and play a role in a variety of cellular functions.[1] Both endogenous and exogenous phospholipases may cause diseases in man. The purpose of this chapter is to review the evidence which implicates various phospholipases in the pathophysiology of human diseases where a clearly defined role has been demonstrated or proposed. The role of phospholipases in normal cellular physiology will not be considered in this chapter since the subject will be covered in several other chapters, such as those by Litosch and Fain and Farese. This general area has also been reviewed recently by van den Bosch.[1] Major topics to be considered include diseases associated with alterations in the activity of endogenous phospholipases (drug-induced lipidosis, ischemic myocardial injury, and pancreatitis) and diseases caused by exogenous phospholipases (gas gangrene and snake bite).

The reactions catalyzed by phospholipases A_1, A_2, and C are shown in schematic form in Figure 1. Phospholipase A_1 hydrolyzes the fatty acyl ester at the *sn*-1 position of phosphatidylcholine producing equimolar amounts of *sn*-2-acylglycerophosphocholine and fatty acid, whereas phospholipase A_2 attacks the *sn*-2 acyl ester producing equimolar amounts of *sn*-1-acylglycerophosphocholine and fatty acid. Phospholipase C hydrolyzes the phosphodiester bond at the *sn*-3 position of phosphatidylcholine forming *sn*-1,2-diacylglycerol and phosphocholine. Properties of the specific phospholipases which are involved in the pathophysiology of human diseases will be discussed in the respective sections below. Readers who are interested in a more detailed classification of the enzymes which degrade phospholipids (phospholipase B, D, lysophospholipase, etc.) may consult the recent review of van den Bosch.[1]

A. Subcellular Localization of Phospholipases

Phospholipases A are ubiquitous components of mammalian cells and they have been found in many different cell types. A detailed discussion of this subject is not within the scope of this chapter and interested readers may consult the recent article by van den Bosch.[1] Rat liver has been studied most extensively with regard to the intracellular distribution of phospholipases and this body of work will be reviewed briefly.

Phospholipase A_2 is found in liver mitochondria[2-4] while microsomes contain phospholipase A_1.[4-6] Both phospholipase A_1 and A_2 activities have been reported in the plasma membrane,[4,6-8] Golgi,[6] and cytosol compartments,[5] while phospholipase A_1 is the major phospholipase of liver lysosomes.[9,10] Mitochondrial phospholipase A_2[11,12] and lysosomal phospholipase A_1[9,10] are the only phospholipases from specific subcellular sties which have been extensively purified and characterized. Their properties are summarized in Table 1 and the data for porcine pancreatic phospholipase A_2 has been included for comparison.[13] Interestingly, mitochondrial phospholipase has a molecular weight and positional specificity nearly identical to that of porcine pancreatic phospholipase A_2 and both enzymes are blocked by *p*-bromophenacylbromide, suggesting that the mitochondrial enzyme may have a histidine residue at the active site[13] as has been demonstrated for the pancreatic enzyme.[14] However, phospholipase A_1 from lysosomes is distinctly different. It has a larger molecular weight

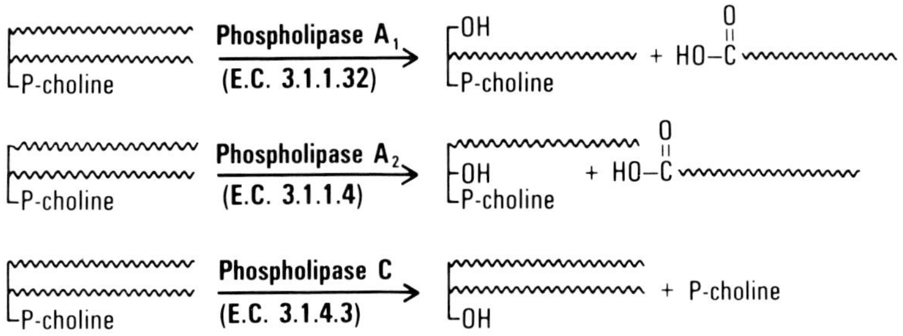

Phospholipase A_1 (E.C. 3.1.1.32)
P-choline → OH, P-choline + HO–C(=O)

Phospholipase A_2 (E.C. 3.1.1.4)
P-choline → OH, P-choline + HO–C(=O)

Phospholipase C (E.C. 3.1.4.3)
P-choline → OH + P-choline

FIGURE 1. Schematic representation of the action of phospholipases A_1, A_2, and C on phosphatidylcholine. The wavy lines denote fatty acyl esters or fatty acid and phosphate oxygens have been omitted for simplicity. The structures shown do not reflect the actual stereoconfiguration of phosphatidylcholine (*sn*-1,2-diacylglycero-3-phosphocholine).

which is two to three times that of mitochondrial phospholipase A_2; it is a glycoprotein which hydrolyzes the *sn*-1 acyl ester (phospholipase A_1) and it does not require Ca^{2+} for activity.[9,10] The pH optimum of 4.0[9,10] is very low compared with that of the phospholipases A_2 (pH 8.4).[11,12] Lysosomal phospholipase A_1 is not affected by *p*-bromophenacylbromide, suggesting the absence of an essential histidine at the active site equivalent to that of pancreatic and mitochondrial phospholipases A_2.[9] An acid phospholipase C which hydrolyzes a variety of phosphoglycerides is present in liver lysosomes[15] and phosphatidylinositol-specific phospholipases C have also been reported in the lysosomes and in the cytosol of liver.[16,17] The latter is presumed to be a key enzyme in the phosphatidylinositol response to various stimuli; this subject is discussed in more detail in the chapters by Litosch and Fain and Farese.

B. Regulation of Phospholipase A

Several mechanisms are known to modulate the activity of intracellular phospholipases. Conversion of a zymogen of phospholipase A_2 to the active form by the action of trypsin has been demonstrated in porcine pancreas by Volwerk and de Haas[18] and will be discussed in Section II.D. Phospholipases A_2 may be regulated by the availability of calcium and their activities may be further affected by cyclic AMP (cAMP), hormones, and changes in membrane structure as noted in the reviews of van den Bosch.[1,19] Certain cationic amphiphilic drugs inhibit intracellular phospholipases and this subject will be discussed in detail in Section II. Recent studies have demonstrated that proteins synthesized in response to glucocorticoids inhibit phospholipase A_2[20,21] and other phospholipases, including phospholipase C and D.[22] The two proteins which have been isolated and characterized are called lipomodulin[22] (mol wt 40,000) and macrocortin[23] (mol wt 15,000), respectively. Finally, three chemotactic factors, some of which are small peptides, were recently reported to stimulate the phospholipase A_2 activity of rabbit neutrophil plasma membranes in vitro; the activation of phospholipase was reported to be receptor mediated.[24]

C. Role of Phospholipases in Cellular Homeostasis

It is not easy to describe or formulate specific functional roles for the intracellular phospholipases, especially when one considers the variety of enzymes present at different subcellular localizations. However, several generalizations are possible. It seems clear that intracellular phospholipases are important in phosphoglyceride deacylation-reacylation reactions (sometimes referred to as remodeling reactions or the Lands cycle) which characterize the ongoing dynamic turnover of phospholipid acyl chains in intracellular membranes.[1,25] In addition, it is clear that phospholipases A and C are important in the generation of

Table 1
PROPERTIES OF PURIFIED INTRACELLULAR PHOSPHOLIPASES A: COMPARISON WITH PURIFIED PANCREATIC PHOSPHOLIPASE A

Source	Mol wt	Positional specificity	pH optimum	Ca^{2+} req.	pBPB inhibition	Glycoprotein	Ref.
Procine pancreas	14,000	A_2	8.4	+	+	0	13
Liver mitochondria	10—15,000	A_2	8.4	+	+	0	11,12
Liver lysosomes	34,000	A_1	3.2—4.0	0	0	+	9
Liver lysosomes	44,000	A_1	3.2—4.0	0	0	+	9

Abbreviations: Mol wt, molecular weight; pBPB, *p*-bromophenacylbromide; A_1, A_2, hydrolyzes the *sn*-1 or *sn*-2 position of *sn*-1,2-diacylglycerophospholipids, respectively.

prostaglandin precursors from the *sn*-2 position of phosphoglycerides (in the case of phospholipase C, this requires the subsequent action of a diacylglycerol lipase). In lysosomes, the phospholipases are important in the turnover of incoming phospholipids which may originate from lipoproteins. These reach the lysosomes by absorptive endocytosis, as well as phospholipids from intracellular organelle membranes, which are degraded after microautophagy. Finally, as discussed in other parts of this book, phosphatidylinositol-specific phospholipase C appears to play an important role in the hormone or neutrotransmitter-induced turnover of phosphatidylinositol and its phosphorylated derivatives. This appears to be involved in several cellular regulatory functions.

II. ROLE OF PHOSPHOLIPASES IN DRUG-INDUCED LIPIDOSIS

In 1971, Yamamoto and co-workers[26,27] in Osaka noted that patients who had been taking the coronary vasodilator, 4,4′bis(diethylaminoethoxy)-α,β-diethyldiphenylethane (4,4′-bis(diethylaminoethoxy)hexestrol; abbreviation, DH), developed an acquired lipid storage disease that resembled Niemann-Pick disease. It was characterized by the enlargement of the liver and spleen and by the appearance of foam cells in the bone marrow and vacuolated lymphocytes in the peripheral blood. A major ultrastructural feature of this disorder is the presence of large numbers of intracellular multilamellar bodies (myelin figures) in the liver and other tissues.[28-30] All classes of phospholipids are substantially increased in liver and other tissues, but sphingomyelin is not increased out of proportion as is the case in Neimann-Pick disease.[31] The disorder can be produced by administration of DH to rats or monkeys, resulting in a marked increase in liver phospholipid content.[31,32] Treatment of experimental animals with chloroquine also causes a similar syndrome characterized by the presence of intracellular multilamellar bodies and increased tissue phospholipids.[33,34]

Since the description of diethylaminoethoxyhexestrol-induced phospholipid fatty liver in man, there have been many reports of drug-induced phospholipidosis. More than 30 agents have now been identified which cause phospholipid storage in man, animals, or cultured cells.[35] Many cells and tissues can be affected including lung, liver, spleen, adrenal, kidney, alveolar macrophages, MDCK cells, and others.[35-39] Although the drugs vary in their principal pharmacological actions, they share cationic and amphiphilic structural characteristics.[35]

A. DH-Induced Phospholipid Fatty Liver

Animal experiments have made it possible to study the mechanisms which lead to human phospholipid fatty liver. Using rats treated with diethylaminoethoxyhexestrol (DH) it has been possible to study drug-induced lipidosis in detail, and it seems likely that the mechanisms elucidated in this system will be relevant to drug-induced phospholipidosis in general.

1. Liver Lipid Analysis in DH Phospholipidosis in Man

Table 2 shows the lipid analyses carried out on liver biopsies from some of the patients studied by Yamamoto and co-workers.[27] Liver phospholipid levels were greatly increased while triglyceride and ester cholesterol levels were essentially normal in most cases; free cholesterol was usually increased.[27] DH was isolated from liver and the structure of its metabolites was determined.[40] It was found that human liver could not hydroxylate the benzene ring of DH, whereas this reaction is very rapid in rats. The lack of hydroxylation of DH in man is proposed to be the cause of the slow metabolism of the drug and may account for the fact that its side effects were not appreciated in the toxicity studies done in animals.[40]

Table 2
EFFECT OF DH ON LIPID CONTENT OF HUMAN LIVER[27a]

	Lipid content, % wet weight				
Lipid	**Normal (5)**	**1**	**2**	**3**	**4[b]**
Phospholipid	2.9 ± 0.3	4.4	4.9	4.6	5.2
Triglyceride	0.5 ± 0.2	0.2	11.2	0.1	—
Cholesterol ester	0.2 ± 0.1	0.2	0.3	0.3	—
Free cholesterol	0.2 ± 0.1	0.5	1.2	0.9	—

[a] Numbers 1—4 represent individual cases of DH-induced lipidosis. Normal represents 5 unaffected control subjects.
[b] Autopsy sample. All others cases represent liver biopsy specimens.

2. Metabolic Basis of Hepatic Phospholipidosis: Studies with DH-Treated Animals

a. Hepatic Changes in DH-Treated Rats

Treatment of rats with 50 to 100 mg DH/kg for 1 week or more resulted in liver enlargement and the appearance of the typical multilamellar bodies in the cell cytoplasm. The structures were shown to contain acid phosphate, a lysosomal marker enzyme. Biochemical studies in rats confirmed that DH caused an increase in the phospholipid content of liver similar to that reported in the human cases.[31,33,34]

b. Subcellular Localization of Phospholipid and Drug in Rat Liver

To determine the intracellular site of phospholipid accumulation, rats were treated with DH 100 mg/kg for 1 week, and the respective subcellular fractions were isolated, characterized for purity using marker enzymes, and the levels of phospholipid were determined.[41] The results are shown in Table 3.

From marker enzyme data, the size of the respective subcellular pools could be determined for protein and phospholipid. Mitochondrial and microsomal phospholipid per milligram of protein did not change significantly, and the total protein and phospholipid pools represented by these fractions did not change. However, lysosomal phospholipid per milligram of protein increased 5.5-fold over that of control lysosomes, and the lysosomal pool of protein and phospholipid per gram of liver increased by 2.9-fold and 15.4-fold, respectively, demonstrating that phospholipid catabolism in lysosomes is impaired to a much greater degree than is protein breakdown. These studies clearly identify the lysosomes as the principal intracellular site of DH effects on phospholipid metabolism.[41]

The levels of DH in purified subcellular fractions are shown in Table 4. Small amounts of DH are present in various membranes and in the cytosol ranging from 2.8 to 6.3 nmol/mg protein. However, lysosomes contain 189 nmol DH/mg protein, representing the bulk of the intracellular drug.[34]

c. Effect of DH on Lysosomal Phospholipid Catabolism

Since lysosomes contain the stored lipid and most of the intracellular DH, we examined the effects of DH on the activity of lysosomal phospholipases. It proved to be a potent inhibitor of both lysosomal pathways of phosphoglyceride degradation as shown in Table 5. For phospholipase A inhibition, the IC_{50}* of DH was 0.23 m*M* vs. 0.20 m*M* for phos-

* The IC is the concentration of a drug required to produce a 50% decrease in the activity of an enzyme, in this case, phospholipase A_1.

Table 3
ESTIMATION OF THE SIZE OF INTRACELLULAR POOLS OF PROTEIN AND PHOSPHOLIPID IN THE LIVER OF DRUG-TREATED RATS[41a]

Treatment	Fraction	Protein (mg/fraction)[b]	Lipid phosphorus (nmol/mg protein)	Total lipid phosphorus/ fraction
Control	Mito	352 ± 59	272 ± 10	96 ± 14
	Micro	269 ± 16	620 ± 22	167 ± 14
	Lyso	26 ± 11	510 ± 54	14 ± 6
DH-treated	Mito	371 ± 44	238 ± 36	87 ± 10
	Micro	268 ± 18	644 ± 39	173 ± 17
	Lyso	76 ± 13[c]	2820 ± 230[d]	215 ± 48[d]

[a] Abbreviations: Mito, mitochondria; Micro, microsomes; Lyso, lysosomes.
[b] Milligram of protein per total fraction assuming 1.0 g total liver protein; μmol is lipid phosphorus per total fraction.
[c] $p < 0.005$.
[d] $p < 0.01$ vs. control (n = 3).

Table 4
INTRACELLULAR LOCALIZATION OF DH IN RAT LIVER[34]

Cell fraction	DH, nmol/mg protein	f[a]
Homogenate	17	1.0
Mitochondria	3	0.2
Lysosomes	189	11.1
Microsomes	7	0.4
Supernatant	4	0.2

[a] Enrichment factor (f) relative to the concentration of DH in the cell homogenate.

pholipase C[42] at pH 4.4. At pH 5.4, the corresponding values for IC_{50} were 0.04 and 6.0 m*M*, respectively. These findings strongly suggest that phospholipase inhibition by DH is an important cause of drug-induced lipidosis. Therefore, it would be of great interest to know the in vivo concentration of DH in liver lysosomes attained after several doses of drug.

d. Measurement of the Concentration of DH in Liver Lysosomes

The data in Table 4 indicate that DH concentrates in lysosomes; however, the drug concentration could not be determined previously due to the lack of a volume (water space) term. This problem has now been overcome by the development of a new method which employs ^3H_2O and [U-^{14}C]sucrose to measure the lysosomal water space. DH is measured in lysosomal extracts by gas liquid chromatography. The results are shown in Table 6. After 2 days of DH (100 mg/kg), the intralysosomal concentration of the drug is 26 m*M*, rising to 69 m*M* at 7 days. Based on the data in Table 5, this level would be sufficient to abolish phospholipase A and C activity in liver lysosomes. Consistent with this observation, lysosomal phospholipid at 2 and 7 days is greatly increased to levels 6.8- and 9.1-fold that of control lysosomes.[43] Thus, DH causes phospholipid fatty liver in man and animals by

Table 5
EFFECT OF DH ON LIVER LYSOSOMAL PHOSPHOLIPASES IN VITRO[42a]

Incubation pH	IC_{50} PLA[b]	IC_{50} PLC[b]
4.4	0.23	0.20
5.4	0.04	6.0

[a] Abbreviations: PLA, phospholipase A; PLC, phospholipase C.
[b] Millimolar concentration of DH required to produce 50% inhibition of the respective lysosomal phospholipase activity.

Table 6
CONCENTRATION OF DH IN LYSOSOMES AND EFFECT OF DH TREATMENT ON LYSOSOMAL PHOSPHOLIPID CONTENT[a]

DH treatment (days)[b]	DH, conc. in lysosomes, m*M*	Lysosomal phospholipid content, nmol (mg/protein)
0	0	309
2	26	2100
7	69	2820

[a] Abbreviation: DH, 4,4′-bis(diethylaminoethoxy)hexestrol.
[b] DH dosage, 100 mg/kg p.o.

inhibiting intralysosomal processing of complex lipids. Inhibition of lysosomal phospholipases may be direct (as shown for DH) or indirect through elevation of the intralysosomal pH.

B. Chloroquine-Induced Phospholipid Fatty Liver

Chloroquine is useful in the treatment of malaria, rheumatoid arthritis, and collagen-vascular disesese.[44] However, its prolonged use is often associated with toxic side effects involving the retina and skeletal muscles. Treatment of animals with chloroquine in large doses for 5 to 7 days or with small doses for longer periods causes phospholipid accumulation in liver, spleen, muscle, and other tissues.[33,34,45] In rat liver, the biochemical changes are very similar to those noted above with DH treatment. After 7 days of chloroquine treatment the liver phospholipid content increases by 50%.[34] The lysosomal fraction contains all of the excess liver phospholipid[34,41] and is highly-enriched in chloroquine.[34]

Two possible mechanisms might explain chloroquine-induced lysosomal phospholipid storage: (1) inhibition of lysosomal phospholipases as shown for DH, or (2) lysosomal depletion of acid phospholipase. The latter mechanism was considered since chloroquine has been shown to cause cultured fibroblasts to secrete lysosomal acid hydrolases into the medium with the result that fibroblast levels of acid hydrolases are reduced.[46,48] Rats were treated with chloroquine in a dosage sufficient to cause a 35% increase in liver phospholipids.

Table 7
EFFECT OF CHLOROQUINE ON RAT LIVER LYSOSOMAL CHLOROQUINE CONCENTRATION

CLQ treatment (hr)	Lysosomal CLQ, m*M*	Lysosomal PL, nmol (mg/protein)
0	0	308 ± 6
12	6.3 ± 0.8	527 ± 29
24	16.5 ± 1.7	794 ± 38
48	61.1 ± 0.8	996 ± 25
72	74.0 ± 5.8	1132 ± 43

Abbreviations: CLQ, chloroquine; PL, phospholipid. Values are mean ± S.D., n = 3.

However, acid phospholipase A levels in liver were found to be increased by chloroquine treatment, essentially ruling out cellular depletion of lysosomal phospholipase in the pathogenesis of the disorder.[49]

Chloroquine concentrates greatly in lysosomes and it is of interest to know the intralysosomal concentrations of this drug in order to evaluate the significance of the in vitro inhibition data.[42] Lysosomes were isolated by free-flow electrophoresis, their water space was determined using $^{3}H_2O$ and [^{14}C]sucrose and chloroquine was measured by fluorescence.[50] As shown in Table 7, intralysosomal chloroquine concentrations of 6.3 m*M* were reached in liver only 12 hr after a single chloroquine dose. Chloroquine rose steadily to 74 m*M* after 3 days of treatment (100 mg/kg) and the increase was accompanied by a nearly fourfold increase in lysosomal phospholipid.

The intralysosomal levels of chloroquine attained are sufficient to account for significant inhibition of acid phospholipase A_1, which has an IC_{50} of 3 to 6 m*M*.[50] Thus, chloroquine produces phospholipidosis through its inhibitory effects on lysosomal catabolism of polar lipids both by direct inhibition[42,50,51] and by virtue of its well-known ability to raise the intralysosomal pH.[52,53] Although the sequence of events has only been established to date in liver and in cultured cells (*vide infra*), it seems likely that the toxic side effects of chloroquine on the retina and muscle may be produced by a similar type of mechanism.

C. Studies with Other Agents Which Cause Phospholipidosis

Additional agents which have been implicated in celluar phospholipidosis, including chlorphentermine, imipramine, chlorpromazine, and 1,7-bis(*p*-amino-phenoxy)heptane, were tested to determine their effects on lysosomal phospholipases. The former two agents have been reported to cause pulmonary phospholipidosis;[54,55] chloropromazine causes phospholipid storage in cultured cells[35] and 1,7-bis(*p*-aminophenoxy)heptane appears to produce phospholipidosis in schistosomes.[56] All were inhibitors of lysosomal phospholipase action as shown in Table 8.[57] The cationic amphiphiles propranolol, amantadine, and tripelennamine also inhibited the phospholipases in vitro. 1,7-bis(*p*-Aminophenoxy)heptane and chlorpromazine were the most active inhibitors of the lysosomal phospholipases with IC_{50} values of 0.01 to 0.07 m*M*, while chlorphentermine, imipramine, DH, and propranolol were intermediate with IC_{50} values of 0.23 to 0.32 m*M* for inhibition of phospholipase A and 0.20 to 0.45 m*M* for phospholipase C, respectively. Chloroquine was a weak inhibitor of phospholipase A at pH 4.4 (Table 8), but at pH 5.4 substantial inhibition was noted as shown in Table 6.[42] Amiodarone, which has also been reported to cause phospholipidosis in the lung[58] is also a potent inhibitor of lysosomal phospholipase A.[59]

Gentamicin and other aminoglycoside antibiotics concentrate in the kidney proximal tubule

Table 8
INHIBITION OF LYSOSOMAL PHOSPHOLIPASES BY CATIONIC AMPHIPHILIC DRUGS IN VITRO[42,57a]

	IC_{50} (m*M*[b])	
Drug	**Phospholipase A**	**Phospholipase C**
bis(*p*-Aminophenoxy)heptane	0.01	0.03
Chlorpromazine	0.03	0.07
Chlorphentermine	0.32	0.45
DH	0.23	0.20
Imipramine	0.23	0.25
Propranolol	0.25	0.38
Tripelennamine	0.50	1.25
Amantadine	3.25	6.75
Chloroquine	10	0.33

[a] DH, 4,4′-bis(diethylaminoethoxy)hexestrol.
[b] IC_{50} millimolar concentration required to produce 50% inhibition of the respective phospholipase; PLA, phospholipase A; PLC, phospholipase C.

cell and cause phospholipid storage and multilamellar bodies accumulation.[60,61] Recently we found that aminoglycosides inhibit lysosomal phospholipase A_1[62,63] and these findings have been extended by Laurent et al.[64] and Carlier et al.[65] Aminoglycosides also inhibit the renal phosphatidylinositol-specific phospholipase C.[66] Thus, it appears that one important property of the agents which have been reported to cause phospholipidosis is their common ability to inhibit the phospholipases of lysosomes.

D. Possible Mechanisms of Drug Inhibition of Lysosomal Phospholipases

Cationic amphiphilic drugs bind to phospholipid and it has been proposed that the resulting complex is resistant to phospholipase action.[35,36,67,68] The tendency of these drugs to bind to phospholipid is well documented; acidic phospholipids such as phosphatidylinositol and phosphatidylserine bind cationic amphiphilic drugs more readily than do neutral phosphoglycerides such as phosphatidylcholine and phosphatidylethanolamine.[67,68]

We recently examined the effect of several cationic amphiphilic agents on lysosomal phospholipase A hydrolysis of phosphatidylinositol, phosphatidylcholine, and phosphatidylethanolamine.[69] The results are shown in Table 9. Conversion of these three phospholipids to their respective lysoderivatives was inhibited by imipramine, propranolol, and chlorpromazine. However, 5 to 25 times more inhibitor was required to inhibit by 50% the conversion of phosphatidylinositol to its lysocompound than that needed to give a similar reduction in the rate of conversion of the neutral phosphoglycerides. This suggests the possibility that enhanced complex formation between the drugs and acidic phospholipids may not be the only factor in the inhibition of lysosomal phospholipases. More detailed studies with purified enzymes will be required to resolve this point.

Lysosomal phospholipase A_1 has recently been purified from rat liver lysosomes. Multiple isoenzymes were isolated which differ in their molecular weights, isoelectric points, and apparent K_m for phosphatidylcholine.[9] Using the major isoenzyme, the inhibitory properties of propranolol and other β-adrenergic blockers were examined recently. There was a wide variation in the ability of various β-adrenergic blockers to inhibit purified lysosomal phospholipase A. In addition, there was a strong negative correlation between the octanol/water partition coefficients and the IC_{50} for phospholipase A_1 inhibition ($r = -0.91$).[70] Thus, hydrophobicity appears to be a very important determinant of the phospholipase inhibitory potential of a cationic amphiphilic drug.

Table 9
CATIONIC AMPHIPHILIC DRUG INHIBITION OF PHOSPHOLIPASE A HYDROLYSIS OF ACIDIC AND NEUTRAL PHOSPHOGLYCERIDE SUBSTRATES[69]

	IC_{50}[b]			
Substrate	**Imipramine**	**Propranolol**	**DH**	**Chlorpromazine**
PE	0.10	0.32	0.11	0.10
PC	0.36	0.71	0.41	0.10
PI	1.8	3.6	2.8	0.50

[a] Abbreviations: PE, phosphatidylethanolamine; PC, phosphatidylcholine, PI, phosphatidylinositol, DH, 4,4′-bis(diethylaminoethoxy)hexestrol.
[b] IC_{50}, the m*M* concentration of drug required to produce 50% inhibition of lysophosphoglyceride formation.

E. Drug-Induced Phospholipid Storage in Cultured MDCK Cells

Phospholipidosis experiments with cationic amphiphilic agents in animals may be complicated by variable tissue distribution of the drugs and further metabolism and excretion of the agents. Amantadine is an anti-influenza A agent previously shown to be a weak inhibitor of lysosomal phospholipases,[57] which was lysosomotropic in Madin-Darby canine kidney cells (MDCK cells). Fifty-five percent (55%) of the cellular amantadine was found in the lysosomal compartment and most of the remainder (34%) was recovered in the supernatant (cytosol) fraction.[71] Lysosomal phospholipases were isolated from MDCK cells and amantadine was found to inhibit their activity.[39] When MDCK cells were grown in the presence of 0.3 m*M* amantadine, cellular phospholipid increased from 225 nmol/mg protein to 313 nmol/mg protein after 3 days, an increase of 39%. Chloroquine, which is highly lysosomotropic in MDCK cells, also inhibited MDCK lysosomal phospholipases. MDCK cells grown in 0.03 m*M* chloroquine showed a 40% increase in their phospholipid content.[39] Thus, chloroquine was 10 times more effective than amantadine in producing phospholipidosis in MDCK cells. This is probably due to its greater degree of lysosomotropism and its greater effectiveness as an inhibitor of lysosomal phospholipases.[39,42,57] In addition to their direct inhibitory effects, it is very likely that these weakly basic compounds also impair phospholipid catabolism by raising intralysosomal pH as has been shown in macrophages[52] and fibroblasts.[53]

F. Proposed Mechanisms of Cationic Amphiphilic Drug-Induced Phospholipidosis

Based on the studies in rats with DH and chloroquine and in cultured MDCK cells with chloroquine and amantidine, it is hypothesized that cellular phospholipidosis results if three steps take place: (1) free entry of the agent into the cell must occur — cells which exclude the agent would be protected from phospholipidosis; (2) the agent must be lysosomotropic; and (3) the drug must inhibit lysosomal phospholipase activity — inhibition may occur either directly, as shown in our in vitro studies with the phospholipases, or indirectly, by elevation of the intralysosomal pH. To date, evidence for the above sequence of events has been provided only in liver and cultured MDCK cells. Nevertheless, it seems quite likely that this sequence of events will explain cationic drug-induced phospholipidosis in other organs where multilamellar bodies have been shown to accumulate as a toxic side effect.

III. PHOSPHOLIPASES AND ISCHEMIC MYOCARDIAL INJURY

A. Biochemical and Ultrastructural Events

In acute myocardial ischemic injury a number of biochemical events occur, ultimately leading to the failure of muscle contraction and to irreversible cell injury. Carbohydrate metabolism shifts to anaerobic glycolysis with accumulation of lactic acid; the intracellular pH declines and glucose metabolism is inhibited at several points.[72] Fatty acid oxidation is reduced due to the limited supply of oxygen and to inactivation of the mitochondrial carnitine acyltransferase system. Intracellular levels of acyl CoA and acylcarnitine rise, inhibiting adenine nucleotide transferase which is required for ATP transfer from mitochondria to cytosol and lysophospholipase. Cellular ATP and high energy phosphate levels fall and muscle contraction ceases. After 40 to 60 min necrosis occurs.[73,74] Calcium influx into the cells is substantial, especially with reperfusion. From an ultrastructural standpoint, some important features of irreversible ischemic injury are (1) disruption of the plasmalemma of the sarcolemma, (2) mitochondrial swelling, (3) distortion of mitochondrial cristae, and (4) the presence of electron-dense amorphous mitochondrial inclusions (amorphous matrix densities).[73,75]

B. The Importance of Phospholipid Metabolism in Ischemic Myocardial Injury

The biochemical and ultrastructural studies noted above point strongly toward membrane damage as a key early event. As proposed by Singer and Nicolson,[76] the biological membrane consists of a phospholipid bilayer which itself provides the permeability barrier; the phospholipid bilayer also contains integral and peripheral membrane proteins which confer special properties on the membrane. However, biological membranes also contain phospholipases A which can hydrolyze the fatty acid ester bonds of the phospholipid components of the bilayer, producing free fatty acid and lysophospholipid.[1]

1. Evidence for Increased Phospholipase Activity During Ischemia

Several lines of evidence implicate endogenous phospholipase A as an important participant in the membrane damage of ischemic myocardial injury. Studies of ischemic heart tissue generally show that levels of lysophospholipid increase following coronary occlusion, although the subject remains controversial. Initially, Sobel and co-workers[77] reported a 60 to 70% increase in lysophospholipids after 15 to 60 min of ischemia in rabbit heart. Using improved methodology, Shaikh and Downar[78] reported that tissue levels of lysophosphatidylcholine increase by 25 to 58% in ischemic pig heart at 8 to 40 min. Corr et al.[79] found a 53% increase in lysophosphoglycerides in feline heart after 10 min of ischemia. After 24 hr of ischemia in dog heart, myocardial lysophosphatidylcholine and lysophosphatidylethanolamine were increased by 107 and 137%, respectively.[80] However, Chien et al.[81] found no increase in lysophosphoglycerides after 3 hr of ischemia in dog heart. Increased levels of lysophospholipid have also been demonstrated in venous blood draining the ischemic regions.[82]

Lysophospholipids have been implicated in the pathogenesis of malignant ventricular dysrhythmias (*vide infra*) and may be cytotoxic due to their detergent properties such as their long-recognized ability to lyse red blood cells (hence, the name "lyso" phospholipid).[83] Fatty acids, the other product of phospholipid breakdown, and fatty acylCoA and fatty acylcarnitines also accumulate in ischemic areas; these compounds also have detergent properties and may also be toxic to cells.[74,84] Elevated levels of free fatty acid have also been linked to arrhythmias.[85]

A second major line of evidence implicating abnormal phospholipid metabolism in irreversible ischemic injury relates to the fact that myocardial tissue phospholipid levels decline during ischemia. Vasdev et al.[86] showed a 15% decrease in mitochondrial phospholipids in dog heart 3 hr after development of ischemia while sarcolemmal membranes exhibited a 29% decline in phospholipid. However, the purity of these membrane fractions was not evaluated by marker enzyme analysis or by electron microscopy. Shaikh and Downar reported that total phospholipid of ischemic pig heart declined 6 and 14% after 8 and 24 hr, respectively. In their study, lysophospholipid levels were increased by 60% as early as 8 min of ischemia, but did not increase further between 40 min and 24 hr.[78] Chien et al.[87] found an 11% decrease in heart phospholipid at 8 hr and a 32% decrease at 12 hr in rats. In sarcoplasmic reticulum isolated from the ischemic rat heart, the amount of phospholipid per milligram of protein declined by 15, 28, and 29% at 4, 8, and 12 hr, respectively. Chien et al.[87] concluded that the apparent decrease was due to increased phospholipid degradation rather than to reduced phospholipid synthesis. However, in the above studies of isolated membrane preparations, the purity of the subcellular fractions was not investigated and the differences noted could be greatly affected by varying degrees of contamination of the respective membrane fractions with other membranes which are less enriched in phospholipid. Nevertheless, the studies noted above strongly suggest that phospholipase A is activated in ischemia.

Only a few direct measurements of myocardial phospholipases have been made with fractions isolated from ischemic tissue. In ischemic rabbit heart, Corr and co-workers[88] reported that phospholipase A_2 activity in isolated mitochondria and microsomes did not increase in in vitro experiments using direct assays. However, activation of phospholipase A_2 has recently been reported in experimental cerebral ischemia.[89]

2. Regulation of Intracellular Phospholipid Metabolism During Myocardial Ischemia

Figure 2 shows the biochemical pathways of importance to phosphoglyceride and lysophosphoglyceride metabolism during ischemia using phosphatidylcholine as an example. From an examination of the metabolic pathways in Figure 2, it is clear that phospholipid depletion during ischemia will depend strongly on the rates of endogenous membrane phospholipase A activities vs. the rates of resynthesis of phosphoglycerides, either by reacylation of lysophospholipids or by *de novo* synthesis. Whether or not levels of lysophospholipids will rise depends on the relative rates of phospholipase A (PLA), lysophospholipase (LPL), LPL/transacylase, and the reacylation of LPC. Either activation of phospholipase(s) A or inhibition of lysophospholipase would tend to increase tissue levels of lysophosphatidylcholine. On the other hand, levels of lysophosphatidylcholine would be reduced if there is activation of lysophospholipase, increased reacylation of LPC or increased activity of LPC/transacylase.

In the heart, phospholipases A have been reported in all subcellular membrane fractions including mitochondria (A_2)[90-92] lysosomes (A_1),[92,93] microsomes (A_1, A_2),[90,92] and sarcolemma (A_1,A_2).[94-96] Recently, a cytosolic phospholipase A_1 has been described in rat heart.[92] Little is presently known about the regulation of these enzymes. Ca^{2+} appears to be required for mitochondrial phospholipase A_2 activity[90-92] and is required for optimum activity with the sarcolemmal enzyme,[94-96] but has little effect on the microsomal and cytosolic phospholipases A.[92] A calcium-dependent phospholipase A_2 has been isolated from rabbit heart and chick embryo myocytes, but its subcellular origin is unclear.[97] Isoproterenol has been reported to stimulate sarcolemmal phospholipase A activity at concentrations of 1 to 10 μM.[98] Myocardial phospholipases from specific intracellular sites have not yet been purified or extensively characterized.

Lysophospholipases are present in mitochondria,[92] microsomes,[92,99] and in the cytosol.[92,99-102] The cytosolic lysophospholipase has been purified and characterized.[100] It has a molecular weight of 23,000 and, interestingly, it is inhibited by palmitoylcarnitine and low pH (pH 6.5). Recently it was demonstrated that the cytosolic lysophospholipase may be identical to palmitoylCoA hydrolase based on co-purification and mutual inhibition by the respective substrates.[101] An interesting nonenergy-requiring enzyme which resynthesizes phosphatidylcholine from 2 mol of lysophosphatidylcholine by transacylation has also been purified and characterized.[103] This enzyme, which has a molecular weight of 63,000 also has lysophospholipase activity and is inhibited by palmitoylcarnitine. Based on the above

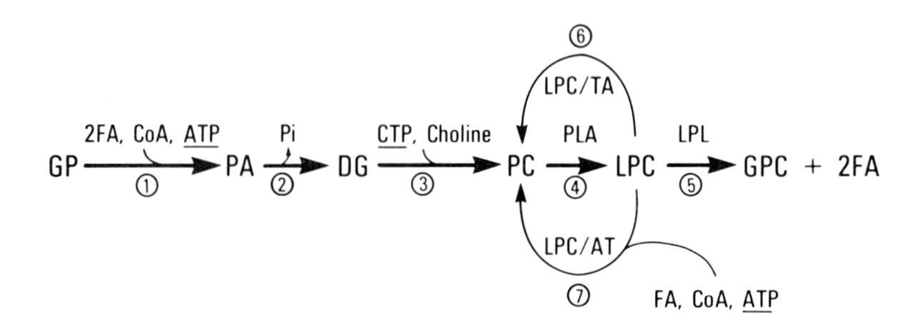

FIGURE 2. Metabolic pathway important in the regulation of the levels of phosphoglycerides and lysophosphoglycerides in ischemia. Enzyme designations: (1) acyl CoA: glycerol-3-phosphate acyltransferase, E.C. 2.3.1.15; (2) phosphatidate phosphohydrolase, E.C. 3.1.3.4; (3) CTP: cholinephosphate cytidylyltransferase, E.C. 2.7.7.14 and CDPcholine: 1,2-diacylglycerol choline phosphotransferase, E.C. 2.7.8.2; (4) phospholipase A_1 E.C. 3.1.1. 32 or phospholipase A_2, E.C. 3.1.1.4; (5) lysophospholipases, E.C. 3.1.1.5; (6) lysophosphatidylcholine transacylase; (7) lysolecithin acyltransferase, E.C. 2.3.1.23. Abbreviations: FA, fatty acid; CoA, coenzyme A; ATP, adenosine triphosphate; CTP, cytidine triphosphate; PC, phosphatidylcholine; LPC, lysophosphatidylcholine; GPC, glycerophosphocholine; PLA_1, phospholipase A_1; PLA_2, phospholipase A_2; LPL, lysophospholipase, Pi, inorganic phosphate. Reactions which require nucleotide triphosphates (underlined) are indicated and other cofactors are noted.

it has been suggested that the accumulation of lysophospholipids during myocardial ischemia may be due in part to inhibition of lysophospholipases by acylcarnitine and low pH.

In myocardial ischemia, there is a marked lack of correspondence between the loss of tissue phosphoglycerides and the accumulation of lysophospholipids. That is, the rise in lysophospholipids during ischemia is not equal to the loss of phospholipids. For example, after 24 hr of ischemia in pig heart, the level of phosphatidylcholine declines by 41 μg lipid P/g wet weight, but lysophosphatidylcholine increases by only 0.6 μg lipid P/g wet weight. This is doubtlessly due to rapid removal of lysophosphatidylcholine by myocardial lysophospholipases.[92,99-102] Recently it has been demonstrated that the observed Vmax for lysophospholipase in rat heart mitochondria, microsomes, and cytosol is 3.6 to 4.5 times greater than the Vmax observed for phospholipase A in the respective cell fractions.[92]

The absence of one-to-one stochiometry involving phospholipid decline and lysophospholipid increase in the myocardium during ischemia is further supported by the finding of a progressive increase in nonesterified arachidonic acid and other long-chain fatty acids during ischemia in dog heart.[81,104,105] The increase in myocardial free fatty acids is directly proportional to the reduction in blood flow as determined using radiolabeled microspheres.[105] The increase in tissue free fatty acid levels as a result of the sequential action of phospholipases A and lysophospholipase might be a better indicator of ischemic injury than the increase in lysophosphoglycerides.[81,105]

Thus, it appears that phospholipase activation is an important early result of ischemia; the phosphoglyceride deacylation-reacylation cycle is greatly increased,[81] followed later by a decline in the levels of phospholipid as nucleotide triphosphate levels become depleted and reacylation of lysophosphatides, acylation of glycerol phosphate, and formation of CDP-choline for *de novo* synthesis can no longer proceed. This is due to the inability of the cell to activate long-chain fatty acids and choline since these reactions require ATP and CTP (see Figure 2).

3. Effects of Lysophospholipids on Cardiac Metabolism

In 1977, Downar et al.[106] found that electrophysiological depressant effects are noted when strips of normal pig myocardium are superfused with "ischemic" venous blood obtained from the coronary vein of pigs following experimental coronary occlusion. The authors[106] concluded that these effects could not be accounted for by increased potassium, hypoxia, acidosis, or hypoglycemia, and they proposed that the release of unidentified factors from ischemic tissue might mediate the electrophysiological effects.

Subsequently, lysophospholipids were shown to be elevated in the ischemic myocardium and in venous blood draining the ischemic areas. These compounds were implicated by many investigators as mediators of the electrophysiological abnormalities which result in the production of malignant cardiac dysrhythmias.[79,80,82,107-111] In addition, lysophospholipids have been shown to cause an increase in myocardial levels of cAMP[112] and to promote Ca^{2+} entry into cardiac myocytes.[113] Increased levels of lysophosphatidylcholine in vitro reduce mitochondrial state 3 respiration and abolish state 4 respiration.[88] Thus, lysophospholipids appear to produce a variety of effects on cardiac metabolism and electrophysiology which may be of great significance in the pathogenesis of ischemic myocardial injury.

The effects of lysophosphoglyceride on cardiac arrhythmias noted above are of particular interest in view of their possible role in ventricular fibrillation, a serious and life-threatening complication of myocardial infarction.[84,88] It should be noted that many of the studies which have been done employed perfusion mixtures containing lysophosphoglycerides in concentrations which are assumed to be equivalent to that present in the ischemic myocardium or venous blood. For example, lysophosphatidylcholine (6 to 3000 μM) is dissolved in Tyrodes solution or Krebs-Henseleit buffer and used in studies of cardiac electrophysiology. However, lysophosphatidylcholine is formed in vivo by phospholipase A action on phospholipid bilayers and it is well known that the products, including lysophosphatidylcholine, may be retained in the lipid bilayer rather than being released into solution.[114,115] In fact, natural biological membranes and artifical phospholipid bilayer membranes may contain rather large amounts of lysophospholipids, representing as much as 17 to 35% of total lipid phosphorus, without loss of their membrane bilayer structure.[116-118] Thus, the use of solutions of lysophosphatidylcholine in concentrations calculated from data based on the amount of lysophospholipid per gram of wet weight of myocardial tissue is questionable in studies where the mixture is administered externally (extracellularly) to obtain electrophysiological data. In other words, one cannot expect the properties of a solution of lysophosphatidylcholine in buffer to be similar to the properties of an equivalent amount of lysophospholipid which is not free in solution, but is bound to proteins, lipoproteins, or present as a component of a phospholipid bilayer. Thus, it seems likely that the problem (i.e., how to test the hypothesis that elevated levels of lysophosphatidylcholine cause a variety of deleterious cardiac electrophysiological effects) will require approaches which rigorously account for both free and bound lysophosphatidylcoline, the latter including that associated with plasma proteins, lipoproteins, and that present both in myocardial cell phospholipid bilayer membranes as well as that in extramyocardial membranes (including erythrocytes and leukocytes).

To date, several studies have attempted to answer the question of the role of bound vs. free lysophosphatidylcholine in the altered electrophysiology. Snyder et al.[82] found that 370 μM lysophosphatidylcholine in feline plasma did not produce electrophysiological changes at pH 7.4, but marked electrophysiological effects were noted with acidotic plasma (pH 6.7) containing 370 μM lysophosphatidylcholine. Man and Choy[110] suggested that bound lysophosphatidylcholine is not responsible for the production of arrhythmias and in their studies, free lysophosphatidylcholine levels above 20 μM were noted to be associated with the production of arrhythmias in perfused hamster hearts.

In summary, the lysophospholipid hypothesis, although controversial, is an attractive and important hypothesis which may be able to explain electrophysiological phenomena which occur during ischemia. If correct, it could lead to pharmacological approaches that could modify the effects on cardiac metabolism and electrophysiology.

4. Pharmacological Interventions to Reduce Phospholipase A Activity

Pharmacological interventions designed to reduce the activity of phospholipases A, and hence, the degradation of membrane phospholipids have been reported in liver, kidney, and heart. Chlorpromazine was reported to prevent liver cell death produced by up to 3 hr of ischemia and in liver it prevented the 40 to 55% decrease in phospholipid levels which were found in nontreated controls.[119] In rat heart, chlorpromazine pretreatment was reported to reduce the phospholipid depletion associated with 12 hr of ischemia[87] and with isoproterenol-induced myocardial damage.[120] Thus, it seems possible that in the future, phospholipase A inhibitors might be used to protect the ischemic myocardium from phospholipase A-induced membrane damage.

IV. ROLE OF PHOSPHOLIPASE A_2 IN PANCREATITIS

The pancreas is the source of various digestive enzymes which are secreted into the small bowel in response to humoral and neural stimuli. Lipolytic enzymes secreted include lipase, phospholipase A_2, cholesterol esterase, and lysophospholipase. When the pancreas becomes inflamed in response to a variety of different stimuli, these enzymes are released locally into the pancreatic tissue and into the circulation. Proteolytic and lipolytic enzymes of the pancreas cause extensive local tissue destruction. Serum levels of pancreatic enzymes, such as amylase and lipase, rise markedly and form the biochemical basis upon which the diagnosis of pancreatitis is made.[121] In severe cases of pancreatitis, fat necrosis, hemorrhagic necrosis, pulmonary dysfunction, and death may occur.[121]

Initially it was thought that the tissue damage was due principally to the action of trypsin and other proteolytic enzymes. In 1969 Schmidt and Creutzfeldt[122] questioned the role of proteolytic enzymes in pancreatitis; they suspected a role for phospholipase A_2 and its breakdown product, lysolecithin. They found large amounts of lysophosphatidylcholine in the necrotic pancreatic tissue of three patients who died of acute pancreatitis and suggested that activation of phospholipase A_2 might be an important factor. They[122] showed that the injection of purified phospholipase A_2 into rat pancreas in the presence of bile acids had a necrotizing effect on pancreas not unlike that seen in the human disease.

Phospholipase A_2 is present in the pancreas as an inactive proenzyme which is converted to the active form by cleavage of a heptapetide from the N-terminus of the polypeptide chain. Prophospholipase A_2 can act only on monomers of phosphatidylcholine because its interface recognition site is exposed only after trypsin cleavage.[18] Thus, one could imagine that minor degrees of pancreatic inflammation and damage could be magnified greatly if phospholipase A_2 is converted by trypsin to the active form which can bind to and attack membrane bilayer phospholipids.

Phospholipase A_2 levels increase in serum during pancreatitis and evidence has been provided showing that the clinical severity of the disease correlates with the increase in serum phospholipase A_2 activity.[123,124] In addition, it has been suggested that increased serum levels of phospholipase A_2 may provoke pulmonary dysfunction by degrading pulmonary surfactant which is comprised principally of two phospholipids: dipalmitoylphosphatidylcholine and dipalmitoylphosphatidylglycerol.[124,125] The pulmonary abnormalities of pancreatitis and the role of phospholipase A_2 have recently been reviewed.[126-128] Interestingly, it has been suggested that two inhibitors of phospholipase, xylocaine, and chlorpromazine, may have beneficial effects in animal models of acute pancreatitis.[129,130] Trasylol (aprotinin) has been reported to be effective in the treatment of pancreatitis, although the results are controversial.[127] It would presumably exert its effects on phospholipase A_2 by inhibiting conversion of prophospholipase A to its active form.

V. ROLE OF EXOGENOUS PHOSPHOLIPASES IN HUMAN DISEASES

A. Microbial Phospholipases

1. Gas Gangrene and Phospholipase C

Clostridium perfringens causes gas gangrene, a serious and life-threatening infection of deep or penetrating wounds. The organism produces numerous toxins, including an α-toxin, the first bacterial toxin shown to be an enzyme.[131] α-Toxin is a potent phospholipase C which hydrolyzes phosphatidylcholine to diglyceride and phosphocholine. The enzyme hydrolyzes most naturally occurring membrane phosphoglycerides. Phospholipase C is considered to be a major factor in gas gangrene because the purified enzyme, when given to guinea pigs, causes dermonecrosis and death, effects similar to those seen in the human gas gangrene.[132] Intramuscular injection of phospholipase C in guinea pigs causes myonecrosis apparently by attacking the muscle cell plasma membrane first,[133] followed later by damage to other intracellular membrane structures, including nuclei, sarcoplasmic reticulum, and mitochondria.[133,134] The toxic effects can often be prevented in animals by prior active or passive immunization against phospholipase C (α-toxin), while immunization against other clostridial toxins does not prevent death.[132] However, the role of α-toxin in gas gangrene has been questioned since α-antitoxin does not provide good protection if extensive muscle necrosis has already occurred.[135] There are alternative explanations for the failure of α-antitoxin to protect once extensive tissue damage has occurred, including release from lysosomes or other intracellular sites of other potent phospholipases A or C which do not react with alpha antitoxin. Toxicity might result from high levels of fatty acids, partial glycerides, or lysophospholipids formed by cellular phospholipases A, but little data appears to be available relative to this question. Antitoxin may not reach the sites of damage when necrosis and tissue injury is already substantial. Another significant question about the role of clostridial phospholipase C is the lack of good correlation between virulence and the ability of a particular strain to produce phospholipase C.[136] Thus, the precise role of clostridial phospholipase C (α-toxin) in gas gangrene remains controversial,[137] although its biochemical effects have been shown to mimic many aspects of the disease in animals.

2. Phospholipase A and Various Other Infectious Diseases

Phospholipases A have been reported to be present in most bacteria, but it is not clear that these enzymes are involved in the pathogenesis of bacterial diseases of man. This subject has been extensively reviewed by Möllby[138] and more recently by Arbuthnott.[139] However, it has recently been proposed that premature labor may be initiated by intrauterine infection with strains of bacteria which produce phospholipase A_2.[140] This results in release of arachadonic acid from the amnion which has been suggested to lead to increased synthesis of prostaglandins known to be involved in the initiation of labor.

Trypanosomes contain phospholipase A_1 which has been reported to accumulate in tissue fluid during experimental trypanosomiasis in the rabbit. Phospholipase A_1 activity appears in the tissue fluid in parallel with the appearance of trypanosomes.[141] It has been proposed that phospholipases A_1 may play a role in the pathophysiology of trypanosomaiasis.[142] At present, evidence to support an important role for phospholipase A is sparse and it would appear that further studies will be needed to establish a role for phospholipase A in the pathophysiology of this disease.

Amoebae of the genus *Acanthamoeba* cause acanthamebiasis, a fatal protozoan infection of the brain. The agent enters through the nasal mucosa, skin, or eye and causes a fulminant meningoencephalitis for which there is no known treatment.[143] Very little is known about the pathophysiology of this disease. Interestingly, cell-free supernatants from two strains of *Acanthamoeba castellanii* (Singh and Lilly strains) were shown to secrete a phospholipase

A into the culture medium. There was a correlation between the cytopathic effects of the supernatants on cultured muscle cells and their phospholipase A content, suggesting a possible role for phospholipase A in the pathogenesis of acanthamebiasis.[144]

In rats infected with *Pneumocystis carinii* phospholipase A levels have been reported to be increased 3.5-fold in broncheoalveolar lavage fluid. Conversely, phospholipid levels in the fluid, representing pulmonary surfactant, are decreased by 40% in experimental pneumocystis pneumonia.[145] However, the nature and origin of the putative phospholipase A is not clear since the authors indicate that they were not able to measure phospholipase A in *Peumocystis carinii* itself.[145] Nevertheless, the studies are interesting since respiratory distress, which may be associated with low levels of surfactant phospholipid, is a major clinical feature of pneumonia caused by pneumocystis.[146] Interestingly, this situation resembles some cases of acute pancreatitis where high plasma levels of pancreatic phospholipase A_2 are thought to cause respiratory distress syndrome by degrading pulmonary surfactant phospholipids. [124-126]

B. Poisonous Snakes and Phospholipase A

Five families of snakes cause injury to man by means of venoms containing complex mixtures of toxins. They include the Elapidae (cobras, mambas, kraits, and coral snakes), the Hydrophidae (sea snakes), the Colubridae (bird snakes and tree snakes), and the Crotalidae (pit vipers). In the U.S. it is estimated that of 8000 individuals bitten, approximately 20 persons die yearly as a result of snake bite; most are victims of the various species of rattlesnake. In the world the annual death rate from snake bite is estimated to be 30,000 to 40,000 deaths yearly, with particularly high death rates found in Brazil and in the Far East.[147]

The signs and symptoms of snake bite vary with the species. By way of example, envenomation by pit vipers is followed by severe local pain, edema, and hemorrhage into the skin. Severe tissue destruction and gangrene often occur. Systemic symptoms include hypotension, shock, hemolysis, and myonecrosis. Respiratory arrest may occur in severe cases of envenomation as a result of blocked neuromuscular transmission.[148]

The venom of poisonous snakes contains many components which have specific enzymatic activities or physical functions. For example, rattlesnake venom contains phospholipase A_2, phosphodiesterase and phosphomonoesterase, numerous proteolytic enzymes, L-amino acid oxidase, neurotoxins, myotoxins, and hemorrhagic toxins.[149] The venom of all five families of poisonous snakes contains phospholipase A_2.[150] In many cases these enzymes have been isolated, purified, and in some cases, the primary structure has been determined. Interestingly, there is a strong sequence homology between pancreatic phospholipase A_2 and venom phospholipases A_2.[150,151] By studying the effects of purified venom compounds in animals, it has been possible to clarify to some degree the role of phospholipase A_2 in the pathophysiology of envenomation. However, the events which result in tissue injury following envenomation are extremely complex and in many cases the precise role of phospholipase A_2 in the overall process is not clear.

One of the most important effects of phospholipase A_2 components of venoms is the blockade of neuromuscular transmission. The crotoxin complex isolated from *Crotalus durissus terrificus* has been studied very thoroughly. It contains a basic phospholipase A_2 and an acidic peptide, crotapotin, which lacks enzymatic activity. These components associate to form a complex, crototoxin, which has pronounced phospholipase A_2 activity against phospholipid bilayer membranes.[152,153] Crotoxin blocks neuromuscular transmission in animals and decreases the contractile response of the diaphragm to phrenic nerve stimulation; the site of action is at the presynaptic motor nerve terminals.[154-157] The phospholipase A_2 activity of crototoxin is believed to be the cause of the blocked nerve transmission since pretreatment of the phospholipase A_2 with *p*-bromophenyacyl bromide (blocks the phospholipase A activity by modifying the active site histidine residue) abolishes the postsynaptic neurotoxicity.[157] In addition, this modification has no effect on the ability of the phospholipase A_2 to complex with crotapotin.[152] Replacement of Ca^{2+} by Sr^{2+} blocks or inhibits both the postsynaptic effects of crotoxin and its phospholipase A_2 activity.[152,155,157] Other toxins which produce neuromuscular blockade also have strong phospholipase A_2 activities.[157,158]

Hypotension and shock are also important features of envenomation. It has been suggested that this may be partly due to the fact that phospholipase A_2 of venom releases arachidonic acid from membrane phospholipids. Arachidonic acid is converted to prostaglandins which lower the blood pressure.[158] The phospholipase A_2 from cobra venom (*Naja naja*) has been reported to increase the formation of prostaglandin-like activity from rat peritoneal cells and perfused guinea pig lungs.[159] However, evidence for increased circulating levels of prostaglandins in vivo during venom shock has not yet been provided.[160]

Myotoxic effects have been reported frequently in humans and can be demonstrated in animal studies with venom phospholipases.[161-163] Myoglobinuria is an especially frequent finding in envenomation by the Malayan sea snake (*Enhydrinia schistosa*) and the Australian tiger snake (*Notechsis scutatus*). In both cases phospholipase A_2 has been identified as the agent responsible for rhabdomyolysis and myoglobinuria.[161,162] Similar findings were reported for the Australian elapid snake, *Pseudechis colletti*.[163] Muscle damage is presumably a result of hydrolysis of the phospholipids of the membranes of myocytes. The presence of increased amounts of lysophospholipids have been shown in toxin-damaged muscles.[162] Even more compelling proof for the role of phospholipase has been provided by experiments showing that modification of the histidine at the phospholipase A_2 catalytic site with *p*-bromophenacylbromide abolishes both the phospholipase A activity and myoglobinuria in mice at a dose representing 50 times the LD_{50}.[161]

Finally, destruction of red blood cells with liberation of hemoglobin (hemolysis) is a long-recognized feature of envenomation, but hemolysis itself is not usually an important factor in cases with a lethal outcome. The lysis of red blood cells is due to the action of phospholipase A_2 with production of detergent-like lysophosphoglycerides. Since hemolysis can be blocked by antibodies to the crude venom, to crototoxin, and to phospholipase A_2 itself, it appears that phospholipase A_2 activity is required for hemolysis.[164] While some venoms appear to possess a direct lytic factor which is not a phospholipase, a wide variety of venom phospholipases A_2 have been shown to cause hemolysis.[165]

In summary, phospholipases A_2 of snake venoms are important in the local and systemic toxicity which occurs with envenomation. They play a role in hypotension and shock, blockade of neuromuscular transmission, hemolysis, myotoxicity, and myoglobinuria, and represent one of the most thoroughly studied groups of toxins and enzymes. Several books have appeared regarding the action of venom toxins and the pathophysiology and treatment of envenomation,[149,150] and a number of excellent reviews about the structure and function of phospholipases A_2 are available.[1,18,19,151,166]

VI. CONCLUDING REMARKS

This chapter has attempted to summarize current knowledge of the role of phospholipases in diseases of man. Of course, much of the detailed knowledge of human diseases is derived from experiments done in animals. Diseases which are caused by modulation of endogenous cellular phospholipases include drug-induced lipidosis and ischemic myocardial injury. The former is due to inhibition of lysosomal phospholipid catabolism while the role of phospholipases in ischemic myocardial injury is due to presumed activation of phospholipase with generation of toxic lysophospholipids and loss of membrane phospholipids in myocardial tissue. Pancreatitis is thought to be caused in part by activation of pancreatic prophospholipase A_2 by trypsin with elevation of phospholipase A_2 levels in the circulation, as well as locally in the pancreatic region. This leads to distant effects including depletion of pulmonary surfactant and to local tissue destruction. I have classified pancreatitis as a disease caused in part by an endogenous phospholipase. Pancreatic prophospholipase A_2 is an endogenous phospholipase which is synthesized and stored in the exocrine cells of the pancreas. However, it is secreted (exogenously) into the GI tract where it plays an important role in digestion. Thus, the involvement of phospholipase A_2 in pancreatitis could also be argued to be an example of an exogenous phospholipase causing disease, since the enzyme does not normally function as an endogenous enzyme nor does it normally appear in the circulation. Exogenous phospholipases play a clear role in the pathophysiology of gas gangrene (*Clostridium perfringens* phospholipase C or α-toxin) and poisonous snake bite injury (various venom phospholipases A_2).

In the past 15 years there has been an explosion in our knowledge of phospholipase structure, function, mechanisms, distribution in nature, and involvement in normal cellular physiology. In terms of the role of phospholipases in the pathophysiology of human disease, our level of understanding is presently very limited. In the coming years, it is likely that our ability to relate the basic biochemistry of phospholipases to the pathophysiology of disease will continue to advance rapidly since it is becoming clear that phospholipases are important factors in drug toxicity, ischemic tissue injury, the inflammatory response, platelet activation, and in the action of various and hormones and other biochemical modulators.

ACKNOWLEDGMENTS

The author's research is supported by NIH Grants AM 32159 and GM 24979 and by the Research Division of the Veterans Administration Medical Center, San Diego, Calif. During the preparation of this chapter, the author was a recipient of a Clinical Investigatorship from the Veterans Administration. Dr. E. J. Victoria reviewed the manuscript and provided valuable commentary and Mr. Cleon Tate assisted in the preparation of the manuscript.

REFERENCES

1. **van den Bosch, H.,** Intracellular phospholipases A, *Biochim. Biophys. Acta*, 604, 191, 1980.
2. **Scherphof, G. L. and van Deenen, L. L. M.,** Phospholipase A activity of rat liver mitochondria, *Biochim. Biophys. Acta*, 98, 204, 1965.
3. **Scherphof, G. L., Waite, M., and van Deenen, L. L. M.,** Formation of lysophosphatidylethanolamines in cell fractions of rat liver, *Biochim. Biophys. Acta*, 125, 406, 1966.
4. **Nachbaur, J., Colbeau, A., and Vignais, P. M.,** Distribution of membrane-confined phospholipases A in the rat hepatocyte, *Biochim. Biophys. Acta*, 274, 426, 1972.
5. **Waite, M. and van Deenen, L. L. M.,** Hydrolysis of phospholipids and glycerides by rat liver preparations, *Biochim. Biophys. Acta*, 137, 498, 1967.
6. **van Golde, L. M. G., Fleischer, B., and Fleischer, S.,** Some studies on the metabolism of phospholipids in Golgi complex from bovine and rat liver in comparison to other subcellular fractions, *Biochim. Biophys. Acta*, 249, 318, 1971.
7. **Newkirk, J. D. and Waite, M.,** Identification of a phospholipase A_1 in plasma membranes of rat liver, *Biochim. Biophys. Acta*, 225, 224, 1971.
8. **Victoria, E. J., van Golde, L. M. G., Hostetler, K. Y., Scherphof, G. L., and van Deenen, L. L. M.,** *Biochim. Biophys. Acta*, 249, 318, 1971.
9. **Hostetler, K. Y., Yazaki, P. J., and van den Bosch, H.,** Purification of lysosomal phospholipase A: evidence for multiple isoenzymes in rat liver, *J. Biol. Chem.*, 257, 13367, 1982.

10. **Robinson, M. and Waite, M.,** Physical-chemical requirements for the catalysis of substrates by lysosomal phospholipase A_1, *J. Biol. Chem.*, 258, 14371, 1983.
11. **de Winter, J. M., Vianen, G. M., and van den Bosch, H.,** Purification of rat liver mitochondrial phospholipase A_2, *Biochim. Biophys. Acta*, 712, 332, 1982.
12. **Natori, Y., Karasawa, K., Arai, H., Tamori-Natori, Y., and Nojima, S.,** Partial purification and properties of phospholipase A_2 from rat liver mitochondria, *J. Biochem.*, 93, 631, 1983.
13. **de Haas, G. H., Postema, N. M., Niewenhuizen, W., and van Deenen, L. L. M.,** Purification and properties of phospholipase A from porcine pancreas, *Biochim. Biophys. Acta*, 159, 103, 1968.
14. **Volwerk, J. J., Pieterson, W. A., and de Haas, G. H.,** Histidine at the active site of phospholipase A_2, *Biochemistry*, 13, 1446, 1974.
15. **Matsuzawa, Y. and Hostetler, K. Y.,** Properties of phospholipase C isolated from rat liver lysosomes, *J. Biol. Chem.*, 255, 646, 1980.
16. **Irvine, R. F., Hemington, N., and Dawson, R. M. C.,** Hydrolysis of phosphatidylinositol by lysosomal enzymes of rat liver and brain, *Biochem. J.*, 176, 475, 1978.
17. **Irvine, R. F., Hemington, N., and Dawson, R. M. C.,** Phosphatidylinositol-degrading enzymes in liver lysosomes, *Biochem. J.*, 164, 277, 1977.
18. **Volwerk, J. J. and de Haas, G.,** Pancreatic phospholipase A_2: a model for membrane-bound enzymes? in *Lipid-Protein Interactions*, Vol. 1, Jost, P. C. and Griffith, O. H., Eds., Academic Press, New York, 1982, chap. 3.
19. **van den Bosch, H.** Phospholipases, in *Phospholipids*, Hawthorne, J. N. and Ansell, G. B., Eds., Elsevier, Amsterdam, 1982, chap. 9.
20. **Flower, R. J. and Blackwell, G. J.,** Anti-inflammatory steroids induce biosynthesis of a phospholipase A_2 inhibitor which prevents prostaglandin generation, *Nature, (London)*, 278, 456, 1979.
21. **Hirata, F., Schiffmann, E., Venkatasubramanian, K., Solomon, D., and Axelrod, J.,** A phospholipase A_2 inhibitory protein in rabbit neutrophils induced by glucocorticoids, *Proc. Natl. Acad. Sci. U.S.A.*, 77, 2533, 1980.
22. **Hirata, F.,** The regulation of lipomodulin, a phospholipase inhibitory protein, in rabbit neutrophils by phosphorylation, *J. Biol. Chem.*, 256, 7730, 1981.
23. **Blackwell, G. J., Carnuccio, R., DiRosa, M., Flower, R. J., Parante, L., and Persico, P.,** Macrocortin: a polypeptide causing the antiphospholipase effect of glucocorticoids, *Nature (London)*, 287, 147, 1980.
24. **Bormann, B. J., Huang, C.-K., Mackin, W. M., and Becker, E. L.,** Receptor-mediated activation of a phospholipase A_2 in rabbit neutrophil plasma membranes, *Proc. Natl. Acad. Sci. U.S.A.*, 81, 767, 1984.
25. **Hill, E. H. and Lands, W. E. M.,** Phospholipid metabolism, in *Lipid Metabolism*, Wakil, S. J., Ed., Academic Press, New York, 1970, chap. 6.
26. **Yamamoto, A., Adachi, S., Kitani, T., Shinji, T., Seki, K.-I., Nasu, T., and Nishikawa, M.,** Drug-induced lipidosis in human cases and in animal experiments, *J. Biochem.*, 69, 613, 1971.
27. **Yamamoto, A., Adachi, S., Ishikawa, K., Yokomura, T., Kitani, T., Nasu, T., Imoto, T., and Nishikawa, M.,** Studies on drug-induced lipidosis: lipid composition of the liver and other tissues in clinical cases of "Niemann-Pick-like syndrome" induced by 4,4′-diethylaminoethoxyhexestrol, *J. Biochem.*, 70, 775, 1971.
28. **Seki, K., Shimji, Y., and Nishikawa, M.,** Studies on drug-induced lipidosis, light and electron microscopic observations on the liver biopsy specimens, *Acta Hepatol. Jpn.*, 12, 226, 1971.
29. **de la Iglesia, F. G., Takada, A., and Matsuda, Y.,** Morphologic studies on secondary phospholipidosis in human liver, *Lab. Invest.*, 30, 539, 1974.
30. **Tashiro, Y.,** An electron microscopic observation on the cytological changes in the experimental drug-induced lipidosis, *Keio J. Med.*, 24, 115, 1975.
31. **Adachi, S., Matsuzawa, Y., Yokomura, T., Ishikawa, K., Uhara, S., Yamamoto, A., and Nishikawa, M.,** Studies on drug-induced lipidosis. V. Changes in the lipid composition of rat liver following the administration of 4,4′-diethylaminoethoxyhexestrol, *Lipids*, 7, 1, 1972.
32. **Matsuzawa, Y., Yamamoto, A., Adachi, S., and Nishikawa, M.,** Studies on drug-induced lipidosis. VIII. Correlation between drug accumulation and acidic phospholipids, *J. Biochem.*, 82, 1369, 1977.
33. **Yamamoto, A., Adachi, S., Matsuzawa, Y., Kitani, T., Hiraoka, A., and Seki, K.-I.,** Studies on drug-induced lipidosis. VII. Effects of bis-α-β diethylaminoether of hexestrol, chloroquine, homochlorocyclizine, prenylamine and diazacholesterol on the lipid composition of rat liver and kidney, *Lipids*, 11, 616, 1976.
34. **Matsuzawa, Y. and Hostetler, K. Y.,** Studies on drug-induced lipidosis: subcellular localization of phospholipid and cholesterol in the liver of rats treated with chloroquine or 4,4′-bis(diethylaminoethoxy)-α, β-diethyldiphenylethane, *J. Lipid Res.*, 21, 202, 1980.
35. **Lüllmann, H., Lüllman-Rauch, R., and Wassermann, D.,** Lipidosis induced by amphiphilic cationic drugs, *Biochem. Pharmacol.*, 27, 1103, 1978.

36. **Lülmann-Rauch, R.,** Drug-induced lysosomal storage disorders, in *Lysosomes in Applied Biology and Therapeutics,* Vol. 6, Dingle, J. T., Jacques, P. J., and Shaw, I. H., Eds., North Holland, Amsterdam, 1979, chap. 3.
37. **Kacew, S. and Reasor, M. J.,** Chlorphentermine-induced alterations in pulmonary phospholipid content in rats, *Biochem. Pharmacol.,* 18, 2683, 1983.
38. **Tjiong, H. B., Lepthin, J., and Debuch, H.,** Lysosomal phospholipids from rat liver after treatment with different drugs, *Hoppe-Seyler's Z. Physiol. Chem.,* 359, 63, 1978.
39. **Hostetler, K. Y. and Richman, D. D.,** Studies on the mechanism of phospholipid storage induced by amantadine and chloroquine in Madin-Darby canine kidney cells, *Biochem. Pharmacol.,* 31, 3795, 1982.
40. **Matsuzawa, Y., Yokomura, T., Ishikawa, K., Adachi, S., and Yamamoto, A.,** Studies on drug-induced lipidosis. VI. Identification and determination of the drug and its metabolite in lipidosis induced by 4,4′-diethylaminoethoxyhexestrol, *J. Biochem.,* 72, 615, 1972.
41. **Matsuzawa, Y. and Hostetler, K. Y.,** Effects of chloroquine and 4,4′-bis(deithylaminoethoxy) α,β-diethyldiphenylethane on the incorporation of [^{3}H]glycerol into the phospholipids of rat liver lysosomes and other subcellular fractions, *in vivo, Biochim. Biophys. Acta,* 620, 592, 1980.
42. **Matsuzawa, Y. and Hostetler, K. Y.,** Inhibition of lysosomal phospholipase A and phospholipase C by chloroquine and 4,4′-bis(diethylamino-ethoxy)hexestrol, *J. Biol. Chem.,* 255, 5190, 1980.
43. **Kubo, M., Yazaki, P. J., and Hostetler, K. Y.,** Unpublished observations, 1984.
44. **Rollo, I. M.,** Chloroquine, in *The Pharmacologic Basis of Therapeutics,* 6th ed., Gilman, A. G., Goodman, L. S., and Gilman, A., Eds., Macmillan, New York, 1980, 1042.
45. **Nilsson, O., Fredman, P., Klinghardt, G. W., Dreyfus, H., and Svennerholm, K.,** Chloroquine-induced accumulation of gangliosides and phospholipids in skeletal muscles. Quantitative determination and characterization of stored lipids, *Eur. J. Biochem.,* 116, 565, 1981.
46. **Wiesmann, U. N., Didonato, S., and Herschkowitz, N. H.,** Effect of chloroquine on cultured fibroblasts: release of lysosomal hydrolases and inhibition of their uptake, *Biochem. Biophys. Res. Commun.,* 66, 1338, 1975.
47. **Hasilik, A. and Neufeld, E. F.,** Biosynthesis of lysosomal enzymes in fibroblasts, *J. Biol. Chem.,* 255, 4937, 1980.
48. **Gonzalez-Noriega, A., Grubb, J. H., Talhad, V., and Sly, W. S.,** Chloroquine inhibits lysosomal enzyme pinocytosis and enhances lysosomal enzyme secretion by impairing receptor recycling, *J. Cell. Biol.,* 85, 839, 1980.
49. **Reasor, M. J. and Hostetler, K. Y.,** Chloroquine does not cause phospholipid storage by depleting rat liver lysosomes of acid phospholipase A, *Biochim. Biophys. Acta,* 793, 497, 1984.
50. **Hostetler, K. Y., Reasor, M. J., and Yazaki, P. J.,** Chloraquine-induced phospholipid fatty liver: measurement of drug and lipid concentrations in rat liver lyosomes, *J. Biol. Chem.,* 260, 215, 1985.
51. **Kunze, H., Hesse, B., and Bohn, E.,** Effects of antimalarial drugs on several rat-liver lysosomal enzymes involved in phosphatidylethanolamine catabolism, *Biochim. Biophys. Acta,* 713, 112, 1982.
52. **Ohkuma, S. and Poole, B.,** Fluorescence probe measurement of the intralysosomal pH in living cells and the perturbation of pH by various agents, *Proc. Natl. Acad. Sci. U.S.A.,* 75, 3327, 1978.
53. **Hollemans, M., Oude-Elferink, R., de Groot, P. G., Strijland, A., and Tager, J. M.,** Accumulation of weak bases in relation to intralysosomal pH in cultured human skin fibroblasts, *Biochim. Biophys. Acta,* 643, 140, 1981.
54. **Reasor, M. J. and Koshut, R. A.,** Augmentation in antioxidant defuse mechanism in rat alveolar macrophages following induction of phospholipidosis with chlorophentemine, *Toxicol. Appl. Pharmacol.,* 55, 334, 1980.
55. **Lüllmann-Rauch, R. and Scheid, D.,** Intraalveolar foam cells associated with lipidosis-like alterations in lung and liver of rats treated with tricyclic psychotropic drugs, *Virchows Arch. B:,* 19, 225, 1975.
56. **Watts, S. D. M., Orpin, A., and MacCormick, C.,** Lysosomes and treatment pathology in the chemotherapy of schistosomiasis with 1,7-bis(*p*-aminophenoxy) heptane (152C51), *Parasitology,* 78, 287, 1979.
57. **Hostetler, K. Y. and Matsuzawa, Y.** Studies on the mechanisms of drug-induced lipidosis. Cationic amphiphilic drug of lysosomal phospholipases A and C, *Biochem. Pharmacol.,* 30, 1121, 1981.
58. **Marchlinski, F. E., Gausler, T. S., Waxman, H. L., and Josephson, M. E.,** Amiodarone pulmonary toxicity, *Ann. Intern. Med.,* 97, 839, 1982.
59. **Hostetler, K. Y., Reasor, M. J., and Frazee, B. W.,** Mechanisms of amiodarone pulmonary toxicity, *Clin. Res.,* 32, 528A, 1984.
60. **Morin, J. P., Viotte, G., Vandewalle, A., van Hoof, F., Tulkens, P., and Filastre, J. P.,** Gentamicin-induced nephrotoxicity: a cell biology approach, *Kidney Int.,* 18, 583, 1980.
61. **Feldman, S., Wang, M.-Y., and Kaloyanides, G. J.,** Aminoglycosides induce a phospholipidosis in the renal cortex of the rat: an early manifestation of nephrotoxicity, *J. Pharmacol. Exp. Ther.,* 220, 514, 1982.
62. **Hostetler, K. Y. and Hall, L. B.,** Aminoglycoside antibiotics inhibit lysosomal phospholipase A and C from rat liver, *in vitro, Biochim. Biophys. Acta,* 710, 506, 1982.

63. **Hostetler, K. Y. and Hall, L. B.,** Inhibition of kidney lysosomal phospholipases A and C by aminoglycoside antibiotics: a possible mechanism of aminoglycoside toxicity, *Proc. Natl. Acad. Sci. U.S.A.*, 79, 1663, 1982.
64. **Laurent, G., Carlier, M.-B., Rollman, B., van Hoof, F., and Tulkens, P. M.,** Mechanisms of aminoglycoside-induced lysosomal phospholipidosis: *in vitro* and *in vivo* studies with gentamicin and amikacin, *Biochem. Pharmacol.*, 31, 3861, 1982.
65. **Carlier, M.-B., Laurent, G., Claes, P. J., Vanderhaeghe, H. J., and Tulkens, P. M.,** Inhibition of lysosomal phospholipases by aminoglycoside antibiotics: *in vitro* comparative studies, *Antimicrob. Agents Chemother.*, 23, 440, 1983.
66. **Lipsky, J. J. and Lietman, P. S.,** Aminoglycoside inhibition of a renal phosphatidylinositol phospholipase C, *J. Pharmacol. Exp. Ther.*, 220, 287, 1982.
67. **Lüllmann, H. and Wehling, M.,** The binding of drugs to different polar lipids *in vitro, Biochem. Pharmacol.*, 28, 3409, 1979.
68. **Lüllmann, H., Plosch, H., and Ziegler,A.,** Ca replacement by cationic amphiphlic drugs from lipid monolayers, *Biochem. Pharmacol.*, 29, 2969, 1980.
69. **Pappu, A. and Hostetler, K. Y.,** Effect of cationic amphiphilic drugs on the hydrolysis of acidic and neutral phospholipids by liver lysosomal phospholipase A, *Biochem. Pharmacol.*, 33, 1639, 1984.
70. **Pappu, A., Yazaki, P. J., and Hostetler, K. Y.,** Inhibition of purified lysosomal phospholipase A_1 by beta adrenergic blockers, *Clin. Res.*, 31, 252A, 1983.
71. **Richman, D. D., Yazaki, P. J., and Hostetler, K. Y.,** The intracellular distribution and antiviral activity of amantadine, *Virology*, 111, 81, 1981.
72. **Braunwald, E. and Sobel, B. E.,** Coronary blood flow and myocardial ischemia, in *Heart Disease: A Textbook of Cardiovascular Medicine*, Braunwald, E., Ed., W. B. Saunders, Philadelphia, 1980, chap. 35.
73. **Jennings, R. B. and Reimer, K. A.,** Lethal myocardial injury, *Am. J. Pathol.*, 102, 241, 1981.
74. **Katz, A. M. and Messineo, F. C.,** Lipid-membrane interactions and the pathogenesis of ischemic damage in the myocardium, *Circ. Res.*, 48, 1, 1981.
75. **Taylor, I. M., Shaikh, N. A., and Downar, E.,** Ultrastructural changes of ischemic injury due to coronary artery occlusion in the porcine heart, *J. Mol. Cell Cardiol.*, 16, 79, 1984.
76. **Singer, S. J. and Nicolson, G.** The fluid mosaic model of the structure of cell membranes, *Science*, 175, 720, 1972.
77. **Sobel, B. E., Corr, P. B., Robison, A. K., Goldstein, R. A., Witkowski, F. X., and Klein, M. S.,** Accumulation of lysophosphatides with arrhythmogenic properties in ischemic myocardium, *J. Clin. Invest.*, 62, 546, 1978.
78. **Shaikh, N. A. and Downar, E.,** Time course of changes in porcine myocardial phospholipid levels during ischemia. A reassessment of the lysolipid hypothesis, *Circ Res.*, 49, 316, 1981.
79. **Corr, P. B., Snyder, D. W., Lee, B. I., Gross, R. W., Keim, C. R., and Sobel, B. E.,** Pathophysiological concentrations of lysophosphatides and the slow response, *Am. J. Physiol.*, 243, H187, 1982.
80. **Mann, R. Y. K., Slater, T. L., Pelletier, M. P., and Choy, P. C.,** Alterations of phospholipids in ischemic canine myocardium during acute arrhythmia, *Lipids*, 18, 677, 1983.
81. **Chien, K. R., Han, A., Sen, A., Buja, M., and Willerson, J. T.,** Accumulation of unesterified arachidonic acid in ischemic canine myocardium. Relationship to a phosphatidylcholine deacylation-reacylation cycle and the depletion of membrane phospholipids, *Circ. Res.*, 54, 313, 1984.
82. **Snyder, D. W., Crafford, W. A., Glashow, J. L., Rankin, D., Sobel, B. E., and Corr, P. B.,** Lysophosphoglycerides in ischemic myocardium effluents and potentiation of their arrhythmogenic effects, *Am. J. Physiol.*, 241, H700, 1981.
83. **Weltzien, H. U.,** Cytolytic and membrane-perturbing properties of lysophosphatidylcholine, *Biochim. Biophys. Acta*, 559, 259, 1979.
84. **Corr, P. B. and Sobel, B. E.,** Arrhythmogenic properties of phospholipid metabolites associated with myocardial ischemia, *Fed. Proc.*, 42, 2454, 1983.
85. **Oliver, M. F.and Yates, P. A.,** Induction of ventricular arrhythmias by elevation of arterial free fatty acids in experimental myocardial infection, *Cardiology*, 57, 359, 1971.
86. **Vasdev, S. C., Biro, G. P., Narbaitz, R., and Kako, K. J.,** Membrane changes induced by early myocardial ischemia in the dog, *Can. J. Biochem.*, 58, 1112, 1980.
87. **Chien, K. R., Pfau, R. G., and Farber, J. L.,** Ishcemic myocardial cell injury. Prevention by chlorpromazine of an accelerated phospholipid degradation and associated membrane dysfunction, *Am. J. Pathol.*, 97, 505, 1979.
88. **Corr. P. B., Ahumada, G. G., and Sobel, B. E.,** Membrane active metabolites: potential mediators of sudden death, in *Cardiovascular Medicine*, Vol. 1, Vogel, J. H. K., Ed., Raven Press, New York, 1982, 161.

89. **Edgar, A. D., Strosznajder, J., and Horrocks, L. A.,** Activation of ethanolamine phospholipase A_2 in brain during ischemia, *J. Neurochem.*, 39, 1111, 1982.
90. **Weglicki, W. B., Waite, M., Sisson, P., and Shohet, S. B.,** Myocardial phospholipase A of microsomal and mitochondrial fractions, *Biochim. Biophys. Acta*, 231, 512, 1971.
91. **Palmer, J. W., Schmid, P. C., Pfeiffer, D. R., and Schmid, H. H. O.,** Lipids and lipolytic enzyme activities of rat heart mitochondria, *Arch. Biochem. Biophys.*, 211, 674, 1981.
92. **Nalbone, G. and Hostetler, K. Y.,** Subcellular localization of the phospholipases of rat heart: evidence for a cytosolic phospholipase A_1, *J. Lipid Res.*, 26, 104, 1985.
93. **Franson, R., Waite, M., and Weglicki, W.,** Phospholipase A activity of lysosomes of rat myocardial tissue, *Biochemistry*, 11, 472, 1972.
94. **Weglicki, W. B., Waite, M., and Stam, A. C.,** Association of phospholipase A with a myocardial membrane preparation containing the $(Na^+-K^+)-Mg^{2+}$-ATPase, *J. Mol. Cell Cardiol.*, 4, 195, 1972.
95. **Franson, R. C., Pang, D. C., Towle, D. W., and Weglicki, W. B.,** Phospholipase A activity of highly enriched preparations of cardiac sarcolemma from hamster and dog, *J. Mol. Cell Cardiol.*, 10, 921, 1978.
96. **Owens, K., Pang, D. C., and Weglicki, W. B.,** Production of lypophospholipids and free fatty acids by a sarcolemmal fraction from canine myocardium, *Biochem. Biophys. Res. Commun.*, 89, 368, 1979.
97. **Franson, R. C., Weir, D. L., and Thakkar, J.,** Solubilization and characterization of a neutral-active, calcium-dependent, phospholipase A_2 from rabbit heart and isolated chick embryo myocytes, *J. Mol. Cell Cardiol.*, 15, 189, 1983.
98. **Franson, R. C., Pang, D. C., and Weglicki, W. B.,** Modulation of lipolytic activity in isolated canine cardiac sarcolemma by isoproterenol and propranolol, *Biochem. Biophys. Res. Commun.*, 90, 956, 1979.
99. **Gross, R. W. and Sobel, B. E.,** Lysophosphatidylcholine metabolism in the rabbit heart: characterization of metabolic pathway and partial purification of myocardial lysophospholipase-transacylase, *J. Biol. Chem.*, 257, 6702, 1982.
100. **Gross, R. W. and Sobel, B. E.,** Rabbit myocardial cytosolic lysophospholipase: purification, characterization and competitive inhibition by L-palmitoyl carnitine, *J. Biol. Chem.*, 258, 5221, 1983.
101. **Gross, R. W.,** Purification of rabbit myocardial cytosolic acyl-CoA hydrolase, identity with lysophospholipase and modulation of enzymic activity by endogenous cardiac amphiphiles, *Biochemistry*, 22, 5641, 1983.
102. **Gross, R. W., Ahumada, G. G., and Sobel, B. E.,** Cytosolic lysophospholipase in cardiac myocytes and its inhibition by L-palmitoylcarnitine, *Am. J. Physiol.*, 246, C266, 1984.
103. **Gross, R. W., Drisdel, R. C., and Sobel, B. E.,** Rabbit myocardial lysophospholipase-transacylase: purification, characterization and inhibition by endogenous cardiac amphiphiles, *J. Biol. Chem.*, 258, 15165, 1983.
104. **Weglicki, W. B., Owens, K., Urschel, C. W., Serur, J. R., and Sonnenblick, E. H.,** Hydrolysis of myocardial lipids during acidosis and ischemia, *Rec. Adv. Stud. Cardiac Struct. Metab.*, 3, 781, 1974.
105. **Van der Vusse, G. J., Roemen, Th. H. M., Pruizen, F. W., Conmans, W. A., and Reneman, R. S.,** Uptake and tissue content of fatty acids in dog myocardium under normoxic and ischemic conditions, *Circ. Res.*, 50, 538, 1982.
106. **Downar, E., Janse, M. J., and Durrer, D.,** The effect of "ischemic" blood on transmembrane potentials of normal porcine ventricular myocardium, *Circ. Res.*, 55, 455, 1977.
107. **Corr, P. B., Cain, M. E., Witkowski, F. X., Price, D. A., and Sobel, B. E.,** Potential arrhythmogenic electrophysiological derangements in canine Purkinje fibers induced by lysophosphoglycerides, *Circ. Res.*, 44, 822, 1979.
108. **Arnsdorf, M. F. and Sawicki, G. J.,** Effects of lysophosphatidylcholine, a toxic metabolite of ischemia, on the components of cardiac excitability in sheep Purkinje fibers, *Circ. Res.*, 49, 16, 1981.
109. **Bergman, S. R., Ferguson, J. B., and Sobel, B. E.,** Effects of amphiphiles on erythrocytes, coronary arteries and perfused hearts, *Am. J. Physiol.*, 240, H229, 1981.
110. **Man, R. Y. K., and Choy, P. C.** Lysophosphatidylcholine causes cardiac arrhythmia, *J. Mol. Cell Cardiol.*, 14, 173, 1982.
111. **Man, R. Y. K., Wong, T., and Choy, P. C.,** Effects of lysophosphoglycerides on cardiac arrhythmias, *Life Sci.*, 32, 1325, 1983.
112. **Ahumada, G. G., Bergmann, S. R., Carlson, E., Corr, P. B., and Sobel, B. E.,** Augmentation of cyclic AMP content induced by lysophosphatidylcholine in rabbit hearts, *Cardiovasc. Res.*, 13, 377, 1979.
113. **Sedlis, S. P., Corr, P. B., Sobel, B. E., and Ahumada, G. G.,** Lysophosphatidylcholine potentiates Ca^{2+} accumulation in rat cardiac myocytes, *Am. J. Physiol.*, 244, H32, 1983.
114. **Van Meer, G., de Kruijff, B., op den Kamp, J. A. F., and van Deenen, L. L. M.,** Preservation of bilayer structure in human erythrocytes and erythrocyte ghosts after phospholipase treatment: a ^{31}P-NMR study, *Biochim. Biophys. Acta*, 596, 1, 1980.

115. **Jain, M. K., van Echteld, C. J. A., Ramirez, F., de Gier, J., de Haas, G., and van Deenen, L. L. M.,** Association of lysophosphatidylcholine with fatty acids in aqueous phase to form bilayers, *Nature (London),* 284, 486, 1980.
116. **De Oliveira Filgueiras, O. M., van den Besselaar, A. M. H. P., and van den Bosch, H.,** Localization of lysophosphatidylcholine in bovine chromaffin granules, *Biochim. Biophys. Acta,* 558, 73, 1979.
117. **De Oliveira Filgueiras, O. M., van den Bosch, H., Johnson, R. G., Carty, S. E., and Scarpa, A.,** Phospholipid composition of some amine storage granules, *FEBS Lett.,* 129, 309, 1981.
118. **Van Echteld, C. J. A., de Kruijff, B., Mandersloot, J., and de Gier, J.,** Effects of lysophosphatidylcholines on phosphatidycholine and phosphatidylcholine/cholesterol lipososome systems as revealed by ^{31}P-NMR, electron microscopy and permeability studies, *Biochim. Biophys. Acta,* 649, 211, 1981.
119. **Chien, K. R., Abrams, J., Serroni, A., Martin, J. T., and Farber, J. L.,** Accelerated phospholipid degradation and associated membrane dyfunction in irreversible, ischemic liver cell injury, *J. Biol. Chem.,* 253, 4809, 1978.
120. **Okumara, K., Ogawa, K., and Satake, T.,** Pretreatment with chloropromazine prevents phospholipid degradation and creatine kinase depletion in isoproterenol-induced myocardial damage in rats, *J. Cardiovasc. Pharmacol.,* 5, 983, 1983.
121. **Soergel, K. H.,** Acute pancreatitis, in *Gastrointestinal Disease: Pathophysiology, Diagnosis and Management,* 3rd ed., Sleisenger, M. H. and Fordtran, J. S., Eds., W. B. Saunders, Philadelphia, 1983, chap. 91.
122. **Schmidt, H. and Creutzfeldt, W.,** The possible role of phospholipase A in the pathogenesis of acute pancreatitis, *Scand. J. Gastroenterol.,* 4, 39, 1969.
123. **Schröder, T., Kivilaakso, E., Kinnunen, P. K. J., and Lempinen, M.,** Serum phospholipase A_2 in human acute pancreatitis, *Scand. J. Gastroenterol.,* 15, 633, 1980.
124. **Schröder, T., Lempinen, M., Kivilaakso, E., and Nikki, P.,** Serum phospholipase A_2 and pulmonary changes in acute fulminant pancreatitis, *Resuscitation,* 10, 79, 1982.
125. **Morgan, A. P., Jenny, M. E., and Haessler, H.,** Phospholipids, acute pancreatitis, and the lungs: effect of lecithinase infusion on pulmonary surface activity in dogs, *Ann. Surg.,* 3, 329, 1968.
126. **Malik, A. B.,** Pulmonary edema after pancreatitis: role of humoral factors, *Circ. Shock,* 10, 71, 1983.
127. **Nevalainen, T. J.,** The role of phospholipase A in acute pancreatitis, *Scand. J. Gastroenterol.,* 15, 641, 1980.
128. **Schmidt, H. and Creutzfeldt, W.,** Etiology and pathogenesis of pancreatisis, in *Gastroenterology,* Vol. 3, Bullock, H. L., Ed., W. B. Saunders, Philadelphia, 1976, chap. 136.
129. **Schröder, T., Kinnunen, P. K. J., and Lempinen, M.,** Xylocaine treatment in experimental pancreatitis in pigs, *Scand. J. Gastroenterol.,* 13, 863, 1978.
130. **Schröder, T., Lempinen, M., Nordling, S., and Kinnunen, P. K. J.,** Chlorpromazine treatment of experimental acute fulminant pancreatitis in pigs, *Eur. Surg. Res.,* 13, 143, 1981.
131. **MacFarlane, M. G. and Knight, B. C. J. G.,** The biochemistry of the bacterial toxins. I. The lecithinase activity of *Cl. Welchii* toxins, *Biochem. J.,* 35, 884, 1941.
132. **Ispolatovskaya, M. V.,** Type A *Clostridium perfringens* toxin, in *Microbial Toxins,* Vol. 2A, Kadis, S., Montie, T. C., and Ajl, S. J., Eds., Academic Press, New York, 1971, chap. 3.
133. **Strunk, S. W., Smith, C. W., and Blumberg, J. M.,** Ultrastructural studies on the lesion produced in skeletal muscle fibers by crude type A *Clostridium perfringens* toxin and its pruified alpha fraction, *Am. J. Pathol.,* 50, 89, 1967.
134. **Grossmar, I. W., Heitkamp, D. H., and Sacktor, B. C.,** Morphologic and biochemical effects of *Clostridium perfringens* alpha toxin on intact and isolated skeletal muscle mitochondria, *Am. J. Pathol.,* 50, 77, 1967.
135. **Bullen, J. J.,** Role of toxins in host-parasite relationships, in *Microbial Toxins,* Vol. 1, Ajl, S. J., Kadis, S., and Montie, T. C., Eds., Academic Press, New York, 1970, chap. 7.
136. **Möllby, R., Holme, T., Nord, C.-E., Smyth, C. J., and Wadström, T.,** Production of phospholipase C (alpha toxin), haemolysins and lethal toxins by *Clostridium perfringens,* Type A to D, *J. Gen. Microbiol.,* 96, 137, 1976.
137. **Van Heyningen, W. E.,** Bacterial exotoxins, in *Medical Microbiology and Infectious Disease,* Brande, A. I., Davis, C. E., and Fierer, J., Eds., W. B. Saunders, Philadelphia, 1981, chap. 5.
138. **Möllby, R.,** Bacterial phospholipases, in *Bacterial Toxins and Cell Membranes,* Jeljaszewski, J. and Wadström, T., Eds., Academic Press, New York, 1978, chap. 11.
139. **Arbuthnott, J. P.,** Bacterial cytolysins (membrane-damaging toxins), in *Molecular Action of Toxins and Viruses,* Vol. 2, Cohen, P. and van Heyningen, S., Eds., Elsevier, Amsterdam, 1982, chap. 4.
140. **Bejar, R., Curbelo, V., Davis, C., and Gluck, L.,** Premature labor. II. Bacterial sources of phospholipase, *Obstet. Gynecol.,* 57, 479, 1981.
141. **Hambrey, P. N., Tizard, J. R., and Mellors, A.,** Accumulation of phospholipase A_1 in tissue fluid of rabbits infected with *Trypanosomia brucei, Tropenmed. Parasitol.,* 31, 439, 1980.

142. **Hambrey, P. N., Mellors, A., and Tizard, J. R.,** The phospholipases of pathogenetic and non-pathogenetic *Trypanosomia* species, *Mol. Biochem. Parasitol.*, 2, 177, 1981.
143. **Cerva, L.,** Amebic memingoencephalitis, in *Medical Microbiology and Infectious Disease*, Braude, A. I., Davis, C. E., and Fierer, J., Eds., W. B. Saunders, Philadelphia, 1981, chap. 162.
144. **Visvesvara, G. S. and Balamuth, W.,** Comparative studies on related free-living and pathogenic amebae with special reference to *Acanthamoeba*, *J. Protozool.*, 22, 245, 1975.
145. **Kernbaum, S., Masliah, J., Alaindor, L. G., Bouton, C., and Christol, D.,** Phospholipase activities of bronchoalveolar lavage fluid in rat *Pneumocystis carinii* pneumonia, *Br. J. Exp. Pathol.*, 64, 75, 1983.
146. **Cohen, S. N.,** Infections with *Pneumocystis carinii, in Medical Microbiology and Infectious Disease*, Braude, A. I., Davis, C. E., and Fierer, J., Eds., W. B. Saunders, Philadelphia, 1981, chap. 124.
147. **Wallace, J. F.,** Disorders caused by venoms, bites and stings, in *Principles of Internal Medicine*, 9th ed., Isselbacher, K. L., Adams, R. D., Braunwald, E., Petersdorf, R. G., and Wilson, J. D., McGraw-Hill, New York, 1980, chap. 215.
148. **Warrell, D. A.,** Venomous and poisonous animals, in *Tropical and Geographical Medicine*, Warren, K. S. and Mahmond, A. A. F., Eds., McGraw-Hill, New York, 1984, chap. 58.
149. **Tu, A. T.,** Chemistry of rattlesnake venous, in *Rattlesnake Venoms: Their Actions and Treatment*, Tu, A. T., Ed., Marcel Dekker, New York, 1982, chap. 5.
150. **Tu, A. T.,** *Venoms: Chemistry and Molecular Biology*, John Wiley & Sons, New York, 1977, chap. 2.
151. **Slotboom, A. J., Verheij, H. M., and de Haas, G. H.,** On the mechanism of phospholipase A_2, in *Phospholipids*, Hawthorne, J. N. and Ansell, G. B., Eds., Elsevier, Amsterdam, 1982, chap. 10.
152. **Haberman, H. and Breithaupt, H.** The crotoxin complex — an example of biochemical and pharmacological protein complementation, *Toxicon*, 16, 19, 1978.
153. **Hendon, R. A. and Fraenke-Conrat, H.,** The role of complex formation in the neurotoxicity of crotoxin components A and B, *Toxicon*, 14, 283, 1976.
154. **Breithaupt, H.,** Neurotoxic and myotoxic effects of crotalus phospholipase A and its complex with crotapotin, *Naunyn-Schmiedebergs Arch. Pharmacol.*, 292, 271, 1976.
155. **Hawgood, B. J. and Smith, J. W.,** The mode of action at the mouse neuromuscular junction of the phospholipase A — crotapotin complex isolated from the venom of the South American rattlesnake, *Br. J. Pharmacol.*, 61, 597, 1977.
156. **Chang, C. C. and Lu, J. D.,** Crotoxin, the neurotoxin of South American rattlesnake venom, is a presynaptic toxin acting like α-Bungarotoxin, *Naunyn-Schmiedebergs Arch. Pharmacol.*, 296, 159, 1977.
157. **Chang, C. C., Su, M. J., Lee, J. D., and Eaker, D.,** Effects of Sr^{2+} and Mg^{2+} on the phospholipase A and the prosynaptic neuromuscular blocking action of α-Bungarotoxin, crotoxin and taipoxin, *Naunyn-Schmiedebergs Arch. Pharmacol.*, 299, 155, 1977.
158. **Hawgood, B. J.,** Physiological and pharmacological effects of rattlesnake venoms, in *Rattlesnake Venoms: Their Actions and Treatment*, Tu, A. T., Ed., Marcel Dekker, New York, 1982, chap. 2.
159. **Damerau, B., Lege, L., Oldigs, H.-D., and Vogt, W.,** Histamine release, formation of prostaglandin-like activity (SRC-C) and mast cell degranulation by direct lytic factor (DLF) and phospholipase A of cobra venom, *Naunyn-Schmiedebergs Arch. Pharmacol.*, 287, 151, 1975.
160. **Gopalakrishnakone, P., Hawgood, B. J., Holbrooke, S. E., Marsh, N. A., Santana de Sa, S., and Tu, A. T.,** Site of action of Mojave toxin isolated from the venom of the Mojave rattlesnake, *Br. J. Pharmacol.*, 69, 421, 1980.
161. **Fohlman, J. and Eaker, D.,** Isolation and characterization of a lethal myotoxic phospholipase A from the venom of the common sea snake. *Enhydrina schistosa* causing myoglobinuria in mice, *Toxicon*, 15, 385, 1977.
162. **Harris, J. B. and MacDonell, C. A.** Phospholipase A_2 activity of notexin and its role in muscle damage, *Toxicon*, 19, 419, 1981.
163. **Mebs, D. and Samejima, Y.,** Myotoxic phospholipases from snake venom, *Pseudechis colletti* producing myoglobunuria in mice, *Experientia*, 36, 868, 1980.
164. **Hanashiro, M. A., de Silva, M. H., and Bier, O. G.,** Neutralization of crotoxin and crude venom by rabbit antiserum to crotalus phospholipase A, *Immunochemistry*, 15, 745, 1978.
165. **Sosa, B. P., Alagon, A. C., Possani, L. D.,and Julia, J. Z.** Comparison of phospholipase activity with direct and indirect lytic effects of animal venoms upon human red cells, *Comp. Biochem. Physiol.*, 64B, 231, 1979.
166. **Dennis, E. A.,** Phospholipases, *Enzymes*, 16, 307, 1983.

Chapter 7

DE NOVO PHOSPHOLIPID SYNTHESIS AS AN INTERCELLULAR MEDIATOR SYSTEM

Robert V. Farese

TABLE OF CONTENTS

I. INTRODUCTION

It is widely appreciated that phospholipids, by virtue of their hydrophobic fatty acid chains and hydrophilic head groups, have the innate ability to form thermodynamically stable, bilayered membranes when dispersed in an aqueous environment; through this property, phospholipids serve as the major "structural" substance in all cellular membranes. On the other hand, there is generally less appreciation for the fact that many phospholipids have very high metabolic turnover rates, and because of their intimate relationship with membrane-associated proteins (receptors, transporters, enzymes, etc.), phospholipids have great potential for regulating a wide variety of metabolic processes in living cells.

The possibility that hormones, neurotransmitters, and other agents may employ phospholipids as intracellular mediators received its first major impetus from the work of Hokin and Hokin.[1] These investigators demonstrated that acetylcholin provokes a rapid increase in the ^{32}P-labeling of phosphatidic acid and phosphatidylinositol during stimulation of zymogen secretion in the exocrine pancreas. Since that original description, ^{32}P-labeling effects involving phosphatidic acid and phosphatidylinositol have been observed during the action of a wide variety of agents (see References 2 to 8). The increase in ^{32}P-labeling of phosphatidic acid and phosphatidylinositol was in some cases (e.g., see Reference 9) found to be associated with a decrease in the mass of phosphatidylinositol, coupled with an increase in the mass of phosphatidic acid. It was therefore thought that the initial event involved hydrolysis of phosphatidylinositol by phospholipase C, followed by generation of diacylglycerol[10] and inositol-1-phosphate, subsequent conversion of diacylglyerol to phosphatidic acid, and ultimately, resynthesis of phosphatidylinositol. More recently, this hypothesis has been modified somewhat in that hydrolysis of triphosphoinositide [phosphatidylinositol-4,5-$(PO_4)_2$] is thought to be the main initial phospholipase C-mediated event, followed by depletion of phosphatidylinositol, either because of conversion to diphosphoinositide (phosphatidylinositol-4-PO_4) and triphosphoinositide, or subsequent hydrolysis of phosphatidylinositol, either by a phospholipase C or phospholipase D (see below).

Increased ^{32}P-labeling of phosphatidic acid and phosphatidylinositol has also been observed in the action of agonists which do not appear to increase hydrolysis of either phosphatidylinositol or triphosphoinositide. Indeed, in the actions of ACTH,[11,12] luteinizing hormone,[13,14] and insulin,[11,15-19] increases in ^{32}P-labeling of phosphatidic acid and phosphatidylinositol appear to be associated with increases in the mass, not only of phosphatidic acid, but also of phosphatidylinositol (rather than decreases, as expected with simple hydrolysis). Clearly, there appear to be two separate mechanisms whereby ^{32}P-labeling of phosphatidic acid and phosphatidylinositol can be increased, viz., increased hydrolysis of phosphoinositides, or increased *de novo* synthesis of phosphatidic acid and phosphoinositides (see Figure 1).

In addition to changes in phosphatidic acid and phosphoinositide metabolism, phospholipid metabolism may be perturbed in other ways. These include methylation of phosphatidylethanolamine to form phosphatidylcholine and phospholipase A_2-mediated hydrolysis of phosphoinositides and other phospholipids, thus liberating arachidonic acid for subsequent conversion to prostaglandins, thromboxanes, and other prostanoids. The mechanisms for perturbing phospholipid metabolism are reviewed elsewhere in this book.

II. THE PHOSPHOINOSITIDE HYDROLYSIS EFFECT(S)

This effect will be reviewed more extensively in another chapter. However, since the *de novo* phospholipid effect and the phosphoinositide hydrolysis effect perturb the same phospholipid cycle so dramatically, since both effects may share common properties, and since some agonists appear to provoke both phospholipid effects simultaneously, some of the

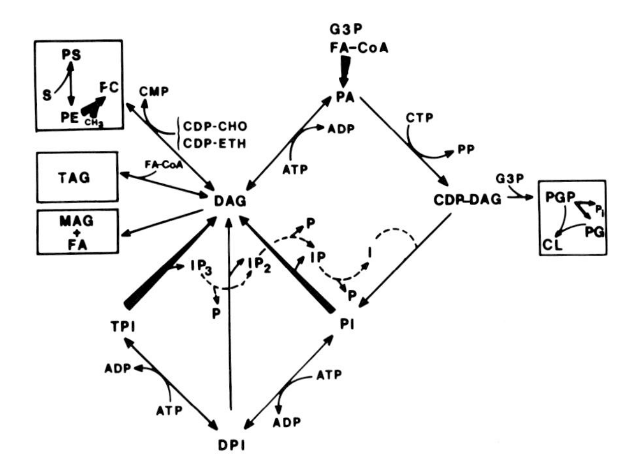

FIGURE 1. The phosphatidate-phosphoinositide pathway. Abbreviations: G3P = glycerol-3′-PO_4; FA-CoA = fatty acid-coenzyme A; PA = phosphatidic acid; CTP = cytidine triphosphate; PP = pyrophosphate; CDP-DAG = cytidine diphosphate-diacylglycerol; PGP = phosphatidylglycerol-PO_4; PG = phosphatidylglycerol; CL = cardiolipin; I = inositol; PI = phosphatidylinositol; DPI = diphosphoinositide; TPI = triphosphoinositide; ATP = adenosine triphosphate; ADP = adenosine diphosphate; IP_3 = inositol-1,4,5-$(PO_4)_3$; IP_2 = inositol-1,4-$(PO_4)_2$; IP = inositol-4-PO_4; DAG = diacylglycerol; TAG = triacylglycerol; MAG = monoacylglycerol; CH_3 = methyl groups; CMP = cytidine monophosphate; CDP-Cho = cytidine diphosphate-choline; CDP-Eth = cytidine diphosphate-ethanolamine; PC = phosphatidylcholine; PE = phosphatidylethanolamine; S = serine; and PS = phosphatidylserine; FA = fatty acid.

more recent developments which concern the phosphoinositide hydrolysis effect will be reviewed in brief.

As alluded to above, decreases in phosphatidylinositol mass, which were observed in conjunction with increases in the labeling of phosphatidic acid and phosphatidylinostol by $^{32}PO_4$, initially suggested that phosphatidylinositol hydrolysis by phospholipase C was the initiating event for the ^{32}P-labeling response to agonists such as acetylcholine or carbachol. This indeed may be the case for some tissues, but in other cases, e.g., the rat pancreas[20] and rat submaxillary gland,[21] the decrease in phosphatidylinositol mass has been shown to be fully dissociated from the ^{32}P-labeling response. The decrease in phosphatidylinositol mass is Ca^{++}-dependent, can be provoked by Ca^{++} ionophores and agonists which do not increase ^{32}P-labeling,[20-23] and in the rat pancreas[102] occurs at lower carbachol concentrations in conjunction with increases in zymogen secretion. On the other hand, the increase in ^{32}P-labeling of phosphatidic acid and phosphatidylinositol is fully independent of Ca^{++}, occurs at considerably higher carbachol concentrations, and occurs only with agonists which mobilize Ca^{++} via activation of cell surface receptors.[102] The ^{32}P-labeling response in the rat pancreas, moreover, appears to be better correlated with Ca^{++}-independent triphosphoinositide (rather than phosphatidylinositol) hydrolysis.[24] Indeed, in many tissues[24-34] it has now been clearly shown that hydrolysis of triphosphoinositide occurs extremely fast, preceding phosphatidylinositol depletion (or hydrolysis) and increases in the ^{32}P-labeling of phosphatidic acid and phosphatidylinositol; it therefore seems likely that the initial biochemical event after receptor activation may be triphosphoinositide hydrolysis by phospholipase C, with concomitant generation of diacylglycerol and inositol-1,4,5-triphosphate, followed by synthesis of phosphatidic acid and phosphatidylinositol and conversion of phosphatidylinositol to diphosphoinositide and then to triphosphoinositide.

Where actual decreases in phosphatidylinositol mass have been demonstrated, the responsible mechanism has not yet been ascertained. In cases in which phosphatidylinositol depletion seems to be linked to Ca^{++}-independent hydrolysis of triphosphoinositide, the decreases in phosphatidylinositol may reflect concomitant hydrolysis or conversion to diphosphoinositide and then to triphosphoinositide, in order to provide a continuous source for the latter as well as generation of diacylglycerol and inositol-triphosphate. In other cases such as the rat pancreas, however, the decrease in phosphoinositide mass appears to be unrelated to triphosphoinositide hydrolysis,[24] and other mechanisms appear to be operative. We have recently obtained evidence[102] that Ca^{++}-dependent activation of both phospholipase C and D may be involved in the decrease in phosphatidylinositol mass in the stimulated rat pancreas. Moreover, no evidence has been found to suggest that phospholipase A_2 action is responsible for the decrease in phosphatidylinositol mass in this organ, although this apparently occurs in platelets.[35,36] Obviously, decreases in phosphatidylinositol may occur by several independent mechanisms. It should be emphasized that in cases where Ca^{++}-dependence for phosphatidylinositol hydrolysis has been observed, it should not be construed that increases in Ca^{++} are in fact responsible for the decreases in phosphatidylinositol. This may indeed be the case, but it is also possible that Ca^{++} is simply required permissively.

Although there were some experimental findings (see Reference 8) which seemed at odds with Michell's original hypothesis that phosphatidylinositol hydrolysis was Ca^{++}-independent, and may thus be the cause for Ca^{++} mobilization from internal or external stores during the action of agonists which activate cell surface receptors,[4] the possibility that hydrolysis of triphosphoinositide may function as the initial, post-receptor, biochemical event which mobilizes Ca^{++} seems to rest on a more solid footing. In virtually all instances,[24-34] phospholipase C-mediated triphosphoinositide hydrolysis was found to be fully independent of Ca^{++}, and occurred rapidly enough to account for changes in Ca^{++} mobilization in the action of various agonists that activate cell surface receptors. Moreover, in the rat pancreas, we have found[24,102] that hydrolysis of triphosphoinositide is better correlated with receptor occupancy than phosphatidylinositol hydrolysis (which largely occurs at lower levels of receptor occupancy). In addition, as indicated above, triphosphoinositide hydrolysis also seems to be closely related to the ^{32}P-labeling of phosphatidylinositol in dose-response studies. The latter observation is important since it indicates that at least in the rat pancreas, the ^{32}P-labeling effect involving phosphatidic acid and phosphatidylinositol is actually more reflective of triphosphoinositide (rather than large-scale phosphatidylinositol) hydrolysis.

As an immediate consequence of triphosphoinositide hydrolysis by phospholipase C, there are rapid increases in diacylglycerol and inositol-triphosphate. The latter substance has recently been shown to mobilize Ca^{++} in pancreatic acinar cells[37] and may therefore function as a natural Ca^{++} ionophore. The increase in diacylglycerol will, of itself, lead to activation of the phospholipid- and Ca^{++}-dependent protein kinase-C through an increase in affinity for binding Ca^{++} at physiological concentrations (0.01 to 10 μM).[38,39] Activation of protein kinase-C in turn apparently leads to the phosphorylation of a number of potentially important proteins. Increases in diacylglycerol will also lead to rapid increases in phosphatidic acid, and thus substance may, like inositol-triphosphate, function as an endogenous Ca^{++} ionophore[40] and mobilize Ca^{++}.

It is becoming increasingly clear that phospholipase-C-mediated triphosphoinositide hydrolysis occurs in the action of all agonists which mobilize Ca^{++} and operate via cell surface receptors. As alluded to above, this hydrolysis of triphosphoinositide appears to be Ca^{++}-independent and may therefore serve as the initial biochemical event which leads to mobilization of Ca^{++}. There have been occasional instances, however, in which triphosphoinositide hydrolysis has been found to be dependent upon Ca^{++}. A notable example is the pancreatic islet, where glucose has been shown to provoke Ca^{++}-dependent triphosphoinositide hydrolysis.[41] The mechanism for the latter is enigmatic, since glucose does not seem to activate a cell surface receptor.[42] It seems more likely that triphosphoinositide hydrolysis in this case is a consequence of glucose metabolism, but it is uncertain how this could occur. This point is discussed further below.

III. THE *DE NOVO* PHOSPHOLIPID SYNTHESIS EFFECT

A. General Description

The possibility that hormones, neurotransmitters, and other agents may perturb the phosphatidate-phosphoinositide cycle by an increase in *de novo* synthesis was given little attention until several years ago. At that time, it was observed that ACTH treatment, either in vitro or in vivo, provoked 50 to 100% increases in adrenal concentrations of phosphatidic acid, phosphatidylinositol, diphosphoinositide, triphosphoinositide, and phosphatidylglycerol.[11,12,43-45] These effects were found to be nearly maximal within 1 to 15 min, the fastest response involving phosphatidic acid, which increased nearly twofold within 1 to 2 min after ACTH treatment in vivo. These increases in phospholipid mass (measured by colorimetric determination of inorganic phosphorus) were quite unexpected, as previous reports suggested that agonists, if anything, simply diminished phosphatidylinositol levels through activation of phospholipase C. Obviously, with a widespread increase in the levels of the above-mentioned phospholipids, it appeared likely that ACTH increased *de novo* synthesis of phosphatidic acid, with rapid conversion of the latter to its derivatives, the phosphoinositides, and the glycerophospholipids (see Figure 1). That these phospholipids were not increased at the expense of other phospholipids became clear in other studies in which it was shown that ACTH also provoked increases in diacylglycerol[46] and lesser relative increases (approximately 25%) in phosphatidylcholine and phosphatidylethanolamine.[47] In fact, the increases in phosphatidylcholine and phosphatidylethanolamine, although less on a relative basis, were nonetheless quite sizeable on an absolute basis.

With confirmation of results observed in the adrenal studies, we have now found[103] that insulin rapidly increases phosphatidic acid, phosphoinositides, diacylglycerol, and phosphatidylcholine + phosphatidylethanolamine in BC3H-1 cultured myocytes to virtually the same degree as that provoked by ACTH in the adrenal cortex. Moreover, in the myocyte culture system, this conclusion has been reached by entirely different methodology. In the latter case, phospholipids and intracellular phosphate pools were labeled to constant specific activity by preincubation for 3 days in the presence of $^{32}PO_4$; under these circumstances, simple changes in radioactivity can be equated with changes in phospholipid mass. In addition, in the myocyte system it has been possible to perform prolonged pulse-chase experiments. From these it is clear that insulin-induced increases in phospholipids are due to increased *de novo* synthesis rather than diminished degradation of preexisting phospholipids. In fact, insulin had no effect on the fractional rate of degradation of preexisting phospholipids, and as with ACTH, there was no evidence to suggest that triphosphoinositide or phosphatidylinositol hydrolysis was provoked (even prior to observed increases in these phospholipids). In addition, it may be surmised that since the fractional rate of degradation is the same in control and insulin-treated myocytes and since the mass of phosphoinositides is increased by insulin, the total rate of phosphoinositide degradation (determined by multiplying the fractional degradation rate by the phosphoinositide concentration) may be increased. The latter possibility is important, since an increase in total phosphoinositide degradation may generate increased amounts of inositol-phosphates and diacylglycerol; this may lead to activation of protein kinase-C and Ca^{++} mobilization (see above).

In most tissues, increases in the above-mentioned phospholipids have not been attended by increases in sphingomyelin or phosphatidylserine. However, insulin appears to increase phosphatidylserine levels, both in rat adipose tissue[48] and in cultured, ^{32}P-prelabeled BC3H-1 myocytes. The increase in phosphatidylserine might be due to an increase in *de novo* synthesis of its immediate precursor, phosphatidylethanolamine. However, it is also possible that insulin provokes the Ca^{++}-dependent exchange of serine for ethanolamine in pre-existing phosphatidylethanolamine. We had, in fact, favored the former explanation, since cycloheximide inhibited insulin-induced increases in both phosphatidylserine and phosphoinositides.[49] However, that postulation reflected our initial belief that cycloheximide diminished phospholipids only by interfering with their synthesis *de novo*. It is now clear that cyclohexamide has a number of effects on phospholipid metabolism, one being the rapid degradation of pre-existing phospholipids in hormone-treated, but not in control, tissues[50] (see below). Thus, cyclohexamide inhibition is nonspecific, and its occurrence does not necessarily imply that increases in phosphatidylserine and other phospholipids are reflective of the *de novo* synthesis effect.

As alluded to above, increases in phosphatidylcholine and phosphatidylethanolamine have been relatively small (0 to 30%), as compared to increases in phosphatidic acid and phosphoinositides (40 to 100%). On an absolute basis, however, the relatively smaller increases in phosphatidylcholine and phosphatidylethanolamine nevertheless reflect large absolute increases in these phospholipids. Although it may be argued that the generalized increase in phospholipid levels in the *de novo* effect is nonspecific, it should be emphasized that relative increases in phosphatidic acid and phosphoinositides are much more dramatic, particularly during the early minutes of stimulation. As a result, there will be a considerable enrichment of phosphatidate, phosphoinositides, glycerophospholipids, and diacylglycerol concentrations in cellular membranes, and it should be further realized that the increases in these substances will be even greater in the vicinity of the enzymes which are responsible for their synthesis.

B. Methodology for Demonstrating the De Novo Phospholipid Effect

In general, it has been easier to gain evidence for phosphoinositide hydrolysis and turnover, since this effect usually leads to sizeable, rapid increases in $^{32}PO_4$ incorporation into phosphatidic acid (via synthesis from diacylglycerol and ATP) and eventually phosphatidylinositol in acute labeling experiments. On the other hand, as discussed below, it has been much more difficult in some systems to study the *de novo* phospholipid synthesis effect.

Prior to realization that the *de novo* phospholipid synthesis effect existed, it had frequently been tacitly assumed that increased $^{32}PO_4$ incorporation into phosphatidic acid and phosphatidylinositol reflected phosphoinositide hydrolysis and subsequent resynthesis (i.e., "turnover"). Clearly, this can no longer be assumed, since *de novo* phospholipid synthesis can provoke similar increases in the acute labeling of these phospholipids. An excellent example of the latter is insulin, which provokes sizeable increases in $^{32}PO_4$ incorporation into phosphatidic acid and phosphoinositides in several target tissues;[16-19,48] however, unlike the classical ^{32}P-labeling effect secondary to phosphoinositide hydrolysis, additional increases in $^{32}PO_4$ incorporation into phosphatidylcholine and phosphatidylethanolamine have been observed. Indeed, the latter, if observed, may well be an indication that the *de novo* effect, rather than phosphoinositide hydrolysis effect, is in operation.

In some systems where the *de novo* effect is provoked, effects of agonists on $^{32}PO_4$ incorporation have been difficult to demonstrate. For example, in the case of ACTH in the adrenal cortex, increases in $^{32}PO_4$ incorporation were observed at lower but not higher ACTH concentrations.[12] The latter biphasic effect of ACTH appeared to be due to a decrease in specific activity of ATP observed at the higher ACTH concentrations.[12] The decrease in ATP specific activity may have been due to large increases in cyclic AMP (cAMP) generation.[12]

There is a second difficulty with the ^{32}P-labeling experiments that may be encountered, especially when agonists and $^{32}PO_4$ are added simultaneously, viz., ATP labeling may lag initially at a time when *de novo* synthesis is occurring extremely rapidly. For example, in our myocytes experiments,[103] it appears that there is an initial insulin-induced burst of phospholipid synthesis during the first 10 to 30 min of treatment, followed by a prolonged plateau period in which phospholipids are simply maintained at a higher level. The phospholipid synthesis rate during the latter plateau period is much slower than that observed in the initial "burst" period; thus, if there is a lag in ATP labeling, phospholipid labeling may only reflect the synthetic rate of the plateau period and may thus appear to be less dramatic than expected. If this is the case, more dramatic increases in ^{32}P-labeling may be observed if ATP and phospholipid precursors are prelabeled during a short prior incubation with $^{32}PO_4$. However, if prelabeling continues too long, isotopic equilibrium will be approached, phospholipids in basal controls will already be heavily labeled, and, under these circumstances, any new labeling will only reflect the changes in phospholipid mass rather than the much greater rate of synthesis (relative to control) during the initial "burst" period.

Although the initial synthetic "burst" may be less dramatic with more prolonged prelabeling periods, this approach is very useful in some cases. For example, if cAMP and $^{32}PO_4$ are added simultaneously to adrenal sections, phospholipid labeling may be decreased despite increases in mass.[12] If, however, adrenal sections are prelabeled for 2 hr with $^{32}PO_4$, and then treated with cAMP, there is an increase in the ^{32}P content of phosphatidic acid and phosphoinositides during the first 2 to 5 min of stimulation, but this may be rapidly replaced by a decrease in labeling over the next 15 to 30 min.[104] The initial increase in ^{32}P-labeled phospholipids presumably reflects the cAMP-induced increase in phospholipid synthesis. The later decrease presumably reflects a decrease in precursor specific activity (and continued rapid turnover of these phospholipids) since phospholipid mass remains elevated. A virtually identical sequence of events has been observed during stimulation of dog kidney proximal tubules with parathyroid hormone:[51] PTH increased phosphoinositide mass over the entire 30-min study period, but increased $^{32}PO_4$ incorporation was only observed in the first 5 min of stimulation following a 30-min preincubation with $^{32}PO_4$.

In the case of higher ACTH concentrations (see above), we have been able to show that the decline in ^{32}P-labeling of phospholipids can be explained by a decline in ^{32}P-labeling of ATP.[12] With cAMP, however, we have not regularly observed decreases in overall ATP specific activity. On the other hand, since it is clear that the specific activity of phospholipids is diminished by cAMP,[12] it appears that simple measurement of overall ATP specific activity does not provide the information that gives insight into the discrepancy. Conceivably, there may be multiple ATP or other precursor pools of varying specific activity throughout the cell, or there may be differences in labeling of the three phosphates of the ATP molecule. Further work will be required to resolve this question.

As discussed above, simple measurement of acute $^{32}PO_4$ incorporation into phosphatidic acid and phosphoinositides cannot distinguish between turnover or increases in *de novo* synthesis. To be certain of the latter, it is necessary to demonstrate absolute increases in phospholipid mass. Obviously, this will not be possible if degradation keeps pace with synthesis, resulting in no change in phospholipid levels. Fortunately, however, in many cases, increases in phospholipid mass have been demonstrable by one of two techniques: direct measurement of the phosphorus content of purified phospholipids, or increases in the ^{32}P content of phospholipids after prelabeling of phospholipids and all relevant cellular phosphate pools to constant specific activity. The latter approach has been applied in two systems (submitted for publication), continuous cultures of BC3H-1 myocytes[103] and primary cultures of rat islets,[105] and this was effected by labeling for 3 days in the presence of $^{32}PO_4$. Upon addition of agonists to these prelabeled cultures, very rapid increases in ^{32}P content (reflecting mass, rather than hydrolysis-induced turnover, which would result in decreased ^{32}P content) of phospholipids were observed, and the increases (^{32}P content) in phosphatidate and phosphoinositides observed in these systems were virtually identical in relative magnitude to those observed after colorimetric analyses. The increases in phosphatidate and phosphoinositides occurred extremely rapidly and preceded lesser relative increases (^{32}P content) in phosphatidylcholine and phosphatidylethanolamine. In short, the ^{32}P prelabeled culture system has confirmed what seemed to be apparent by simple measurement of phosphorus content of purified phospholipids. Obviously, the culture-prelabeling approach allows simple measurement of ^{32}P to reflect phospholipid mass and can be applied to relatively small amounts of tissue because of enhanced sensitivity of this measurement (relative to the more crude phosphorus determination). Another advatage of the culture-prelabeling method is that only viable cells will be well labeled, and these are more likely to be responsive to agonists; this obviously will minimize the damping effects observed when total phospholipid contents of active and inactive cells are measured simultaneously.

We have had mixed success in using labeled inositol or glycerol to reflect increases in *de novo* synthesis. In BC3H-1 myocytes, insulin clearly increases [^{3}H]inositol content of phosphoinositides after the cultures have been prelabeled for 3 days with [^{3}H]inositol. However, in both rat adrenal cells (or sections) and rat epididimyl fat pads, ACTH and insulin, respectively, have not provoked increases in [^{3}H]inositol incorporation into phospholipids during acute incubations (despite increasing phospholipid mass and $^{32}PO_4$ incorporation). The poor ability for labeled-inositol to reflect the *de novo* effect in some circumstances may be explained either by production of nonlabeled-inositol during agonist stimulation of glycolysis and metabolism along the hexose monophosphate pathway (subsequently diluting the specific activity of labeled inositol) or the incorporation of label may reflect a simple exchange reaction which is limited by factors other than the concentration of phosphoinositides. Although we have observed that ACTH may increase [^{3}H]glycerol incorporation into adrenal phospholipids,[52] this is not always observed with agonists which provoke a *de novo* phospholipid synthesis effect. This may reflect the fact that the labeled glycerol will be diluted by unlabeled glycerol-3-phosphate which is frequently increased through enhanced carbohydrate metabolism during the action of many agonists that evoke the *de novo* phospholipid effect. Another problem which may limit the usefulness of labeled glycerol is that the glycerol kinase step may be rate limiting and may be the sole determinant for the rate of incorporation of labeled glycerol into phospholipids.

To summarize, the only reliable methods for demonstration of the *de novo* phospholipid synthesis effect are to measure: (a) phosphorus mass of purified phospholipids, or (b) ^{32}P content of purified phospholipids after prelabeling to constant specific activity.

C. Requirements for the De Novo Phospholipid Synthesis Effect

The synthesis of phospholipids is an energy requiring process, and whenever agents have been present which concomitantly inhibit energy production (as evidenced by diminished $^{32}PO_4$ incorporation into phospholipids and/or labeled amino acid incorporation into protein), we have failed to find increases in phospholipids after agonist stimulation. This, for example, has been the case with A23187 in concentrations exceeding 1 μM in rat adrenal cells or sections.[53] In rat epididymal fat pads, it is necessary to have an external energy source in the form of glucose or fructose to maintain overall $^{32}PO_4$ incorporation into phospholipids, and to observe full effects of insulin on *de novo* phospholipid synthesis.[49] Small effects of insulin can be observed in the absence of an external fuel supply in this tissue,[19,49] and this is compatible with the idea that the effects of insulin are not simply explained by an increase in glucose uptake. Along the latter lines, in myocytes an external carbohydrate source is not needed, at least during a 30 to 60 min incubation period; presumably these cells have adequate internal fuel sources to support phospholipid synthesis in this time span.

In both the rat adrenal cortex[52] and rabbit kidney cortical tubules,[54] Ca^{++} has been required for demonstration of effects of ACTH and parathyroid hormone on *de novo* phospholipid synthesis. To show this requirement for Ca^{++}, it has been necessary to deplete cellular Ca^{++} stores, by either employing EGTA during incubation or both preincubating and incubating the tissues in the absence of extracellular Ca^{++} (without large amounts of EGTA).

Simple removal of Ca^{++} from the incubation media at the time of addition of agonist addition failed to inhibit the *de novo* synthesis effect of ACTH.[52] It seems clear that internal (but not external) Ca^{++} stores are required for expression of the *de novo* effect. The reason for this requirement is not readily apparent. Cycloheximide and puromycin, which are protein synthesis inhibitors, have been found to inhibit the inductive effects of ACTH,[11,12,43-47,50] insulin,[48] luteinizing hormone,[55] parathyroid hormone,[56] and angiotensin-II[57] on *de novo* phospholipid synthesis in their respective target tissues. In the absence of co-added agonists (i.e., in control tissues), protein synthesis inhibitors oddly enough provoke mild increases in *de novo* phospholipid synthesis, possibly via increases in glycogenolysis and presumably increased availability of glycerol-3-phosphate for phosphatidate synthesis. Clearly, the effects of protein synthesis inhibitors vary, depending upon whether or not the tissue is being concomitantly stimulated by certain agonists. The inhibitory effects of protein synthesis inhibitors on agonist-induced increases in phospholipids at first suggested to us that a labile protein may be required for provoking the *de novo* phospholipid effect. However, we have now clearly shown that cycloheximide treatment causes pre-existing, agonist-induced increases in phospholipids to be degraded at a rate which greatly exceeds the degradation rate of phospholipids in control or agonist-stimulated tissues incubated without cycloheximide. The rapid increase in phospholipid degradation is presumably secondary to activation of phospholipase(s), and a nonspecific phospholipase C may be involved, as we have found in both the rat adrenal gland[12,50] and rat epididymal fat pad[12] that cycloheximide-induced decreases in phospholipid mass are associated with increases in $^{32}PO_4$ incorporation. This paradoxical increase in acute ^{32}P-labeling is best explained by degradation of unlabeled phospholipids and partial resynthesis of their headgroups with $^{32}PO_4$, a pattern characteristic of phospholipase C-mediated turnover. Unlike phospholipase C activation, which is induced by more naturally occurring agonists, the hydrolysis and partial resynthesis of phosphatidylcholine and phosphatidylethanolamine accompanies that of phosphoinositides during cycloheximide treatment. These diverging effects of protein synthesis inhibitors on ^{32}P-labeling and mass of phospholipids underscore the fact that it is not possible to use ^{32}P-labeling to determine changes in *de novo* phospholipid synthesis during cycloheximide or puromycin treatment. It is only possible to examine the effects of protein synthesis inhibitors if phospholipid mass is determined directly by phosphorus measurement or indirectly by $^{32}PO_4$ determination, after labeling of phospholipids to constant specific activity (see above).

Another difficulty with studies involving the use of inhibitors of protein synthesis to inhibit the *de novo* phospholipid synthesis effect is that phospholipase C-mediated degradation of phospholipids may lead to lesser or fleeting, but nevertheless effective increases in diacylglycerol and phosphatidic acid[50] which may continue to provoke metabolic changes if the latter are dependent upon these substances rather than increased levels of other phospholipids.

In addition to increases in phospholipid degradation, protein synthesis inhibitors also partially inhibit diacylglycerol kinase,[46] and furthermore, probably inhibit glycerol-3-phosphate acyltransferase in intact tissue, as we have observed inhibiton of [^3H]glycerol incorporation into phospholipids and neutral lipids in hormone-stimulated tissues.[50] Just how cycloheximide provokes all of these changes in phospholipid metabolism in the agonist-stimulated state is not readily apparent. It is possible that an increase in phospholipase activity may promote the appearance of a variety of substances in cellular membranes which may secondarily inhibit a variety of phospholipid-synthesizing enzymes.

It should be stressed again that these phospholipid inhibitory effects of protein synthesis inhibitors occur only (or to a much greater extent) with concomitant stimulation by agonists which provoke the *de novo* phospholipid synthesis effect. Protein synthesis inhibitors have no (or lesser) inhibitory effects on phospholipid metabolism in control tissues, and furthermore, do not inhibit the phosphoinositide hydrolysis effect(s). Since alterations of phospholipid metabolism are observed with both cycloheximide and puromycin (although effects on cycloheximide seem more prominent), it seems likely that inhibition of protein synthesis per se somehow leads to an altered state of metabolism (accumulation of a metabolite?) that is conducive to phospholipid degradation, especially if a *de novo* synthesis-provoking agonist is present. Perhaps the latter interaction reflects metabolic instability of phospholipid pools that are expanded during the *de novo* effect.

Obviously, the effects of protein synthesis inhibitors on phospholipid metabolism are very complex and difficult to interpret correctly. For example, although cycloheximide sensitivity of hormonal effects may suggest involvement of phospholipids, failure to observe cycloheximide sensitivity does not necessarily rule out phospholipid involvement. This is because phospholipase C-mediated phospholipid degradation may continue to generate diacylglycerol and phosphatidic acid, although possibly to a lesser degree. In addition, phospholipase degradation may yield noxious substances that may nonspecifically inhibit cellular processes which require intact membranes. As a result, our enthusiasm for use of protein synthesis inhibitors to gain insight into the question of whether the *de novo* phospholipid synthesis effect is important for metabolic actions of agonists has diminished. On the other hand, our studies clearly show that the demonstration of sensitivity of hormonal- or agonist-provoked metabolic effects to protein synthesis inhibitors cannot simply be attributed to direct involvement of a labile protein therein, as the effects may be more dependent on phospholipid metabolism. The latter, in fact, seems likely to be the case in both agonist-stimulated steroidogenesis and insulin effects on pyruvate dehydrogenase. In both cases, the decreases in enzyme activity occur extremely rapidly, along with equally rapid decreases in phospholipids, and it seems more logical to impute altered phospholipid metabolism, rather than protein synthesis, as the reason for inhibition.

D. Possible Mechanisms for Provoking the *De Novo* Phospholipid Synthesis Effect

There appear to be at least two mechanisms for provoking the *de novo* phospholipid synthesis effect. Since glucose provokes this effect in pancreatic islets,[104] it seems likely that direct provision of glycolytic intermediates, such as glycerol-3-phosphate and dihydroxyacetone-phosphate, may lead to increases in *de novo* synthesis of phospholipids. Further proof for this possibility stems from the fact that glyceraldehyde, presumably after its phosphorylation, also increases *de novo* phospholipid synthesis in this tissue. It therefore appears that substrate availability is limiting, and substrate provision through enhanced glycolysis can influence the rate of *de novo* phosphatidic acid synthesis.

Other agonists may increase phosphatidate synthesis through similar mechanisms, e.g., ACTH may increase substrate availability through glycogenolysis[58] and glycolysis,[59] insulin may increase glucose uptake, and parathyroid hormone[60] and angiotensin[61] may also provoke glycogenolysis. In the case of insulin, however, a simple increase in glucose uptake cannot explain the insulin effects in BC3H-1 myocytes. In the latter system, insulin increases the levels of phosphatidic acid and phosphoinositides in the absence of extracellular glucose, and there must be an additional effect of insulin. Along these lines, insulin has been shown to provoke increases in the activity of glycerol-3′-PO_4 acyltransferase in both liver and fat,[62,63] and this may also be the case in muscle. Obviously, further work is required to answer this question.

The mechanisms whereby glycerol-3-PO_4 acyltransferase activity may be regulated is not clear, but activity of this enzyme in adipose tissue is reported to be increased by phosphatase treatment and decreased by phosphorylation provoked by cAMP-dependent protein kinase.[64] It is therefore possible that insulin, which increases phosphatase activity and thereby increases the activity of various other enzymes,[65,66] may also regulate phospholipid synthesis through a similar mechanism.

E. Effects of "Second Messengers" on *De Novo* Phospholipid Synthesis

We have observed increases in phosphatidate and phosphoinositide levels after cAMP treatment in the rat adrenal cortex,[11,12,44,52] rabbit kidney cortical tubules,[54,56] rat adrenal zona glomerulosa,[57] and rat Leydig cells.[55] Thus, with hormones such as ACTH, luteinizing hormone, and parathyroid hormone (which apparently operate via cAMP), the *de novo* phospholipid effect may be provoked by agonist-induced increases in tissue cAMP levels. This point, however, is not fully settled, since cAMP has not provoked the same effects on $^{32}PO_4$ incorporation into phospholipids which are observed during hormone treatment, e.g., as with luteinizing hormone in ovarian tissues[106] and ACTH in the adrenal cortex.[12] Whether the latter discrepancy reflects isotopic problems (see above) encountered with addition of large amounts of cAMP or its derivatives, or whether the hormones increase *de novo* phospholipid synthesis by a cAMP-independent mechanism, remains to be determined.

In the case of agonists which operate via Ca^{++} instead of cyclicAMP, e.g., angiotensin-II and K^+ in adrenal glomerulosa, the observed *de novo* effects[57] obviously must be provoked by a cAMP-independent mechanism. Ca^{++} is required for the *de novo* phospholipid effect[57] (see above text and Reference 57), and we have also observed a *de novo* effect with A23187 treatment in rabbit kidney cortex tubules (unpublished); it is therefore possible that Ca^{++} may also provoke a *de novo* phospholipid effect during the actions of certain agonists. Along these lines, both cAMP and Ca^{++} may increase glycogenolysis and thus provide glycerol-3-PO_4 for phosphatidate synthesis.

In the case of insulin, its "second messenger" has not been defined.[65,66] Insulin could increase *de novo* phospholipid synthesis through increases in glucose uptake and/or activation of glycerol-3′-PO_4 acyltransferase (see above). Conceivably, the "second messenger" which may activate the acyltransferase could be an unidentified "chemical mediator"[67,68] (causing activation via its reported phosphatase activity?), Ca^{++} or cGMP,[65] or other undefined substances.

Finally, it may be questioned whether increases in phosphatidic acid and phosphoinositides may be directly provoked by agonist action at the level of the plasma membrane (possibly even without an intermediate "second messenger"), instead of in the endoplasmic reticulum and outer mitochondrial membrane. Synthesis of these phospholipids, especially the polyphosphoinositides, appears to occur in the plasma membrane. ACTH has been found to influence polyphosphoinositide synthesis in cell-free preparations,[69] and this might occur in intact tissue as well; however, increases in polyphosphoinositides seemed to follow phosphatidate synthesis in adrenal studies.[11,45] Phosphatidic acid is also synthesized in plasma membranes,[70] and it is conceivable that this may be directly affected by hormone-receptor complexes or some other local change in the plasma membrane. Obviously, there is more work to be done to clarify this tissue.

F. General Consequences of Increases in *De Novo* Phospholipid Synthesis

The large, nearly twofold increases in phosphatidic acid, diacylglycerol, and phosphoinositides during the *de novo* synthesis effect may be responsible for other wide-ranging effects. Phosphatidic acid has been suggested to be an endogenous Ca^{++} ionophore,[33] and intracellular Ca^{++} translocations may be effected by increases in this phospholipid. Increases in phosphoinositides (phosphatidylinositol, diphosphositide, and triphosphoinositide) may result in increases in inositol-phosphate generation (if the fractional rate of degradation of these phospholipids remains the same while their mass increases). We have, however, in preliminary experiments, not observed increases in the levels of [^{3}H]inositol-phosphates (mono-, di-, and tri-) in several systems which we have evaluated; in these cases, however, agonists failed to increase [^{3}H]inositol incorporation into phosphoinositides, despite provoking a *de novo* phospholipid synthesis effect. It seems likely that increased generation of labeled inositol-phosphates will only be observed if labeled inositol incorporation into phospholipids is increased. Along the latter lines, it is unfortunately more difficult to measure the absolute levels of the inositol-phosphates, as opposed to simple appearance of labeled inositol therein, and this may be necessary to be certain of whether or not inositol-phosphate levels are altered by agonists which provoke the *de novo* phospholipid synthesis effect without concomitantly increasing phosphoinositide hydrolysis. If there are increases in inositol-phosphates, particularly inositol-triphosphate, which may be an endogenous Ca^{++} ionophore,[37] this too may be an important mechanism for mobilizing intracellular Ca^{++}.

Increases in the level of triphosphoinositides, particularly at its primary location in the plasma membrane, may lead to increases in the binding of cations, such as Ca^{++} and Mg^{++}. This may alter a variety of metabolic processes which are controlled by receptors, enzymes, and transporters therein.

Increases in diacylglycerol during the *de novo* phospholipid effect (now documented both with ACTH[46] and insulin treatment)[103] may activate protein kinase-C (see above), [38,39] and this may provide a cascading mechanism to regulate a variety of metabolic events through changes in phosphorylation-dephosphorylation.

In addition to the above indirect effects, increases in specific phospholipids may directly change the activity of membrane-bound enzymes, transporters, and receptors. Obviously, there may be multiple mechanisms which are regulated or modulated by a single change in *de novo* synthesis of phosphatidic acid.

In addition to specific effects on membranes and membrane-associated proteins, a generalized increase in phospholipid synthesis would lead to expansion of cellular membranes, and this may be a forerunner to cellular hypertrophy. The latter is particularly evident in the actions of ACTH and insulin in adrenal and fat tissue, respectively.

G. Comparison of the *De Novo* Phospholipid Synthesis Effect to the Phosphoinositide Hydrolysis Effect(s)

At first glance, it may appear that phosphoinositide hydrolysis and *de novo* phospholipid synthesis are opposing effects; this is true in that phosphoinositide levels decrease with phosphoinositide hydrolysis and increase with *de novo* phospholipid synthesis. However, as alluded to above, in both cases there will be increases in potentially important signalling substances such as diacylglycerol and phosphatidic acid. It should be noted, however, that these increases may be more prominent in plasma membranes with phosphoinositide hydrolysis, and more prominent in interior membranes (endoplasmic reticulum and outer mitochondrial membranes) with *de novo* phospholipid synthesis. Increases in inositol-triphosphate, which occurs in the triphosphoinositide hydrolysis effect, may conceivably occur in the *de novo* synthesis effect (see above), but there is no definitive information presently available on this point.

It should be noted that several agonists appear to provoke both phosphoinositide hydrolysis and *de novo* phospholipid synthesis effects simultaneously. This is clearly evident during angiotensin-II action in adrenal glomerulosa tissue.[34,57,71] In addition, glucose provokes *de novo* phospholipid synthesis in rat pancreatic islets, while simultaneously increasing the hydrolysis of phosphatidylinositol[73] and triphosphoinositide.[41] It is possible that the simultaneous occurrence of both effects leads to intensified generation of potentially important intracellular messengers such as phosphatidic acid, diacylglycerol, and inositol-phosphates. In other words, there may be synergistic effects of phosphoinositide hydrolysis and *de novo* phospholipid synthesis. Although some agents apparently provoke both effects, other agents appear to provoke only the phosphoinositide hydrolysis effect (e.g., acetylcholine and cholecystokinin), and some agents appear to provoke only the *de novo* phospholipid synthesis effect (e.g., insulin and ACTH).

IV. ROLE OF PHOSPHOLIPID METABOLISM IN THE ACTION OF ACTH AND OTHER STEROIDOGENIC AGENTS

ACTH has been found to provoke the *de novo* phospholipid synthesis effect in the adrenal cortex,[11,12,43-50] and we have also observed a similar effect of ACTH in preliminary studies in rat epididymal fat pads. To date, we have found no evidence to indicate that ACTH increases the hydrolysis of phosphoinositides. Although it is presently uncertain what role the *de novo* effect serves during ACTH action, it seems likely that this effect is important in the stimulation of steroidogenesis. In support of the latter possibility, are the following: (a) increases in phosphatidate and phosphoinositide synthesis precede or accompany increases in steroidogenesis during ACTH and cAMP action;[11,45] (b) there is reasonably good correlation between increases in phospholipids and increases in steroidogenesis in ACTH-dose-response studies,[12,44,53] (c) there is good correlation of phospholipid metabolism and steroidogenesis in conditions of varying Ca^{++} deficiency;[52] (d) the rapid reversal of ACTH effects on adrenal steroidogenesis observed in vivo is preceded by equally or more rapid decreases in phospholipids after cycloheximide treatment;[45] (e) the relatively slow reversal of ACTH action after cycloheximide treatment of adrenal sections in vitro is accompanied by an equally slow decrease in phospholipid levels;*[74] and (f) the *de novo* phospholipid synthesis effect has been observed in the action of every tested steroidogenic agonist, although not under all experimental conditions, including (see below) luteinizing hormone, angiotensin-II, K^+, serotonin, and luteinizing hormone-releasing hormone.

While it seems likely that *de novo* phospholipid synthesis effect is important for steroidogenesis, it also seems likely that other factors such as availability of free, unesterified cholesterol are important in determining the total steroid production rate.[76]

The mechanism whereby phospholipids may be important in steroidogenesis is presently obscure. Polyphosphoinositides and cardiolipin have been found to increase cholesterol side chain cleavage when added to adrenal mitochondria,[77,78] and the increased levels of diphosphoinositide observed in crude adrenal mitochondrial fractions after ACTH treatment[44] could explain this observation. However, it should be noted that it is uncertain whether diphosphoinositide is truly present in the mitochondrion itself. Attempts to obtain adrenal mitochondrial preparations free of plasma membrane contamination in our laboratory have so far been unsuccessful. It is therefore uncertain whether increases in diphosphoinositide would be meaningful to stimulation of mitochondrial cholesterol side chain cleavage in the intact cell. In most tissues, triphosphoinositide is primarily localized to plasma membranes, and this phospholipid seems unlikely to be involved in mitochondrial steroidogenesis. Phosphatidylglycerol-phosphate has a polyphosphorylated head group and therefore seems to be another logical candidate for stimulating steroidogenesis.[77,78] This phospholipid is a short-lived, but obligatory precursor to phosphatidylglycerol (its levels are increased acutely by ACTH).[11,45] Moreover, it is likely that both phosphatidylglycerol and phosphatidylglycerol-phosphate are present in mitochondria.[79] Unfortunately, to date there have been no studies on phosphatidylglycerol-phosphate with respect to its steroidogenic effects, or its concentration after ACTH treatment. Cardiolipin has been found to decrease the K_m for the cytochrome P_{450} responsible for cholesterol side chain cleavage,[80] but this phospholipid is relatively inert metabolically and ACTH has failed to influence the levels of this phospholipid in acute studies. Obviously, any of the above phospholipids could theoretically stimulate cholesterol side chain cleavage by altering the activity of the mitochondrial cholesterol side chain

* Half-lives of phospholipids and steroidogenesis in this system in vitro are nearly identical, viz., 25 to 30 min (this contrasts with the much more rapid decrease in protein synthesis in this system,[75] which is maximally inhibited within 10 min of cycloheximide treatment). These results suggest that cycloheximide sensitivity is much better explained by phospholipid changes than changes in protein synthesis per se.

cleavage enzyme complex, or even more likely, promoting the movement of cholesterol from the outer to the inner mitochondrial membrane, the site of the side chain cleavage enzyme complex. Along the latter lines, diphosphoinositide enhances the delivery of cholesterol to active sites of cytochrome P_{450} of the cholesterol side chain cleaving enzyme in reconstituted vesicles.[81] In addition to direct effects of phospholipids on cholesterol side chain cleavage, there are a number of other mechanisms whereby phospholipids could influence steroidogenesis, including (a) phospholipids could promote shifts and/or binding of cations such as Ca^{++}, and thereby alter the movement of cholesterol within the cell to the mitochondrial cholesterol side chain cleavage enzyme complex;[82]; (b) activation of protein kinase-C through increases in diacylglycerol could provoke phosphorylations that may directly or indirectly activate mitochondrial cholesterol side chain cleavage; and (c) phospholipids could interact with carrier proteins and thereby modify or regulate the movement of cholesterol from extra- to intramitochondrial sites.

The recent demonstration of a cycloheximide-sensitive, steroidogenic peptide by Pederson and Brownie[83] also suggests other potential mechanisms whereby phospholipids could be related to steroidogenesis: (a) phospholipids, as the cycloheximide-sensitive step in ACTH action, could precede and provoke (e.g., directly or by protein kinase-C activation, etc.) the formation or release (perhaps by proteolytic action) of the steroidogenic peptide: and (b) both the steroidogenic peptide and phospholipids may operate independently to increase steroid production.

The *de novo* phospholipid synthesis effect could also be involved in other actions of ACTH in the adrenal cortex. Increases in Ca^{++} uptake which are provoked by ACTH are inhibited by cycloheximide,[84] and phospholipid metabolism may therefore be involved in this effect. As alluded to above, the increase in phospholipid synthesis may lead to expansion of intracellular membranes; this might ultimately trigger cellular hypertrophy, which is observed in ACTH action.[85]

V. ROLE OF PHOSPHOLIPID METABOLISM IN THE ACTION OF LUTEINIZING HORMONE

Luteinizing hormone has been found to increase $^{32}PO_4$ incorporation into phospholipids in bovine corpus luteum[13] and rat ovarian granulosa cells.[14] In the latter case, a small increase in phosphoinositide mass has also been demonstrated, suggesting that this effect of luteinizing hormone is reflective of increased *de novo* phospholipid synthesis. In other cases, it has been difficult to show increases in phospholipid mass, but there has been no effect of luteinizing hormone on hydrolysis of either phosphatidylinositol or triphosphoinositide. It seems unlikely that the observed increase in ^{32}P-labeling of phosphatidic acid and phosphatidylinositol is due to phosphoinositide hydrolysis, which is readily demonstrable in the same tissue preparations with other agonists such as luteinizing hormone-releasing hormone.

The mechanism whereby luteinizing hormone increases $^{32}PO_4$ incorporation into phosphatidic acid and phosphatidylinositol in ovarian tissues is presently uncertain. To date, cAMP, its derivatives, and agents which are known to provoke increases in cAMP, such as cholera toxin, have not been found to increase $^{32}PO_4$ incorporation into phospholipids in ovarian preparations.[106,107] As stated above, it is presently uncertain whether the latter failure reflects difficulties in the interpretation of $^{32}PO_4$ incorporation data in situations where cAMP levels are increased, or whether luteinizing hormone alters phospholipid metabolism through a mechanism which is independent of cAMP.

In rat testicular Leydig cells,[55] luteinizing hormone was found to provoke effects on phospholipid metabolism which were similar to those of ACTH in the adrenal cortex. Increases in phosphatidic acid, phosphatidylinositol, polyphosphoinositides, and to a lesser extent, phosphatidylcholine and phosphatidylethanolamine, were observed during in vitro treatment with luteinizing hormone. In addition, as in the ACTH-adrenal studies, 8-bromo-cAMP provoked similar increases in phospholipids, and cycloheximide inhibited the increases in phospholipids which were induced by luteinizing hormone. Unlike adrenal studies, we were unable to show effects of luteinizing hormone on $^{32}PO_4$ incorporation into phospholipids over a 2-hr incubation period. However, in these experiments, luteinizing hormone and ^{32}P were added simultaneously, and these studies need to be repeated with sufficient ^{32}P-pre-labeling (see above).

VI. ROLE OF PHOSPHOLIPID METABOLISM IN THE ACTION OF ANGIOTENSIN, K^+, AND SEROTONIN IN ADRENAL GLOMERULOSA TISSUE

All agonists which increase aldosterone synthesis, including angiotensin-II, K^+, serotonin, ACTH, and cAMP, were found to provoke the *de novo* phospholipid synthesis effect during incubations of rat adrenal glomerulosa preparations in vitro.[57] Dramatic increases in phosphatidic acid and phosphoinositides and lesser relative increases in phosphatidylcholine and phosphatidylethanolamine were also observed[72] in the rat adrenal zona glomerulosa after in vivo treatment with the diuretic, furosemide, which presumably caused maximal stimulation of the renin-angiotensin-adrenal axis. Of all of these agonists, only angiotensin-II concomitantly provoked triphosphoinositide hydrolysis.[34] The latter effect of angiotensin-II occurred extremely rapidly (maximal within 5 sec) and may have been the initial biochemical event leading to mobilization of intracellular Ca^{++} (see above) and other subsequent events, including *de novo* phospholipid synthesis and steroidogenesis. K^+, on the other hand, did not provoke triphosphoinositide hydrolysis, but did increase phosphatidylinositol hydrolysis[57] in rat adrenal glomerulosa cells. Angiotensin-II also provoked phosphatidylinositiol hydrolysis (in addition to triphosphoinositide hydrolysis) in adrenal glomerulosa cell preparations,[57] and the phosphatidylinositol hydrolysis effects of both K^+ and angiotensin-II were dependent upon Ca^{++}. The latter Ca^{++}-dependency contrasts with the apparent independence of angiotensin-II-induced triphosphoinositide hydrolysis upon Ca^{++},[34] and these differences in phosphoinositide and triphosphoinositide hydrolysis are similar to the differences which we have observed in the exocrine pancreas (see above).[20,24] It therefore seems plausible that angiotensin-II utilizes triphosphoinositide hydrolysis to mobilize intracellular Ca^{++}, while K^+ increases Ca^{++} by plasma membrane depolarization. In both cases, the resultant increases in Ca^{++} may then be responsible for provoking increases in phosphatidylinositol hydrolysis, although the latter is not certain and may be provoked by a mechanism which simply requires Ca^{++} permissively.

The increases in *de novo* phospholipid synthesis, which are provoked by angiotensin, K^+, ACTH, and serotonin[57] probably involve different mechanisms, since the latter two agents increase tissue cAMP levels, whereas the former two agonists do not (and presumably operate via Ca^{++}). As discussed above, it seems likely that increases in *de novo* phospholipid synthesis occur secondarily to increases in "second messengers", such as cAMP and Ca^{++}, but this is by no means certain. In any event, the fact that all aldosterone-stimulating agents increase *de novo* phospholipid synthesis and aldosterone production by a cycloheximide-sensitive and Ca^{++}-requiring mechanism is in keeping with the concept that the *de novo* phospholipid synthesis effect has an important post-second messenger role in eliciting this steroidogenic response.

VII. ROLE OF PHOSPHOLIPID METABOLISM IN PARATHYROID HORMONE ACTION

Increases in *de novo* synthesis of phosphatidic acid and phosphoinositides have been observed during parathyroid hormone treatment in both rabbit[54,56,86] and dog[51] kidney cortex tubular preparations in vitro. As with ACTH and other agonists, these effects occur very rapidly, are inhibited by cycloheximide, and require Ca^{++}. Dibutyryl cAMP has been found to provoke similar changes in phospholipid metabolism in rabbit,[56] but not dog[51] kidney preparations. The latter failure to observe an effect may have been due to use of a relatively small dose (1 m*M*) of dibutyryl cAMP (in our rabbit study, 5 m*M* was more effective).

The role of phospholipids in the action of parathyroid hormone is uncertain. The phosphaturic effect of parathyroid hormone was inhibited by cycloheximide,[86] and it is therefore possible that phospholipids may be involved in this effect. Increases in amino acid transport are also inhibited by protein synthesis inhibitors,[87] and this effect may also involve changes in phospholipid metabolism. In view of the relationship of changes in phospholipids to changes in binding and transmembranous movement of ions (see above), it is tempting to suggest that phospholipid metabolism may be involved in the effects of parathyroid hormone on the handling of Ca^{++} and PO_4^{-3} ions in both kidney and bone.

VIII. ROLE OF PHOSPHOLIPID METABOLISM IN THE ACTION OF INSULIN

Insulin has been found to provoke the *de novo* phospholipid synthesis effect in two target tissues, including rat epididymal fat pad[15] and BC3H-1 myocytes[103] In both cases, we have not observed effects of insulin on phosphoinositide hydrolysis. In addition to increasing the concentrations of phosphatidic acid and phosphoinositides in these two tissues, insulin increases the concentration of phosphatidylserine.[48] Insulin also provokes modest increases in the concentrations of phosphatidylcholine and phosphatidylethanolamine in the myocytes, and although measurable concentrations of these substances are not altered appreciably in the rat epididymal fat pad, incorporation of $^{32}PO_4$ into these phospholipids is also increased, presumably reflecting a generalized increase in *de novo* phospholipid synthesis.

The importance of changes in phospholipid metabolism during the action of insulin is presently uncertain. Phosphatidylserine and/or phosphatidic acid have been shown to directly increase the activity of mitochondrial pyruvate dehydrogenase[88] and low K_m cAMP phosphodiesterase,[89] and insulin-induced increases in these phospholipids could play an important role in the direct activation of these enzymes. Increases in phosphatidic acid may also increase Ca^{++} translocations and thereby stimulate pyruvate dehydrogenase.[90] Increases in diacylglycerol may also be important, as we have found[108] that phorbol esters (which, like diacylglycerol, bind to and activate protein kinase-C) also provoke increases in pyruvate dehydrogenase activity.

Increases in triphosphoinositide and diphosphoinositide may play an important role in the observed increases in Ca^{++} binding to plasma membranes after insulin treatment in vitro;[91] changes in Ca^{++} binding could profoundly alter the activity of constituents such as transporting proteins, enzymes, and receptors in the plasma membrane. Increases in phosphatidic acid and possibly inositol-phosphates (see above) also play important roles in the mobilization of intracellular Ca^{++}, as observed in the action of insulin.[92] Changes in Ca^{++} could obviously influence Ca^{++}-dependent protein kinases and thus alter the phosphorylation-dephosphorylation state of important enzymes. Increases in diacylglycerol could also activate protein kinase-C, which may also influence phosphorylation and dephosphorylation. Along these lines, it has been suggested that ribosomal protein S6, a potentially important protein for regulating cytoplasmic protein synthesis, may be activated by a protein kinase-C mechanism.[93] Increases in glucose transport, as well as activation of pyruvate dehydrogenase, have been observed in the action of agents, such as exogenously-added phospholipase C.[94] These agents increase diacylglycerol, phosphatidic acid, and inositol-phosphates, and it is possible that through *de novo* phospholipid synthesis insulin-induced increases in these substances may be important in the activation of glucose transport.

As stated above, it is difficult to interpret cycloheximide inhibitory effects during hormone action. Nevertheless, one cannot ignore the fact that several effects of insulin have been shown to be inhibited by cycloheximide, including activation of pyruvate dehydrogenase,[49,95] increases in amino acid transport,[96] and internalization of the insulin receptor.[97] It is possible that degradation of phospholipids during insulin action is responsible for these inhibitory effects, rather than requirement of specific newly synthesized proteins per se. In fact, the decreases in pyruvate dehydrogenase provoked by cycloheximide occur so rapidly[103] that it is difficult to imagine the existence of a protein with such lability. However, even if these inhibitory effects of cycloheximide are related to changes in phospholipid metabolism rather than protein synthesis, it remains uncertain whether the phospholipids are important intracellular mediators for regulating these processes, or whether the insulin effects are inhibited through nonspecific, deleterious effects on cellular functions dependent on membrane integrity. Against the latter possibility is the fact that cyclohexamide does not inhibit basal (this is actually increased slightly) or cAMP-induced increases in pyruvate dehydrogenase activity.[49,95] With respect to glucose transport, unfortunately, cycloheximide has intrinsic insulin-like activities whereby it provokes considerable increases in 2-deoxyglucose transport, total glucose uptake, and conversion of labeled ^{14}C glucose to $^{14}CO_2$.[109] In some cases, it is possible to observe an effect of insulin on glucose transport or uptake which is additive to that of cycloheximide. This suggests that this effect of insulin is not inhibited by cycloheximide and may therefore not be mediated or provoked through phospholipids as intracellular mediators. However, as pointed out above, the effects of cycloheximide on phospholipid metabolism are exceedingly complex, and there may be continued (albeit lesser) increases in the products of phospholipase C action (diacylglycerol, phosphatidic acid and inositol-phosphates) which may be of importance in provoking this and other effects of insulin.

IX. ROLE OF PHOSPHOLIPID METABOLISM IN THE STIMULATION OF ISLET INSULIN SECRETION IN PANCREATIC ISLETS BY GLUCOSE AND OTHER SECRETAGOGUES

Glucose has been found to provoke very rapid increases in *de novo* phospholipid synthesis in pancreatic islets,[105] presumably by providing glycerol-3′PO_4 as substrate for phosphatidic acid synthesis. Glyceraldehyde, which is converted to glyceraldehyde-3-PO_4 and subsequently converted to glycerol-3′PO_4 and dihydroxyacetone-PO_4, also provokes a *de novo* phospholipid synthesis effect.[105] Since it has been thought that glucose must be metabolized along the glycolytic pathway to increase insulin secretion,[98] it is tempting to infer that glucose metabolism may be linked to insulin secretion through *de novo* phospholipid synthesis and generation of phosphatidic acid, diacylglycerol, phosphoinositides, and inositol-phosphates. Along these lines, it seems likely that exocytotic secretion in platelets[99,100] can be explained by combining protein kinase-C activation with subthreshold increases in Ca^{++} (i.e., as provoked by very low doses of ionophores), as would be expected with increases in diacylglycerol, phosphatidic acid, and inositol-triphosphate (see above). This also seems to be the case for insulin secretion, which can be provoked by combined treatment with phorbol esters (to activate protein kinase-C) and small doses of A23187 (to increase Ca^{++} slightly).[101]

In addition to increasing *de novo* phospholipid synthesis, glucose also reportedly increases phosphoinositide hydrolysis.[41,73] The latter effect would also increase the production of the above-mentioned signalling substances and could obviously contribute to insulin secretion. The mechanism whereby glucose increases triphosphoinositide hydrolysis[41] is presently uncertain. This, unlike receptor-activated triphosphoinositide hydrolysis, is Ca^{++}-dependent, and could result either from increases in Ca^{++} (possibly provoked by the *de novo* phospholipid synthesis effect) or via reported[102] direct activation of phospholipase C by diacylglycerol (which is presumably, as seen above, directly derived from the *de novo* phospholipid synthesis effect). Alternatively, increases in the generation of H^+, NADH, or other cofactors through enhanced glycolysis might somehow increase triphosphoinositide hydrolysis independently of the *de novo* phospholipid synthesis effect. However, this does not seem to be the case in other organs, e.g., during heightened glycolysis which occurs during ACTH and insulin action in their target tissues. In either case, both *de novo* phospholipid synthesis and triphosphoinositide hydrolysis would be expected to synergize in the production of substances which may provoke and maintain the secretion of insulin. It remains for future research to determine which effect, *de novo* phospholipid synthesis or phosphoinositide hydrolysis, is more primary in the effects of glucose on insulin secretion.

X. CONCLUDING REMARKS

I have presented the *de novo* phospholipid synthesis effect herein as a single phospholipid effect, which is distinct from the phosphoinositide hydrolysis effect. It is possible, however, that these are multiple "de novo" effects (or mechanisms to increase the synthesis of specific phospholipids in phosphatidate-phosphoinositide cycle) and increases in *de novo* phosphoinositide synthesis may occur to greater or lesser degrees in the action of agents which are presently recognized to simply increase phosphoinositide hydrolysis (in which case, resultant levels of the phosphoinositides will depend upon whether synthesis or hydrolysis predominates). Indeed, there are at least several cases where increases in polyphosphoinositides do not appear to be adequately explained by increases in phosphatidylinositol and a number of instances where polyphosphoinositide levels have been found to initially decrease and subsequently increase after agonist treatment. Obviously, there is much more to learn about the nature and interrelationship between these phospholipid effects, and their ultimate role in controlling cellular functions.

REFERENCES

1. **Hokin, M. R. and Hokin, L. E.,** Enzyme secretion and the incorporation of P^{32} into phospholipids of pancreas slices, *J. Biol. Chem.*, 203, 967, 1953.
2. **Michell, R. H.,** Inositol phospholipids and cell surface receptor function, *Biochim. Biophys. Acta,* 415, 81, 1975.
3. **Hawthorne, J. N. and White, D. A.,** Myo-inositol lipids, *Vitam. Horm.*, 33, 529, 1975.
4. **Michell, R. H., and Kirk, C. J.,** Why is phosphatidylinositol degraded in response to stimulation of certain receptors? *Trends Pharmacol. Sci.*, 2, 86, 1981.
5. **Berridge, M. J.,** Phosphatidylinositol hydrolysis: a multifunctional transducing mechanism, *Mol. Cell. Endocrinol.*, 24, 115, 1981.
6. **Putney, J. W.,** Recent hypotheses regarding the phosphatidylinositol effect, *Life Sci.*, 29, 1183, 1981.
7. **Fain, J. N.,** Involvement of phosphatidylinositol breakdown in elevation of cytosol Ca^{2+} by hormones and relationship to prostaglandin formation, in *Hormone Receptors,* Kohn, L. D. Ed., John Wiley & Sons, New York, 1982, 237.
8. **Farese, R. V.,** Phosphoinositide metabolism and hormone action, *Endocr. Rev.*, 4, 78, 1983.
9. **Hokin-Neaverson, M.,** Acetylcholine causes a net decrease in phosphatidylinositol and a net increase in phosphatidic acid in mouse pancreas, *Biochem. Biophys. Res. Commun.*, 58, 763, 1974.
10. **Banschback, M. W., Geison, R. L., and Hokin-Neaverson, M.,** Acetylcholine increases the level of diglyceride in mouse pancreas, *Biochem. Biophys. Res. Commun.*, 58, 714, 1974.
11. **Farese, R. V., Sabir, M. A., and Larson, R. E.,** On the mechanism whereby ACTH and cyclic AMP increase adrenal polyphosphoinositides. Rapid stimulation of the synthesis of phosphatidic acid and derivatives of CDP-diacylglycerol, *J. Biol. Chem.*, 255, 7232, 1980.

12. **Farese, R. V., Sabir, M. A., Larson, R. E., Trudeau, W., III.** Further observations on the increases in inositide phospholipids after stimulation by ACTH, cAMP and insulin, and on discrepancies in phosphatidylinositol mass and $^{32}PO_4$-labeling during inhibition of hormonal effects by cyclohexamide, *Cell Calcium*, 4, 195, 1983.
13. **Davis, J. S., Farese, R. V., and Marsh, J. M.,** Stimulation of phospholipid labeling and steroidogenesis by luteinizing hormone in isolated bovine luteal cells, *Endocrinology*, 109, 469, 1981.
14. **Davis, J. S., Farese, R. V., and Clark, M. R.,** Stimulation of phospholipid synthesis by luteinizing hormone in isolated rat granulosa cells, *Endocrinology*, 112, 2212, 1983.
15. **Farese, R. V., Larson, R. E., and Sabir, M. A.,** Insulin acutely increases phospholipids in the phosphatidate-inositide cycle in rat adipose tissue, *J. Biol. Chem.*, 257, 4042, 1982.
16. **Manchester, K. L.,** Stimulation by insulin of incorporation of [^{32}P]phosphate and ^{14}C from acetate into lipid and protein of isolated rat diaphragm, *Biochim. Biophys. Acta*, 70, 208, 1963.
17. **Torrontegui, G. and Berthet, J.,** The action of insulin on the incorporation of [^{32}P]phosphate in the phospholipids of rat adipose tissue, *Biochim. Biophys. Acta*, 116, 477, 1966.
18. **Stein, J. M. and Hales, C. N.,** The effect of insulin on $^{32}P_i$ incorporation into rat fat cell phospholipids, *Biochim. Biophys. Acta*, 337, 41, 1974.
19. **Garcia-Sainz, J. A. and Fain, J. N.,** Effect of insulin, catecholamines and calcium ions on phospholipid metabolism in isolated white fat-cells, *Biochem. J.*, 186, 781, 1980.
20. **Farese, R. V., Larson, R. E, and Sabir, M. A.,** Ca^{++}-dependent and Ca^{++}-independent effects of pancreatic secretagogues on phosphatidylinositol metabolism, *Biochim. Biophys. Acta*, 710, 391, 1982.
21. **Farese, R. V., Larson, R. E., and Sabir, M. A.,** Ca^{++}-dependent and Ca^{++}-independent mechanisms for phosphatidylinositol turnover during cholinergic stimulation of the rat submaxillary gland, *in vitro*, *Arch. Biochem. Biophys.*, 219, 204, 1982.
22. **Farese, R. V., Larson, R. E., and Sabir, M. A.,** Effects of Ca^{++} ionophore A23187 and Ca^{++} deficiency on pancreatic phospholipids and amylase release *in vitro*, *Biochim. Biophys. Acta*, 633, 479, 1980.
23. **Cockroft, S., Bennett, J. P., and Gomperts, B. D.,** Stimulus-secretion coupling in rabbit neutrophils is not mediated by phosphatidylinositol breakdown, *Nature (London)*, 288, 275, 1980.
24. **Orchard, J. L., Davis, J. S., Larson, R. E., and Farese, R. V.,** Effects of carbachol and cholecystokinin on polyphosphoinositide metabolism in the rat pancreas *in vitro*, *Biochem. J.*, 217, 281, 1983.
25. **Downes, P. and Michell, R. H.,** Phosphatidylinositol 4-phosphate and phosphatidylinositol 4,5-bisphosphate: lipids in search of a function, *Cell Calcium*, 3, 467, 1982.
26. **Berridge, M. J.,** 5-Hydroxytryptamine stimulation of phosphatidylinositol hydrolysis and calcium signaling in the blowfly salivary gland, *Cell Calcium*, 3, 385, 1982.
27. **Berridge, M. J., Dawson, R. M. C., Downes, C. P., Heslop, J. P., and Irvine, R. F.** Changes in the levels of inositol phosphates after agonist-dependent hydrolysis of membrane phosphoinositides, *Biochem. J.*, 212, 473, 1983.
28. **Berridge, M. J.,** Rapid accumulation of inositol triphosphate reveals that agonists hydrolyse polyphosphoinositides instead of phosphatidylinositol, *Biochem. J.*, 212, 849, 1983.
29. **Kirk, C. J., Creba, J. A., Downes, C. P., and Michell, R. H.,** Hormone-stimulated metabolism of inositol lipids and its relationship to hepatic receptor function, *Biochem. Soc. Trans.*, 9, 377, 1981.
30. **Billah, M. M. and Lapetina, E. G.,** Rapid decrease of phosphatidylinositol 4,5-bisphosphate in thrombin-stimulated platelets, *J. Biol. Chem.*, 257, 12705, 1982.
31. **Creba, J. A., Downes, C. P., Hawkins, P. T., Brewster, G., Michell, R. H., and Kirk, C. J.,** Rapid breakdown of phosphatidylinositol 4-phosphate and phosphatidylinositol 4,5-bisphosphate in rat hepatocytes stimulated by vasopressin and other Ca^{2+}-mobilizing hormones, *Biochem. J.*, 212, 733, 1983.
32. **Weiss, S. J., McKinney, J. S., and Putney, J. W., Jr.,** Receptor-mediated net breakdown of phosphatidylinositol 4,5-bisphosphate in parotid acinar cells, *Biochem. J.*, 206, 555, 1982.
33. **Putney, J. W., Jr., Burgess, G. M., Hulenda, S. P., McKinney, J. S., and Rubin, R. P.,** Effects of secretagogues on [^{32}P]phosphatidylinositol 4,5-bisphosphate metabolism in the exocrine pancrease, *Biochem. J.*, 212, 483, 1983.
34. **Farese, R. V., Larson, R. E., and Davis, J. S.,** Rapid effects of angiotensin II and dibutyryl cyclic AMP on phosphatidylinositol metabolism, $^{45}Ca^{2+}$ fluxes, and aldosterone synthesis in bovine adrenal glomerulosa cells, *Life Sci.*, 33, 1771, 1983.
35. **Lapetina, E. G., Billah, M. M., and Cuatrecasas, P.,** The initial action of thrombin on platelets. Conversion of phosphatidylinositol to phosphatidic acid preceding the production of arachidonic acid, *J. Biol. Chem.*, 256, 5037, 1981.
36. **Rittenhouse-Simmons, S.,** Differential activation of platelet phospholipases by thrombin and ionosphore A23187, *J. Biol. Chem.*, 256, 4153, 1981.
37. **Streb, H., Irvine, R. F., Berridge, M. J., and Schulz, I.,** Release of Ca^{2+} from a nonmitochondrial store in pancreatic acinar cells by inositol-1,4,5-triphosphate, *Nature (London)*, 306, 67, 1983.

38. **Nishizuka, Y. and Takai, Y.** Calcium and phospholipid turnover in a new receptor function for protein phosphorylation, in *Protein Phosphorylation*, Cold Spring Harbor Conferences on Cell Proliferation, Rosen, O. M. and Krebs, E. G., Eds., Cold Spring Harbor Laboratory, Cold Spring Harbor, N.Y., 1981, 8.
39. **Nishizuka, Y.,** Phospholipid degradation and signal translation for protein phosphorylation, *Trends Biochem. Sci.*, 8, 13, 1983.
40. **Putney, J. W., Jr., Weiss, S. J., Van DeWalle, C. M., and Haddas, R. A.,** Is phosphatidic acid a calcium ionophore under neurohumoral control? *Nature (London)*, 284, 345, 1980.
41. **Laychock, S. G.,** Identification and metabolism of polyphosphoinositides in isolated islets of Langerhans, *Biochem. J.*, 216, 101, 1983.
42. **Wolheim, C. B. and Sharp, G. W. G.,** Regulation of insulin release by calcium, *Physiol. Rev.*, 61, 914, 1981.
43. **Farese, R. V., Sabir, M. A., and Vandor, S. L.,** Adrenocorticotropin acutely increases adrenal polyphosphoinositides, *J. Biol. Chem.*, 254, 6842, 1979.
44. **Farese, R. V., Sabir, A. M., Vandor, S. L., and Larson, R. E.,** Are polyphosphoinositides the cycloheximide-sensitive mediator in the steroidogenic action of adrenocorticotropin and adenosine-3′,5′-monophosphate? *J. Biol. Chem.*, 255, 5728, 1980.
45. **Farese, R. V., Sabir, M. A., and Larson, R. E.,** Kinetic aspects of cycloheximide-induced reversal of adrenocorticotropin effects on steroidogenesis and adrenal phospholipids *in vivo*, *Proc. Natl. Acad. Sci. U.S.A.*, 77, 7189, 1980.
46. **Farese, R. V., Sabir, M. A., and Larson, R. E.,** Effects of adrenocorticopropin and cycloheximide on adrenal diglyceride kinase, *Biochemistry*, 20, 6047, 1981.
47. **Farese, R. V., Sabir, M. A., and Larson, R. E.,** Comparison of changes in inositide and non-inositide phospholipids during acute and prolonged ACTH treatment *in vivo*, *Biochemistry*, 21, 3318, 1982.
48. **Farese, R. V., Sabir, M. A., Larson, R. E., Trudeau, W. L., III,** Insulin treatment acutely increases the concentrations of phosphatidylserine in rat adipose tissue, *Biochim. Biophys. Acta*, 750, 200, 1983.
49. **Farese, R. V., Farese, R. V., Jr., Sabir, M. A., Larson, R. E., Trudeau, W. L., III, and Barnes, D.,** The mechanism of actin of insulin on phospholipid metabolism in rat adipose tissue: requirement for protein synthesis and a carbohydrate source, and relationship to activation of pyruvate dehydrogenase, *Diabetes*, 33, 648, 1984.
50. **Farese, R. V., Sabir, M. A., and Davis, J. S.,** Apparent increases in phospholipid degradation and turnover during combined treatment with protein synthesis inhibitors and adrenocorticotropin, *Biochim. Biophys. Acta*, 793, 317, 1984.
51. **Meltzer, E., Weinreb, S., Bellorin-Font, E., and Hruska, K. A.,** Characterization of the effects of parathyroid hormone on renal phosphoinositide metabolism, *Biochim. Biophys. Acta*, 712, 258, 1982.
52. **Farese, R. V., Sabir, M. A., and Larson, R. E.,** Andrenocorticotropin and adenosine 3′,5′-monophosphate stimulate *de novo* synthesis of adrenal phosphatidic acid by a cycloheximide-sensitive, Ca^{++}-dependent mechanism, *Endocrinology*, 109, 1895, 1981.
53. **Farese, R. V., Sabir, M. A., and Larson, R. E.,** A23187 inhibits adrenal protein synthesis and ACTH effects on steroidogenesis and phospholipid metabolism in rat adrenal cells *in vitro*. Further evidence implicating phospholipids in the steroidogenic action of ACTH, *Endocrinology*, 108, 1243, 1981.
54. **Farese, R. V., Bidot-Lopez, P., Larson, R. E., and Sabir, M. A.,** Effects of parathyroid hormone and cyclic-AMP on renal phospholipid metabolism, in *Biochemistry of Kidney Functions (INSERM Symp. No. 21)*, Morel, F., Ed., Elsevier, Amsterdam, 1982, 205.
55. **Lowitt, S., Farese, R. V., Sabir, M. A., and Root, A. W.,** Ray Leydig cell phospholipid content is increased by luteinizing hormone and 8-bromo-cyclic AMP, *Edocrinology*, 111, 1415, 1982.
56. **Bidot-Lopez, P., Farese, R. V., and Sabir, M. A.,** Parathyroid hormone and adenosine-3′,5′-monophosphate acutely increase phospholipids of the phosphatidate-polyphosphoinositide pathway in rabbit kidney cortex tubules *in vitro* by a cycloheximide-sensitive process, *Endocrinology*, 108, 2078, 1981.
57. **Farese, R. V., Larson, R. E., Sabir, M. A., and Gomez-Sanchez, C. E.,** Effects of angiotensin-II, K^+, adrenocroticotropin, serotonin, cyclic-AMP, cyclic-GMP, A23187 and EGTA on aldosterone synthesis and phospholipid metabolism in the rat adrenal zona glomerulosa, *Endocrinology*, 113, 1377, 1983.
58. **Haynes, R. E. and Berthel, L.,** Studies on the mechanism of action of the adrenocorticotropic hormone, *J. Biol. Chem.*, 225, 115, 1957.
59. **Bell, J., Brooker, G., and Harding, B. W.,** ACTH activation of glycogenolysis in the rat adrenal, *Biochem. Biophys. Res. Commun.*, 41, 938, 1970.
60. **Moxley, A., Bell, N. H., Wagle, S. R., Allen, D. O., and Ashmore, J.,** Parathyroid hormone stimulaion of glucose and urea production in isolated liver cells, *Am. J. Physiol.*, 227, 1058, 1974.
61. **Keppens, S., Vandenheede, J. R., and DeWulf, H.,** On the role of calcium as second messenger in liver for the hormonally induced activation of glycogen phosphorylase, *Biochim. Biophys. Acta*, 496, 448, 1977.

62. **Bates, E. J., Topping, D. L., Sooranna, S. P., Saggerson, D., and Mayes, P. A.,** Acute effects of insulin on glycerol phosphate acyl transferase activity, ketogenesis and serum free fatty acid concentration in perfused rat liver, *FEBS Lett.*, 84, 225, 1977.
63. **Sooranna, S. R. and Saggerson, E. D.,** Interactions of insulin and adrenaline with glycerol phosphate acylation processes in fat-cells from rat, *FEBS Lett.*, 64, 36, 1976.
64. **Nimmo, H. G. and Houston, B.,** Rat adipose-tissue glycerol phosphate acyltransferase can be inactivated by cyclic AMP-dependent protein kinase, *Biochem. J.*, 176, 607, 1978.
65. **Czech, M. P.,** Molecular basis of insulin action, *Annu. Rev. Biochem.*, 46, 359, 1977.
66. **Czech, M. P.,** Insulin action, *Am. J. Med.*, 70, 142, 1981.
67. **Larner, J., Galasko, G., Cheng, K., DePaoli-Roach, A. A., Huang, L., Daggy, P., and Kellogg, J.,** Generation by insulin of a chemical mediator that controls protein phosphorylation and dephosphorylation, *Science*, 206, 1408, 1979.
68. **Jarett, L. and Seals, J. R.,** Pyruvate dehydrogenase activation in adipocyte mitochondria by an insulin-generated mediator from muscle, *Science*, 206, 1407, 1979.
69. **Jolles, J., Zwiers, H., Dekker, A., Wirtz, K. W. A., and Gispen, W. H.,** Corticotropin-(1-24)-tetracosapeptide affects protein phosphorylation and polyphosphoinositide metabolism in rat brain, *Biochem. J.*, 194, 283, 1981.
70. **Bennett, J. P., Cockcroft, S., Casell, A. H., and Gomperts, B. D.,** Plasma-membrane location of phosphatidylinositol hydrolysis in rabbit neutrophils stimulated with formylmethionyl-leucylphenylalanine, *Biochem. J.*, 208, 801, 1982.
71. **Farese, R. V., Sabir, M. A., and Larson, R. E.,** Potassium and angiotensin II increase the concentrations of phosphatidic acid, phosphatidylinositol and polyphosphoinositides in rat adrenal capsules, *in vitro*, *J. Clin. Invest.*, 66, 1428, 1980.
72. **Farese, R. V., Larson, R. E., Sabir, M. A., and Gomez-Sanchez, C.,** Effects of angiotensin II and potassium on phospholipid metabolism in the adrenal zona glomerulosa, *J. Biol. Chem.*, 256, 11093, 1981.
73. **Clements, R. S., Jr., Evans, M. H., and Pace, C. S.,** Substrate requirements for the phosphoinositide response in rat pancreatic islets, *Biochim. Biophys. Acta*, 674, 1, 1981.
74. **Farese, R. V., Sabir, M. A., and Larson, R. E.,** Kinetic aspects of cycloheximide-induced reversal of ACTH effects on steroidogenesis and phospholipid metabolism in rat adrenal sections *in vitro*, *Endocrinology*, 109, 1424, 1981.
75. **Farese, R. V., Linarelli, L. G., Glinsmann, W. H., Ditzion, B. R., Paul, M. I., and Pauk, G. L.,** Persistence of the steroidogenic effect of adenosine-3′,5′-monophosphate *in vitro*: evidence for a third factor during the steroidogenic effect of ACTH, *Endocrinology*, 85, 867, 1969.
76. **Farese, R. V., Ling, N. C., Sabir, M. A., Larson, R. E., Trudeau, W. L., III,** Comparison of effects of ACTH and Lys-V_3-MSH on steroidogenesis, cyclic AMP production and phospholipid metabolism in rat adrenal fasciculata-reticularis cells, *in vitro*, *Endocrinology*, 112, 129, 1983.
77. **Farese, R. V. and Sabir, A. M.,** Polyphosphorylated glycerolipids mimic adrenocorticotropin-induced stimulation of mitochondrial pregnenolone synthesis, *Biochim. Biophys. Acta*, 575, 299, 1979.
78. **Farese, R. V. and Sabir, A. M.,** Polyphosphoinositides: stimulator of mitochondrial cholesterol sidechain cleavage and possible identification as an ACTH-induced, cycloheximide-sensitive, cytosolic, steroidogenic factor, *Endocrinology*, 106, 1869, 1980.
79. **Davidson, J. B. and Stanacev, N. Z.,** Biochemistry of polyglycerophosphatides in central nervous tissue. I. On the biosynthesis, structure, and enzymatic degradation of phosphatidylglycerophosphate and phosphatidylglycerol in isolated sheep brain mitochondria, *Can. J. Biochem.*, 48, 633, 1970.
80. **Lambeth, J. D.,** Cytochrome P-450_{scc}. Cardiolipin as an effector of activity of a mitocondrial cytochrome P-450, *J. Biol. Chem.*, 256, 4757, 1981.
81. **Kowluru, R. A., Genge, R., and Jefcoate, C. R.,** Polyphosphoinositide activation of cholesterol side chain cleavage with purified cytochrome P-450_{scc}, *J. Biol. Chem.*, 258, 8053, 1983.
82. **Hanukoglu, I., Privalle, C. T., and Jefcoate, C. R.,** Mechanisms of ionic activation of adrenal mitochondrial cytochromes P-450_{scc} and $P450_{11}$, *J. Biol. Chem.*, 256, 4329, 1981.
83. **Pedersen, R. C. and Brownie, A. C.,** Cholesterol side-chain cleavage in the rat adrenal cortex: isolation of a cycloheximide-sensitive activator peptide, *Proc. Natl. Acad. Sci. U.S.A.*, 80, 1882, 1983.
84. **Leier, D. J. and Jungmann, R. A.,** Adrenocorticotropic hormone and dibutylryl adenosine cyclic monophosphate-mediated Ca^{2+} uptake by rat adrenal glands. *Biochim. Biophys. Acta*, 329, 196, 1973.
85. **Farese, R. V. and Reddy, W. J.,** Observations on the interrelations between adrenal protein, RNA and DNA during prolonged ACTH administration, *Biochim. Biophys. Acta*, 76, 145, 1963.
86. **Farese, R. V., Bidot-Lopez, P., Sabir, A., Smith, J., Schinbeckler B., and Larson, R.,** Parathyroid hormone acutely increases polyphosphoinositides of the rabbit kidney cortex by a cycloheximide-sensitive process, *J. Clin. Invest.*, 65, 1523, 1980.
87. **Weiss, I. W., Morgan, K., and Phang, J. M.,** Cyclic adenosine monophosphate-stimulated transport of amino acids in kidney cortex, *J. Biol. Chem.*, 247, 760, 1983.

88. **Kiechle, F. L. and Jarett, L.,** Phospholipids and the regulation of pyruvate dehydrogenase from rat adipocyte mitochondria, *Mol. Cell. Biochem.*, 56, 99, 1983.
89. **Macaulay, S. L., Kiechle, F. L., and Jarett, L.,** Comparison of phospholipid effects on insulin-sensitive low K_m cyclic AMP phosphodiesterase in adipocyte plasma membranes and microsomes, *Biochim. Biophys. Acta*, 760, 293, 1983.
90. **Denton, R. M. and Hughes, W. A.,** Pyruvate dehydrogenase and the hormonal regulation of fat synthesis in mammalian tissue, *Int. J. Biochem.*, 9, 545, 1978.
91. **McDonald, J. M., Burns, D. E., and Jarett, L.,** Ability of insulin to increase calcium binding by adipocyte plasma membrane, *Proc. Natl. Acad. Sci., U.S.A.*, 73, 1542, 1976.
92. **McDonald, J. M., Burns, D. E., and Jarett, L.,** The ability of insulin to alter the stable calcium pools of isolated adipocyte subcellular fraction, *Biochem. Biophys. Res. Commun.*, 71, 114, 1976.
93. **Le Peuch, C. J., Ballester, R., and Rosen, O. M.,** Purified rat brain calcium- and phospholipid-dependent protein kinase phosphorylates ribosomal protein S6, *Proc. Natl. Acad. Sci. U.S.A.*, 80, 6858, 1983.
94. **Honeyman, T. W., Strohnsnitter, W., Scheid, C. R., and Schimmel, R. J.,** Phosphatidic acid and phosphatidylinositol labelling in adipose tissue: relationship to the metabolic effects of insulin and insulin-like agents, *Biochem. J.*, 212, 489, 1983.
101. **Dawson, R. M. C., Hemington, N. L., and Irvine, R. F.,** Diacylglycerol potentiates phospholipase attack upon phospholipid bilayers: possible connection with cell stimulation, *Biochem. Biophys. Res. Commun.*, 117, 196, 1983.
102. **Farese, R. V., Orchard, J. L., Larson, R. E., and Davis, J. S.,** Triphosphoinositide hydrolysis and phosphatidylinositol hydrolysis are separable responses during secretagogue action in the rat pancreas, submitted.
103. **Farese, R. V., Barnes, D. E., Davis, J. S., Standaert, M. L., and Pollet, R. J.,** Effects of insulin and protein synthesis inhibitors on phospholipid metabolism, diacylglycerol levels, and pyruvate dehydrogenase activity in BC3H-1 cultured myocytes, *J. Biol. Chem.*, 259, 7094, 1984.
104. **Farese, R. V., Barnes, D. E., Sabir, M. A., Standaert, M. L., and Pollet, R. J.,** Effects of insulin on phospholipid metabolism in cultured BC3H-1 myocytes. Program, presented at The Chilton Conference, Dallas, Texas, January, 1984.
105. **Farese, R. V., DiMarco, P. E., Barnes, D. E., Sabir, M. A., Larson, R. E., Davis, J. S., and Morrison, A. D.,** Glucose-induced *de novo* synthesis of phosphatidic acid, phosphatidylinositol and polyphosphoinositides in pancreatic islets (a possible link between glucose metabolism and insulin secretion). Submitted.
106. **Davis, J. S. and Clark, M. R.,** Phosphatidylinositol labeling in rat granulosa cells: lack of effect of cAMP, *Biol. Reprod.*, 28, 34A, 1983.
107. **Davis, J. S., West, L. A., and Farese, R. V.,** Effects of luteinizing hormone on phosphoinositide metabolism in rat granulosa cells, *J. Biol. Chem.*, submitted.
108. **Farese, R. V., Standaert, M. L., Barnes, D. E., Davis, J. S., and Pollet, R. J.,** Phorbol ester (TPA) provokes insulin-like effects on glucose transport, amino acid uptake and pyruvate dehydrogenase activity in BC3H-1 cultured myocytes, submitted.
109. **Farese, R. V., Standaert, M. L., Pollet, R. J.,** unpublished observations.

INDEX

A

B

C

D

E

F

G

H

I

M

N

O

P